Calorimetry and Thermal Analysis of Polymers

Edited by Vincent B. F. Mathot

Calorimetry and Thermal Analysis of Polymers

Edited by Vincent B. F. Mathot

With contributions by
L. Benoist, H. Berghmans, W. Hemminger,
G. W. H. Höhne, J. A. J. Jansen, V. B. F. Mathot,
M. J. Richardson, R. Riesen, A. Schuijff, M. Wingfield

With 274 Illustrations and 15 Tables

Hanser Publishers, Munich Vienna New York

Hanser/Gardner Publications, Inc., Cincinnati

Editor:
Vincent B. F. Mathot, DSM Research, P.O. Box 18, NL-6160 MD Geleen,
The Netherlands

Distributed in the USA and in Canada by
Hanser/Gardner Publications, Inc.
6600 Clough Pike, Cincinnati, Ohio 45244-4090, USA
Fax: + +1 (513) 527-8950

Distributed in all other countries by
Carl Hanser Verlag
Postfach 86 04 20, 81631 München, Germany
Fax: + +49 (89) 98 48 09

The use of general descriptive names, trademarks, etc., in this publication, even if the former are not especially identified, is not to be taken as a sign that such names, as understood by the Trade Marks and Merchandise Marks Act, may accordingly be used freely by anyone.

While the advice and information in this book are believed to be true and accurate at the date of going to press, neither the authors nor the editor nor the publisher can accept any legal responsibility for any errors or omissions that may be made. The publisher makes no warranty, expressly or implied, with respect to the material contained herein.

Library of Congress Cataloging-in-Publication Data

Calorimetry and thermal analysis of polymers / edited by Vincent B. F.
   Mathot : with contributions by L. Benoist . . . [et al.].
      p.   cm.
   Includes index.
   ISBN 1-56990-126-0
   1. Polymers–Thermal properties.   2. Calorimetry.   I. Mathot,
Vincent B. F.   II. Benoist, L.
QD382.H4C35   1993
620.1'9204296'0287–dc20   93-48558 CIP

Die Deutsche Bibliothek – CIP-Einheitsaufnahme

Calorimetry and thermal analysis of polymers / ed. by Vincent
B. F. Mathot. With contributions by L. Benoist . . . –Munich;
Vienna; New York: Hanser; Cincinnati: Hanser-Gardner,
1994.
   ISBN 3-446-17511-3 (Hanser)
   ISBN 1-56990-126-0 (Hanser-Gardner)
NE: Mathot, Vincent B. F. [Hrsg.]; Benoist, Luc

ISBN: 3-446-17511-3 Carl Hanser Verlag, Munich Vienna New York
ISBN: 1-56990-126-0 Hanser/Gardner Publications, Inc., Cincinnati
Library of Congress Catalog Card Number 93-48558

Typography and production: Charlotte Fabian, Dublin

© 1994 Carl Hanser Verlag, Munich Vienna New York
Print Origination: Datapage International Ltd, Dublin
Printed and bound in Germany by Passavia GmbH, Passau

Preface

In the past few decades, the use of thermal analysis techniques in research, development and production has increased enormously. The number of thermal analysis professionals and users, in the widest sense of the word, has boomed accordingly, as evidenced by the rapid growth of national thermal analysis societies.

In particular the study of the thermal properties of polymers by differential thermal analysis (DTA), and more specifically by differential scanning calorimetry (DSC), has become widespread. The main reason for this trend, apart from the growing interest in materials development, is the increased availability of commercial equipment.

In the case of polymeric materials, DSC can guide the work of development engineers who use relationships between thermal properties and molecular structure, processing, morphology and other properties. These engineers need to be familiar with the thermal phenomena associated with processes such as vitrification, crystallization and melting. Polymer chemists and chemical reaction engineers, too, increasingly rely on thermal analysis tools, specifically at the level of molecular engineering.

Compared with calorimeters of old, the DSCs currently available allow simple and rapid measurement, with reasonably accurate results, which is why they are now in routine use in many areas and for many purposes. A major advantage of the apparatus currently available is that the scanning rate can be varied, which makes it possible to investigate the kinetic determinacy of all kinds of processes that is so characteristic of polymers.

Careful tuning and perfected methods can bring about a further increase in accuracy. Now that peripheral equipment for computerization of the measuring process is available, resulting in greatly expanded possibilities of control and data manipulation, the field of application and the potential of differential calorimeters will be extended even further. The point has already been reached where accuracy is no longer restricted by peripheral apparatus, but only by the calorimeter itself.

At the same time, there is one cause for concern. Equipment users are often given insufficient technical information by equipment manufacturers, sometimes without being aware of this. Their instruments are increasingly becoming black boxes: not only in terms of hardware, with the measuring signal sometimes undergoing untransparent or even unknown processing steps before it enters the recorder or the computer, but also in terms of software, which, if it is transparent at all, is often inaccessible. This is a problem, especially as the software is sometimes substandard and often not flexible enough for research purposes, while in quite a number of cases it has not been sufficiently adapted to applications requiring unattended continuous operation.

A great deal of development work will be necessary over the next ten years or so. In the field of hardware, the range of experimental possibilities should be increased; it should become possible to apply higher, controlled cooling and heating rates (to enable kinetic studies and simulation of production processes), to measure at high pressures, and so on. Software should be made more transparent and updated to current models, researchers should be able to incorporate their own models into the software they use, and it should be possible to couple the software to data-banks. In my opinion, the development of the potential of DSC equipment will stagnate compared with that of other analytical techniques if these conditions are not met.

It is precisely because of the growing popularization of thermal analysis techniques, and the fact that these techniques are increasingly being used by researchers who have little opportunity to familiarize themselves with the instrumental background and the potential of the instrument, that there is a strong need for training facilities.

In this respect, there is an important task ahead for existing organizations in the field of calorimetry and thermal analysis. Many organizations already operate at national level, and the International Confederation for Thermal Analysis and Calorimetry (ICTAC), an international umbrella organization that is an Affiliated Society of the IUPAC, is also actively engaged in training.

This book is intended as an aid in the training of thermal analysts in the fields of calorimetry and thermal analysis on polymers. In addition, I hope it will help experienced analysts to deepen their insight. Some chapters are devoted to the use of DSC in the field of polymers, while others provide background information or show links with other techniques. Of course, I welcome readers' comments and suggestions. The contents of this book are based on a TPS (Transnational Training Project in Polymer Science and Technology) course organized in 1989. I greatly appreciate the fact that several contributors to the course were willing to elaborate their lectures for inclusion in this book. I should also like to thank Ronald Koningsveld for his pleasant and stimulating cooperation during the TPS course and in the preparation of this book. Finally, I am grateful to my wife Mildred and my sons Arjen, Erwin and Tim for their patience during the editing period.

Vincent B. F. Mathot

Front cover illustration:
Transmission electron micrograph showing the lamellar morphology of an ethylene-1-octene co-polymer (a Very Low Density Polyethylene with a density of 888 kg/m³). The spherical entities (compact semi-crystalline domains with a diameter of about 0.3 µm) in the seemingly structureless matrix are thought to be the result of melt demixing prior to crystallization of chains differing in comonomer content and comonomer distribution. The differential scanning calorimetry curves schematically show the crystallization and melting behaviour of an ethylene-1-octene copolymer (a Linear Low Density Polyethylene with a density of 920 kg/m³). For both copolymers, the heterogeneity of comonomer incorporation in the chains is reflected in the material's morphology and thermal behaviour.

Contents

Chapter 1

Thermodynamics

Abraham Schuijff, Ds. Pasmastraat 24, NL-3981 CX Bunnik, The Netherlands

Contents

1.1 Temperature and heat

The use of the concepts of temperature and heat in thermal analysis is so common that it is often not realized how difficult they are and how long it has taken scientists to understand them properly. The definition of temperature in the zeroth law of thermodynamics may seem so obvious that one asks whether it is necessary to state it, and why it is regarded as a law. The notion of heat, on the other hand, is so abstract that many investigators who are involved daily with heat measurements would not find it easy to define correctly.

It is clear to anyone working in the field of thermal analysis that there is an intimate relation between temperature and heat. This relation is a good starting point for understanding the character of both concepts.

Heat is one of the two possible forms of energy transfer, the other of which is work. There are many ways of exerting work on a system, each connected to its own parameters and well defined in classical mechanics. We shall return to this below. There is, however, only one way to supply heat to or extract heat from a system, and that is by a temperature difference. Therefore heat can be defined as an energy transfer caused by a temperature difference. However, this definition is not complete: it makes sense only if temperature can be defined independently of heat. To see this we have to go to the second law of thermodynamics first: this is exactly how it happened in history. The second law states that heat flows from high to low temperature, and not in the opposite direction. It is clear that this statement can apply only if temperature (in particular high and low temperature) is already defined. Obviously the definition of high and low temperature cannot be based on the direction of flow of heat, as this would give a circular argument. An independent definition is therefore necessary, and this definition has become known as the zeroth law. The *zeroth law* defines the property 'temperature' by stating that if two systems (or bodies) are separately in thermal equilibrium with a third system, then they have the same temperature. In addition, 'high' and 'low' are defined by stating that dissipation of work in a pure substance can never cause a temperature decrease (but in general will cause a temperature increase). The character and direction of temperature being defined, the only thing remaining is to fix the unit of temperature. Presupposing that there is an absolute zero for temperature, it is enough to add a number to a fixed point. By convention, the triple point of water is given the value 273.16 degrees Kelvin (K), implying of course that one degree K is equal to one degree centigrade.

1.2 The first and the second law of thermodynamics

As we have seen above, temperature and heat cannot be understood properly without knowledge of the first and the second laws. Although most chemists and physicists are presumably (more or less) familiar with these laws, it may be helpful to review briefly their most important implications.

The *first law* says that the state of a system, which is the object of our interest, can be changed in two ways: by doing *work* on the system (or letting the system do the work) or by allowing the system to exchange *heat* with its surroundings.

Work W and heat Q are the two possible forms of energy transfer, and together they determine the energy change dU of the system:

2

$$dU = Q + W \tag{1.1}$$

Energy, heat and work are all expressed in the same unit, the joule (J).

One may well ask why the energy change dU is written as a differential, while work W and heat Q apparently are not differentials. This is indeed a fundamental point. The energy U of a system is a function of the parameters that describe the state of the system, i.e. a function of state. Therefore it is legitimate to speak of an increase of the energy of the system, and mathematically this is written dU for an infinitesimal increase and ΔU for a finite increase. In contrast, work and heat are not functions. They are both means of energy transfer, or, as is often said, 'energy on the move'. Therefore one cannot speak of the heat of a system or of the work of a system, or consequently of an increase of the heat or of the work of a system. This can easily be demonstrated by the fact that the temperature of a system, say a beaker of water, can be increased by bringing it in contact with a hotter object (heat transfer) or by exerting work on it (stirring). Afterwards it cannot be deduced which kind of process has caused the temperature increase. Therefore, it does not make sense to speak of an increase of the heat of the system. It is the energy of the system that has increased, not the heat or the work.

As mentioned above, the only parameter connected to heat is temperature. This means that heat can only be transferred by a temperature difference and, conversely, without a temperature difference there is no heat. In contrast, there are many forms of work, each connected to two parameters, a force and a coordinate. Most important for thermal analysis is the volume work, $p\,dV$, not because of its extent but because it appears in theory.

Energy is thus a function of the parameters of the system. Given a state of a system, the energy of a system is fixed. Therefore it is called a function of state. Once again, heat and work are forms of energy transfer and not properties of a system. In thermal analysis heat can be supplied to a sample, but after this has been done the sample has undergone an energy increase and does not contain heat. The same applies to work.

The *second law* also deals with heat and work. One of the formulations of this law is that heat cannot flow spontaneously from a system with a certain temperature to a system with a higher temperature. As we have seen above, this statement makes sense only when high and low temperatures are defined independently, and this is done by the zeroth law.

From this formulation of the second law a temperature scale can be derived without reference to a specific system. It is therefore an absolute temperature scale. This derivation is given in thermodynamics textbooks and is not discussed here. For our purpose, what matters is that there exists an absolute temperature scale and that the temperature T in the ideal gas law $pV = nRT$ has the same properties as the absolute temperature. All other thermometers can therefore be scaled on the ideal gas thermometer. As mentioned above, the unit of this so-called Kelvin scale is fixed by designation of the triple point of water at 273.16 degrees Kelvin.

Analogously to the first law, the second law gives rise to a function of state, the *entropy*, which is not derived here. Entropy is related to heat and temperature as follows:

$$dS = \frac{Q_{\text{rev}}}{T} \tag{1.2}$$

This can also be conceived as an axiom. In Eq. 1.2 S is the entropy, dS an infinitesimal increase in the entropy of the system under consideration, T is the absolute temperature and Q_{rev} is an infinitesimal amount of heat transferred to the system reversibly which means, i.e. under equilibrium conditions.

To obtain a finite entropy increase, the Eq. 1.2 must be integrated:

$$\Delta S = \int \frac{Q_{rev}}{T} \tag{1.3}$$

The integral can be calculated when Q_{rev} is expressed in the parameters that act on the system. For instance, when heat is supplied to the system by a temperature difference, Q_{rev} can be written $C_p \, dT$, where C_p is the heat capacity of the system and dT is its temperature increase. Integration from T_1 to T_2 gives

$$\Delta S = \int_1^2 \frac{C_p \, dT}{T} = C_p \ln \frac{T_2}{T_1} \tag{1.4}$$

An entropy change can therefore be calculated only under equilibrium conditions. However, this does not imply that the entropy itself can only be defined when the system is in equilibrium. Entropy is a function of state, and to define entropy the state of the system must be capable of definition. This means that the parameters that describe the system must have a well-defined value. This is not only the case for the equilibrium state, but for instance also for a stationary state and for metastable equilibrium.

At this point we must return to the energy, which, as we have seen, is also a function of state. The character of energy and entropy, however, is quite different. Energy has a very fundamental and unique property: it is conserved under all circumstances. This means that the energy of a system that is isolated from its environment is constant. This conservation of energy can be considered as a law, or as an axiom, or even as the very basis of the energy concept. Sometimes the first law of thermodynamics is formulated as the law of conservation of energy, but on philosophical grounds this is not adopted here. In any case, the equivalence with the first law can easily be shown.

As mentioned above, conservation is a unique property of energy; no other state function in thermodynamics, including entropy, has this property.

The energy of an isolated system is constant. When an isolated system undergoes an internal process, for instance a chemical reaction, its energy does not change but its parameters do, and so does its state. In classical thermodynamics we are only interested in macroscopic systems, and not in the molecular structure of the system. The notion of state implies that we can ascribe a certain value to the macroscopic parameters that describe the system. Therefore it is necessary that these parameter values are time-independent.

This situation will develop when the system is left to itself long enough. We call this state the *equilibrium state*. Obviously this is a preliminary definition of equilibrium. A precise definition, based on the properties of energy and entropy, is derived at a later stage of thermodynamic theory and can be found in textbooks. From the above, it is clear that the equilibrium state is an important concept in thermodynamics. It is a time-independent state for which it is possible to define functions of state unambiguously.

This is the basis for the whole theory of classical thermodynamics, with its powerful methods. Classical thermodynamics is therefore also called equilibrium or reversible thermodynamics. In practice, however, we are often concerned with *processes*, which are clearly not equilibrium situations. Classical thermodynamics deals with this problem by supposing that the process under consideration proceeds so slowly that it can be conceived as a sequence of equilibrium states (a so-called reversible process).

Another, often more realistic, treatment is to allow non-equilibrium processes. In that case reversible thermodynamics has to be extended to the field of irreversible thermodynamics. In this discipline the same state functions are used, but the equilibrium state is replaced by another time-independent state, the stationary state. We shall return to this in section 1.4.

1.3 Enthalpy and heat

Although energy is a fundamental function, it is not convenient for practical use. Energy itself cannot be measured; heat Q and work W are measured. In thermal analysis the quantity to be measured is generally Q: the only advantage of Q is that it can be measured. It 'gives access to' the energy or, more precisely, it gives information about the energy content of the system. However, it is not the only contribution to the energy of the system, the other contribution being W. The ideal quantity for use in thermal analysis (and in thermochemistry in general) would be a state function with the property that its increase in the system is equal to the heat Q supplied to the system. It is evident that this function has to be constructed from the energy in such a way that the work W is eliminated. Because in almost all cases the only work involved is volume work performed by the atmospheric pressure, it is sufficient to eliminate this contribution. This is possible by a very simple mathematical procedure, called a Legendre transformation. Starting with the first law in its infinitesimal form (Eq. 1.1), it follows for the heat

$$Q = dU - W \qquad (1.5)$$

Assuming that only volume work is present, and therefore replacing W by $-p\,dV$

$$Q = dU + p\,dV \qquad (1.6)$$

At constant pressure this can also be written

$$Q_p = dU + p\,dV + V\,dp$$

or

$$Q_p = d(U + pV)$$

The quantity $U + pV$ is also a state function, because U, p and V are state functions: given a certain state of the system, they all have a fixed value.

Introducing the symbol H and the name *enthalpy* for the new state function:

$$Q_p = dH_p \qquad \text{(only volume work)} \qquad (1.7)$$

or for a finite process

$$Q_p = \Delta H_p \qquad (1.8)$$

From this formula, use of the term 'heat content' for the enthalpy is understandable. Although the enthalpy is a function of state like the energy, it is not conserved. This is easy to see when we imagine a system that is isolated when not yet in equilibrium.

5

For this system U and V are constant but p is not necessarily constant, because the pressure may change during the processes the system undergoes on its way to equilibrium. Thus $U + pV$ is not constant for the isolated system. Knowing this, we appreciate enthalpy as a convenient auxiliary function derived from the fundamental energy function.

The usefulness of enthalpy can be demonstrated by considering the heat capacity at constant pressure. This property of a (pure) substance is defined as the temperature increase of a unit of substance as a result of the supply of a unit of heat. It is a very important property of a substance because it gives an indication of how energy is assimilated by the substance. Because of this character it is called a response function. However, from the definition given above it is not clear that it is a function of state, because heat is not a function of state. From Eq. 1.7, however, heat under certain circumstances (p constant, only volume work) is equal to dH and the enthalpy H is a function of state. Therefore the definition of the heat capacity at constant pressure C_p can be written as the partial derivative of H with respect to T

$$C_p = \left(\frac{\partial H}{\partial T}\right)_p \tag{1.9}$$

It can be seen from Eq. 1.6 that if the volume instead of the pressure is kept constant

$$Q_V = \mathrm{d}U_V \tag{1.10}$$

and thus

$$C_V = \left(\frac{\partial U}{\partial T}\right)_V \tag{1.11}$$

Stability theory, an important part of thermodynamics, shows that C_p and C_V of a pure substance in equilibrium *can never be negative*, which means that the supply of heat to a pure substance can never result in a decrease of temperature. It is outside the scope of this chapter to give the proof of this stability condition [1].

The stability condition does not mean that the heat capacity of a substance cannot decrease with increasing temperature, as is for instance the case in water below 4 °C.

1.4 Relation of energy and entropy

As we saw in section 1.2, energy and entropy are both functions of state, yet they are quite different in character. This can be demonstrated elegantly with the concept of a flow. A simple kind of flow is a mass flow. When for example force is exerted on a liquid in a pipe, a flow of liquid develops. Because a liquid is incompressible, the amount of liquid that flows into the pipe per unit time is equal to the amount of liquid that leaves the pipe. When the force on the liquid is kept constant, the same amount of liquid will flow through each cross-section per unit of time. Thus if a certain part of the pipe is considered as the system of interest, it contains at any moment a constant amount of liquid. Although the system is clearly not in equilibrium, its state does not change with time, i.e. it is time-independent. Such a state is called a *stationary state*. In fact its time-independence is based on the law of conservation of mass. For energy there also exists a conservation law. Energy is more difficult to visualize than mass, but we can also generate a flow of energy: one possibility is to apply a temperature difference to a system that was initially in equilibrium. To avoid complications we choose for the system a metal bar placed between two blocks of different but constant temperature (Fig. 1.1).

As we know, a flow of heat will develop through the bar from high to low temperature. Because each block is kept at constant temperature, a situation will arise that will not change with time: a stationary state. The temperature in the bar is then not equable, but decreases when we go from high to low temperature; however, at each point in the bar it is constant. Because heat is a form of energy transfer, this heat flow can be considered to be an *energy flow*. Now energy is a function of state, and because the (stationary) state of the bar is time-independent its energy content must be constant. This means that the same amount of heat that enters the bar at the high temperature side must leave the bar at the low temperature side. The difference

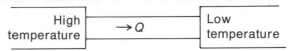

Fig. 1.1. In a metal bar, placed between two blocks of different temperatures, a heat flow develops

between the initial equilibrium state and the developed stationary state is thus that in equilibrium the temperature of the bar is everywhere the same, and in the stationary state there exists a temperature gradient. The notion of energy flow, however, deserves some reflection. Compared to mass flow, where it is clear what is flowing, it is less obvious what is flowing in the case of an energy flow. In fact it is quite unclear. Energy is a function of the parameters of the system, in this case the metal bar. Therefore energy does not flow, but just defines the state of the system. The only fact we know is that on one side heat enters the system, causing an energy increase, while on the other side an equal amount of heat leaves the system, causing an energy decrease. Because this is a continuous process we speak of an energy flow.

Let us now consider the concept of entropy flow. From Eq. 1.2, the entropy change of a system as the result of heat exchange with the environment can easily be calculated. In the case of our system, the metal bar in a stationary state between a high and a low temperature reservoir, a quantity of heat Q flows into the bar at the high temperature side. The entropy *increase* of the bar as a result of the process is

$$\Delta S = \frac{Q}{T_{high}}$$

because the heat is transferred at the high temperature.

At the low temperature side an equal quantity of heat leaves the bar, but at the low temperature. The entropy *decrease* of the bar is therefore

$$\Delta S = \frac{Q}{T_{low}}$$

Because $T_{high} > T_{low}$, the entropy increase is less than the entropy decrease. From this one might conclude that, as a result of the heat flow through the bar, its entropy decreases. This is, however, not the case. Entropy is a state function and the bar is in a stationary, i.e. time-independent, state. The entropy of the bar in this state must therefore be constant. The only conclusion can be that, as a result of the heat flow, entropy is produced in the bar and the entropy produced is removed from the bar immediately.

7

Just as in the case of the energy, we speak here of an *entropy flow*. There is however a striking difference between the two flows. While the energy flow is constant in a stationary state, the entropy flow increases during its passage through the system. Generalizing this conclusion, we can say that in a system in which temperature differences exist heat flows will appear, resulting in entropy production. In the same way it can be shown that concentration or density differences will give rise to mass flows with their resulting entropy production. All flows will go on until equilibrium is attained. If the system is kept isolated from its environment, the produced entropy cannot leave the system and the entropy will ultimately reach a maximum.

The general conclusion, proven more rigorously in thermodynamic theory, is therefore that an isolated system that is not in equilibrium will increase its entropy, attaining a maximum entropy value in the equilibrium state. From this it is clear that entropy is not conserved: entropy is produced inside the system without anything having passed its boundaries.

If we consider only heat flows, which are the most relevant phenomena in thermal analysis, we can summarize the relation between energy and entropy as follows. Entropy is a measure of the distribution of energy in the system. When energy is divided evenly over the system its entropy is high; when it is divided unevenly over the system its entropy is low. In the latter case the entropy of the system will grow on its way towards equilibrium.

The outcome of the discussion about energy, entropy and equilibrium allows us to formulate a more sophisticated definition of equilibrium than that given in section 1.2, where it was defined as the state towards which a system evolves in the course of time.

Now that we know that entropy grows to a maximum in an isolated system, it is clear that it behaves as a concave function when it is plotted against a parameter λ which governs the process occurring in the system, for example a temperature difference (heat flow), a concentration difference (mass flow) or an extent of reaction (chemical reaction). This is illustrated in Fig. 1.2, where the maximum of the curve represents the equilibrium state and all other points on the curve are non-equilibrium states. The system will always develop in the direction of increasing entropy, independently of its actual state (see arrows).

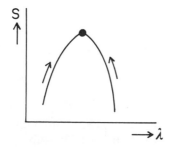

Fig. 1.2. The entropy S of an isolated system in a non-equilibrium state grows to a maximum as a consequence of the processes occurring in the system; λ stands for the parameter governing the process

Obviously the equilibrium state in an isolated system is given by

$$\frac{\partial S}{\partial \lambda} = 0; \qquad \frac{\partial^2 S}{\partial \lambda^2} < 0$$

The condition 'isolated system' is defined by 'no work, no heat', which can be expressed by constancy of the state function energy and of the volume of the system. So we have as the criterion for equilibrium

$$\left(\frac{\partial S}{\partial \lambda}\right)_{UV} = 0; \qquad \left(\frac{\partial^2 S}{\partial \lambda^2}\right)_{UV} < 0 \tag{1.12}$$

which is often shortened to

$$(dS)_{UV} = 0; \qquad \Delta S_{UV} < 0 \tag{1.13}$$

1.5 The absolute entropy: zeroth law

Besides the fact that energy is conserved and entropy is not, there is an important difference between the two state functions. As we have seen from the first law, only energy *differences* of a system can be measured, not the energy itself. As a consequence we have no natural zero point for the energy and for all state functions derived from it, such as the enthalpy H, the Helmholtz energy A and the Gibbs energy G. Therefore a zero point must be defined. By convention the elements at 298.15 K and 1 atmosphere pressure are chosen for this purpose. For the entropy the situation is different. Around 1906, Nernst formulated his heat theorem which later became known as the third law of thermodynamics. It states that in the neighborhood of zero Kelvin the entropy change of a process is zero. Nernst derived this statement from the fact that at room temperature almost all chemical reactions with a positive affinity (negative ΔG) show a positive heat development (negative ΔH).

Planck showed that, at least in the field of chemistry, this is equivalent to the statement that the entropy of all pure substances is zero at zero Kelvin. So obviously there is a natural zero for the entropy. The importance of this fact is indicated by the name 'third law of thermodynamics'.

The existence of a zero level for the entropy gives the opportunity to measure absolute entropies. Therefore the convention used for the zero level of the energy (and all other state functions derived from the energy) does not apply for the entropy. To measure absolute entropies one clearly has to start at zero Kelvin. Because the absolute zero of temperature cannot be attained, the measurements are generally started at about 5 K and the entropy increase from 0 K to 5 K is calculated on theoretical grounds. From 5 K onwards the entropy calculation is based on Eq. 1.3, in the form

$$\Delta S = \frac{Q_{rev}}{T} = \frac{\Delta H}{T}$$

for a phase transition at constant temperature (ΔH = heat of transition), whereas for the ranges between transitions Eq. 1.4 is used. Because of the additivity of the entropy, the contributions can be added to yield the absolute entropy at 298 K. From the above, it will be clear that the measurement of entropies consists of the measurements of heat capacities and of heats of transition, completed with a (small) theoretical contribution below 5 K.

Because of the different zero levels for energy and entropy, the values for the Helmholtz energy $A = U - TS$ and the Gibbs energy $G = U - TS + pV$ obviously cannot be calculated from the separate energy and entropy values. This problem, however, does not arise for ΔA and ΔG because in the equations

$$\Delta A = \Delta U - T\,\Delta S$$

$$\Delta G = \Delta U - T\,\Delta S + p\,\Delta V = \Delta H - T\,\Delta S$$

the zero levels in ΔU and ΔS cancel (just like the difference in height between two mountains that are measured separately from sea level).

1.6 Gibbs energy and Helmholtz energy

From section 1.5, it will be clear that in a thermodynamic description of processes and of equilibrium, both energy and entropy are involved. To facilitate this the two functions are combined to auxiliary functions which, although less fundamental than the basic functions, are more related to experimental conditions and more convenient for treatment of thermodynamic problems.

The two main auxiliary functions are

the Helmholtz energy A, defined as $\qquad A = U - TS$ (1.14)

the Gibbs energy G, defined as $\qquad G = U - TS + PV$ (1.15)

Because U, T, S, P and V all are functions of state, so are A and G. Both functions contain the energy U for which the absolute value cannot be measured; in fact the absolute values of A and G also cannot be measured. However, in classical thermodynamics there is no need to know the absolute values; the interesting quantities are changes in A and G as a consequence of processes. These changes can be calculated straightforwardly from the first and second laws. The first law (Eq. 1.1) is valid for all processes, including reversible processes; in this case it takes the form

$$dU = Q_{rev} + W_{rev}$$

Taking only volume work into account gives Eq. 1.6.

The second law introduces entropy S by Eq. 1.2, or $Q_{rev} = T\,dS$. Combination of Eqs. 1.6 and 1.2 gives

$$dU = T\,dS - p\,dV \qquad (1.16)$$

This equation gives the infinitesimal increase in energy as a result of infinitesimal changes in entropy and volume. Because the entropy is defined only when the state of the system is defined, the equation can only be used when this condition is fulfilled; this is the case not only in the equilibrium state but, under certain additional restrictions, also in other time-independent states such as the stationary state.

One might think from Eq. 1.16 that the energy of a system can be obtained by simple integration; this, however, is possible only when T is known as a function of S and p as a function of V, which is generally not the case. But even if it were possible it would not be very advantageous. It is far more important that use can be made of the fact that U is a function of state and that dU is thus a complete differential. From this property a number of differential equations may be derived which are very useful. It is outside the scope of this survey to show how this is done: it can be found in most books on thermodynamics.

From the definition of the Helmholtz energy $A = U - TS$, together with Eq. 1.16, another differential equation can be derived by a Legendre transformation:

$$dA = d(U - TS) = dU - T\,dS - S\,dT$$

Introducing $dU = T\,dS - p\,dV$ gives

$$dA = -S\,dT - p\,dV \qquad (1.17)$$

Because, as we have seen, A is a function of state, Eq. 1.17 is also a complete differential from which even more useful equations between the parameters of a system can be derived. In exactly the same way as for the Helmholtz energy, the Legendre method can be applied to the Gibbs energy $G = U - TS + pV$, yielding the complete differential

$$dG = -S\,dT + V\,dp \tag{1.18}$$

The basic set of equations is now complete, and a number of differential equations (the Maxwell equations) as well as the relation of the state functions with heat and work can easily be derived. For the Maxwell equations the reader is referred to textbooks on thermodynamics.

The relation of the Helmholtz energy with work follows from Eq. 1.17. It will be remembered that in the derivation use is made of the equation for dU in the form $dU = T\,dS - p\,dV$, which means that only volume work $W = -p\,dV$ is taken into account. The complete equation however is

$$dU = T\,dS + W_{rev} \tag{1.19}$$

If this form of dU is used, Eq. 1.18 becomes

$$dA = -S\,dT + W_{rev} \tag{1.20}$$

For a process at constant temperature, it follows that

$$dA_T = W_{rev} \tag{1.21}$$

From this equation we see that in a reversible process at constant temperature, the increase of the Helmholtz energy is equal to the work done on the system. The fact that the Helmholtz energy can be related to work only under certain conditions is an indication that it is a less fundamental function than, for instance, energy. The same applies for the Gibbs energy, for which it can easily be derived that

$$dG = -S\,dT + V\,dp + W_{eff} \tag{1.22}$$

where

$$W_{eff} = W_{rev} - \text{volume work} \qquad (= \text{useful work})$$

so that

$$dG_{T,p} = W_{eff} \tag{1.23}$$

Summarizing all relations of the state functions with heat and work:

$$\left.\begin{array}{l} dU = Q + W \\ T\,dS = Q_{rev} \end{array}\right\} \quad \text{fundamental functions}$$

$$\left.\begin{array}{l} dH_p = Q \\ dA_T = W_{rev} \\ dG_{T,p} = W_{eff} \end{array}\right\} \quad \text{auxiliary functions}$$

1.7 Irreversible thermodynamics

Because in thermal analysis non-equilibrium situations are more the rule than the exception, we shall briefly show how this is dealt with in thermodynamics. We therefore return to our example of the metal bar which, as the result of a temperature difference, is in a stationary state. We have seen that as a consequence of a heat flow from high to low temperature entropy is produced. For the stationary state this entropy production d_iS can easily be calculated:

$$d_iS = \frac{Q_{rev}}{T_{low}} - \frac{Q_{rev}}{T_{high}} = Q_{rev}\left(\frac{1}{T_{low}} - \frac{1}{T_{high}}\right) = Q_{rev}\,\Delta\left(\frac{1}{T}\right)$$

The subscript 'rev' is usually dropped and Q is written as dQ, so that the entropy production per unit of time P becomes

$$P = \frac{d_iS}{dt} = \frac{dQ}{dt}\cdot\Delta\left(\frac{1}{T}\right)$$

The cause of the entropy production $\Delta(1/T)$ is called the 'force' and given the symbol X_Q. The resultant dQ/dt of the force is called the 'flow' and given the symbol J_Q. Since the entropy production is always positive, we have

$$P = J_Q\cdot X_Q \geq 0 \tag{1.24}$$

If there is another force, e.g. a concentration difference or an electric potential difference, this will also give an entropy production. Because entropy is an extensive property the two contributions must be added:

$$P = \sum_i J_iX_i \geq 0 \tag{1.25}$$

We shall demonstrate the further use of this basic formula for the case of thermo-electric phenomena; in this field irreversible thermodynamics has proved to be very useful.

It can be shown that in the presence of an electric potential the force X_e and the flow J_e are given by

$$X_e = \Delta\left(\frac{\psi}{T}\right) \quad\text{and}\quad J_e = \frac{de}{dt}$$

where ψ is the electric potential and e is the charge. Both quantities are vectors, like X_Q and J_Q, but we shall omit the vector symbols because this property is not important for our purpose.

We now consider a metal wire on which a temperature difference ΔT is imposed; as we have seen, a heat flow J_Q will develop. If at the same time an electric potential is applied to the wire, this will give rise to a flow of electrons. Applying Eq. 1.25 to the system, we have

$$P = J_Q\cdot X_Q + J_e\cdot X_e \geq 0 \tag{1.26}$$

To derive results from this equation it is necessary to know the relations between the flows and the forces. These relations are the well-known phenomenological equations:

Fourier's law: $\quad J_Q = -\alpha\,\Delta T$ \hfill (1.27)

Ohm's law: $\quad J_e = \kappa\,\text{grad}\,\psi$ \hfill (1.28)

where α and κ are constants, the coefficient of heat conduction and the electric conductivity respectively. Both laws are valid only when the force is not too large, i.e. not too far from equilibrium.

In this range the flows are thus linearly dependent on the forces; the notation generally used in thermodynamics is

$$J = LX \tag{1.29}$$

where L is the proportionality constant. We know from theory that heat is (partially)

12

transported by electrons; it is therefore obvious that heat flow and electron flow are not independent. There must be a certain coupling between the two flows. This is reflected by writing

$$J_e = L_{ee}X_e + L_{eQ}X_Q \tag{1.30}$$

$$J_Q = L_{Qe}X_e + L_{QQ}X_Q \tag{1.31}$$

The subscripts of the coefficients L indicate their functions: L_{eQ} for instance is a measure of the dependence of the electron flow on X_Q.

From these equations it is clear that electron flow and heat flow each depend on both forces. It follows from this that a heat flow in a wire as a consequence of a temperature difference will cause an electron flow. When the wire is *not* a closed electric circuit the electron flow will be stopped by the electric potential difference that is built up in the wire. This potential difference is the Seebeck effect, on which the thermocouple is based.

The ratio $\Delta\psi/\Delta T$ at $J_e = 0$ is called the thermo-electric power; it is a property of the material of the wire, and we therefore speak of the thermo-electric power of the metal of which the wire is made. The thermo-electric power cannot be measured directly by a single wire, therefore a thermocouple is a combination of two wires of different materials in the well-known configuration of Fig. 1.3.

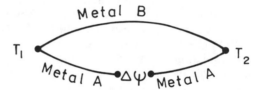

Fig. 1.3. Scheme of a thermocouple consisting of wires of different materials A and B. When temperatures T_1 and T_2 are different, there exists a potential difference $\Delta\psi$

When T_1 and T_2 (Fig. 1.3) are different temperatures, the metal wires A and B show different $\Delta\psi$ values. The thermo-electric power of the thermocouple as a whole is the difference between the two $\Delta\psi$ values, and can be measured. From this the temperature difference $T_1 - T_2$ can be deduced: this is the purpose of the use of a thermocouple.

A formula for the thermo-electric potential can easily be derived from Eq. 1.30. Inserting the expressions for X_e and X_Q gives

$$J_e = L_{ee}\,\Delta\!\left(\frac{\psi}{T}\right) + L_{eQ}\,\Delta\!\left(\frac{1}{T}\right) \tag{1.32}$$

Assuming that $\Delta T = T_1 - T_2$ is small, we can replace $\Delta(1/T)$ by $-(\Delta T/T^2)$ and $\Delta(\psi/T)$ by $\Delta\psi/T$:

$$J_e = L_{ee}\cdot\frac{\Delta\psi}{T} - L_{eQ}\frac{\Delta T}{T^2} \tag{1.33}$$

In an open circuit J_e has to be zero in the stationary (time-independent) state, and therefore Eq. 1.33 yields for each wire separately

$$\left(\frac{\Delta\psi}{\Delta T}\right)_{J_e=0} = \frac{1}{T}\frac{L_{eQ}}{L_{ee}} \tag{1.34}$$

Because the coefficients L are material properties they have different values for the two metals, and the total thermo-electric power for the thermocouple is given by

$$\left(\frac{\Delta\psi}{\Delta T}\right)_{J_e=0} = \frac{1}{T}\left(\frac{L_{eQ}^{A}}{L_{ee}^{A}} - \frac{L_{eQ}^{B}}{L_{ee}^{B}}\right) \tag{1.35}$$

From Eq. 1.35 we see that the thermo-electric power of a thermocouple can be calculated if the coefficients L_{eQ} and L_{ee} for the two metals are known. Unfortunately, however, these quantities are in general not known. It is even difficult, although not impossible, to measure the thermo-electric power of a single metal which, according to Eq. 1.34, would give the quotient of both coefficients. Therefore the thermo-electric power of thermocouples is generally measured directly with the help of a calibrated temperature difference. Tables of the thermo-electric power of different couples as a function of temperature can be found for example in [2], which uses the unit millivolts per Kelvin.

In the derivation of Eq. 1.35 we assumed that ΔT is not large. This is not in agreement with the large temperature ranges for which thermo-electric power values are given in the tables. However, this is not a problem because the tables contain experimentally determined values for which the restriction to small ΔT of course does not apply. Besides, it appears that the applicability range of thermodynamic theory of irreversible processes for the case of thermo-electric phenomena is rather large. A more rigorous and very elegant derivation of a formula for the thermo-electric power is possible by use of a property called 'entropy of transfer', which does not require the restriction to small ΔT; this derivation however falls outside the scope of this short thermodynamic treatment. The aim of this last section is to show the origin of the thermo-electric power and to demonstrate the possibility of using thermodynamics in non-equilibrium situations. For further study the reader is referred to textbooks on irreversible thermodynamics.

Literature

Recommended for further study:
Callen, M. B.: Thermodynamics. John Wiley and Sons, New York, 1985, Second Ed.
Denbigh, K. G.: Chemical Equilibrium. Cambridge University Press, Cambridge, 1989, Fourth Ed.
Maase, R.: Thermodynamik der irreversiblen Prozesse. Steinkopff, Darmstadt, 1963
Katchalsky, A., Curran, P. F.: Nonequilibrium Thermodynamics in Biophysics. Harvard University Press, Cambridge, 1965
Prigogine, J., Defay, R.: Chemical Thermodynamics. Longmans, Green and Co., London, 1954
Zemansky, M. W., Dittmann, R. H.: Heat and Thermodynamics. McGraw-Hill, New York, Tokyo, 1981

On thermocouples in particular:
Callen, M. B.: Physical Review 73, p. 1349 (1948)
Domenicali, C. A.: Rev. of Modern Physics 26, p. 237 (1954)
Handbook of Chemistry and Physics. *Weast, R. C.* (Ed.), CRC Press, Boca Raton

References

1 *Prigogine, J., Defay, R.:* Chemical Thermodynamics. Longmans, Green and Co., London, 1954
2 Handbook of Chemistry and Physics. *Weast, R. C.* (Ed.), CRC Press, Boca Raton

Nomenclature

A	Helmholtz energy
C_p	heat capacity at constant pressure
C_V	heat capacity at constant volume
d	differential
e	electric charge
G	Gibbs energy
grad	gradient
H	enthalpy
J	flow
J_e	electron flow
J_i	flow of i
J_Q	heat flow
L	coefficient in linear equation
L_{ee}	coefficient relating electron flow with electric force
L_{eQ}	coefficient relating electron flow with temperature difference
L_{Qe}	coefficient relating heat flow with electric force
L_{QQ}	coefficient relating heat flow with temperature difference
p	pressure
P	entropy production per unit of time
Q	heat
Q_p	heat at constant pressure
Q_V	heat at constant volume
S	entropy
$d_i S$	entropy production
t	time
T	temperature
U	energy
V	volume
W	work
X	(thermodynamic) force
α	thermal conductivity
Δ	difference
κ	electric conductivity
λ	parameter that governs a process
ψ	electric potential

Chapter 2

Calorimetric Methods

Wolfgang Hemminger, Physikalisch-Technische Bundesanstalt, Bundesallee 100,
D-38116 Braunschweig, Germany

Contents

2.1 Introduction

'Calorimetry' means the measurement of heat. Heat is the amount of energy exchanged in the form of a heat flow between two systems within a period of time. Heat is not a function of state; energy with the particular name 'heat' regrettably does not constitute a total differential. The heat exchanged in a process may therefore vary according to the path of the process. This must be kept in mind in high-pressure DSC, for instance. The history of calorimetry covers various concepts of how to measure heat. Such concepts form the basis for creating a classification system.

2.2 Classification of calorimeters

Calorimeters can be classified according to various criteria. Generally, in any classification system, calorimeters are arranged in groups according to particular characteristics. In spite of the large number of calorimeters available, a simple classification system will be established which is good enough to enable discussion of characteristic features and error sources of groups of instruments commonly used [1]. It is important that the measuring principle and the mode of operation be clearly stated. Even the classification is, at least to a certain extent, a matter of convenience. The aims just mentioned exclude criteria associated with names of persons ('Bunsen ice calorimeter'), or designations such as 'microcalorimeter' because it is unclear whether 'micro' refers to the instrument, the sample, the measured signal or the price. (Of course when we use a short name for a specific calorimeter, e.g. 'Bunsen calorimeter' or 'ice calorimeter' the expert knows which kind of calorimeter is spoken of, but these designations cannot be used for a classification.)

To be excluded are, of course, names that are definitely wrong, e.g. 'differential photocalorimeter'. In this case, neither radiation nor light power is measured: what is concerned is a DSC with an auxiliary device to activate a sample reaction by means of radiation.

The criteria that can be used to classify calorimeters are as follows.
(a) The measuring principle:
  (i) measurement of the energy required for compensation of the measuring effect (compensation calorimeters)
  (ii) measurement of the temperature change of the calorimeter substance due to the measuring effect (temperature-changing calorimeters)
  (iii) measurement of the heat exchange between calorimeter substance and surroundings (heat-exchanging calorimeters).
(b) The mode of operation:
  (i) isothermal
  (ii) isoperibol
  (iii) adiabatic
  (iv) scanning of surroundings
  (v) isoperibol scanning
  (vi) adiabatic scanning.
(c) The principle of construction:
  (i) single calorimeter
  (ii) twin or differential calorimeter.

Most of the calorimeters used nowadays can be classified by the above-mentioned criteria. Of course, not all combinations of (a), (b) and (c) are possible.

18

In the following, a number of calorimeters are presented on the basis of this classification, primarily according to the measuring principle. The classic calorimeters which are of great importance even today are dealt with only briefly; DSCs are treated in closer detail.

2.2.1 Compensation calorimeters

In compensation calorimeters, the effect of the heat to be measured—usually a temperature change—is 'suppressed'. For this purpose, an equally high, known amount of energy of opposite sign is added or eliminated in the form of heat. The possibilities are compensation with the aid of the heat of a phase transition or with electric energy. The advantage of compensation calorimeters lies in the quasi-isothermal conditions during the measurement; heat leaks do not, therefore, represent important error sources.

2.2.1.1 The ice calorimeter (Fig. 2.1)

BUNSEN [2] developed the well-known ice calorimeter of LAVOISIER and LAPLACE into an instrument for the precise measurement of heat. After the sample has been placed in the calorimeter, the sample heat is transferred to an ice/water mixture at 0 °C, and changes the ice/water ratio by melting the ice. This change is detected via the change in the density of the mixture. The inner ice/water mixture is shielded against the surroundings by means of a second ice/water mixture, i.e. an isothermal– adiabatic mode of operation is reached. The drawback of ice calorimeters and generally of all phase-change calorimeters (e.g. boil-off calorimeters) is the restriction to a fixed working temperature. GINNINGS and co-workers carried out the last important measurements with an ice calorimeter in the late 1940s (see e.g. ref. 3).

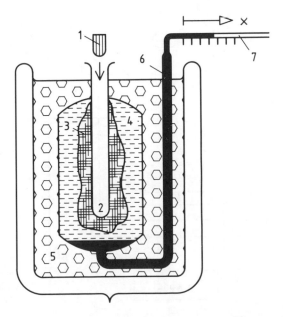

Fig. 2.1. (a) Ice calorimeter (according to BUNSEN [2]): (1) sample, (2) sample container, (3) ice, (4) water, (5) ice–water mixture, (6) mercury, (7) capillary with mercury column (× = displacement of the mercury meniscus)

19

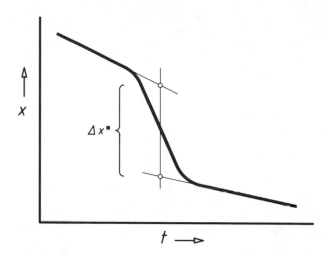

Fig. 2.1. (b) Ice calorimetry, measured curve (exothermic effect): $Q = K \Delta x^*$

2.2.1.2 Compensation by means of electric energy (Fig. 2.2)

A Dewar flask filled with water is equipped with a stirrer, a controllable electric heater and a sensitive thermometer [4]. At a constant temperature, an endothermically dissolving salt is added. The heater is controlled so as to ensure a constant liquid temperature. The supplied energy is then equal to the heat of solution.

To allow measurement of exothermic effects, a constant Peltier cooling power is usually fed into the system, while an equally high controlled Joule's heat is added to keep the temperature constant. To compensate an endothermic effect, the heating power is increased in order to keep the temperature of the calorimeter constant. An

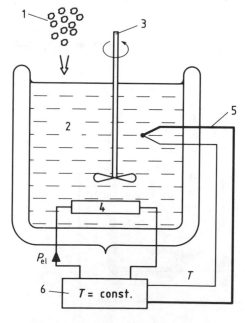

Fig. 2.2. (a) Compensation calorimeter (according to BRÖNSTED [4]): (1) sample (salt), (2) water, (3) stirrer, (4) electric heater, (5) thermometer, (6) controller

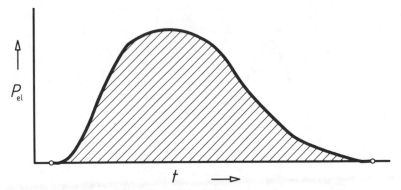

Fig. 2.2. (b) Compensation calorimeter, measured curve (endothermic effect): $Q = \int P_{el}(t) \, dt$

exothermic effect is compensated by decreasing the heating power, with the Peltier power remaining unchanged. The advantage of these quasi-isothermal calorimeters is the simple and very precise measurement of the electric compensation energy.

2.2.1.3 The adiabatic scanning calorimeter (Fig. 2.3)

In adiabatic scanning calorimeters, the temperature program can be preset. Electric heating power is supplied to the sample according to the given temperature program. In practice, the heating power is often given (this is imperative in the case of heat of

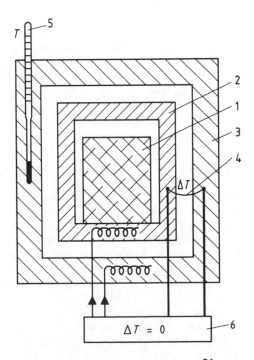

Fig. 2.3. (a) Adiabatic scanning calorimeter: (1) sample, (2) sample container with heater, (3) adiabatic jacket, (4) temperature difference sensor, (5) thermometer, (6) controller and programmer

21

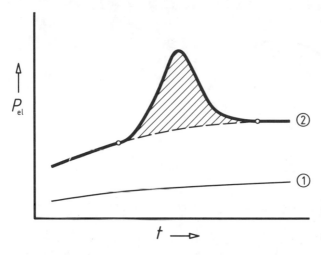

Fig. 2.3. (*b*) Adiabatic scanning calorimeter, measured curve: ① run without sample, ② run with sample

fusion measurements) and the resulting heating rate is measured. Heat losses are minimized by an adiabatic jacket. Calorimeters of this type allow the heat capacity to be measured with a very high accuracy (uncertainty $\leq 0.1\%$). Heats of transition can also be measured.

These calorimeters are used for the accurate, absolute and direct measurement of heat capacities and heats of fusion. Reference materials for calibrating DSCs via heat capacities and heats of fusion can be measured with calorimeters of this type.

2.2.1.4 The power-compensation DSC (Fig. 2.4)

Here, the temperature of the sample surroundings remains constant (i.e isoperibolic operation). The calorimeter comprises two identical measuring systems according to the twin principle: two microfurnaces, each equipped with a temperature sensor [5]. One system contains the sample, the other the reference. The temperature difference between the two systems is measured. The individual microfurnaces are heated separately so that they comply with the given temperature–time program. In the case of ideal thermal symmetry between the two measuring systems, the same heating power is required for sample and reference. An exothermic or endothermic process in the sample leads to a temperature deviation between the microfurnaces. This activates a proportional controller so that the electric heat supplied to the sample is decreased or increased by just the amount that has been generated or consumed during transition. It is important to note that the measured signal is the remaining temperature difference between the two microfurnaces. The individual temperatures cannot be made exactly equal because the proportional controller is activated only if there is a temperature difference. In other words, the temperature difference between sample and reference is at the same time the deviation from the set value (given by the program) which activates the control circuit and the measured signal to which the compensation heating power is proportional. This means that even in power-compensation DSCs a temperature difference exists between sample and reference supports, just as is the case with the heat-flux DSCs. But in contrast to the heat-flux DSCs this temperature difference is reduced by means of the controller; the typical errors of the heat-flux DSCs, which are due to this temperature difference, are then reduced to a

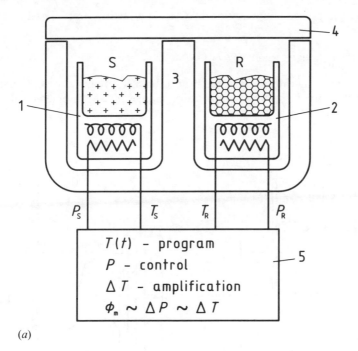

(a)

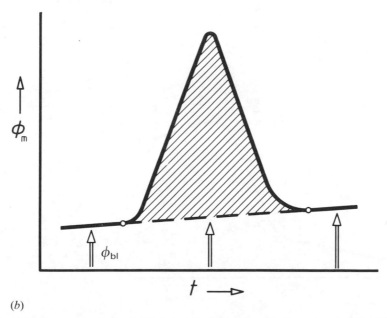

(b)

Fig. 2.4. (a) Power-compensation DSC (isoperibol): S = sample, R = reference, (1), (2) micro-furnaces, (3) surroundings (isoperibol), (4) lid, (5) controller and programmer; (b) measured curve: $Q = K_Q \int (\phi_m - \phi_{bl}) \, dt$

small amount. The power-compensation DSC is described in more detail in section 2.3.2.

2.2.2 Temperature-changing calorimeters

In temperature-changing calorimeters, the heat to be measured is not 'suppressed' by compensation, but gives rise to a temperature change in a 'calorimeter substance'

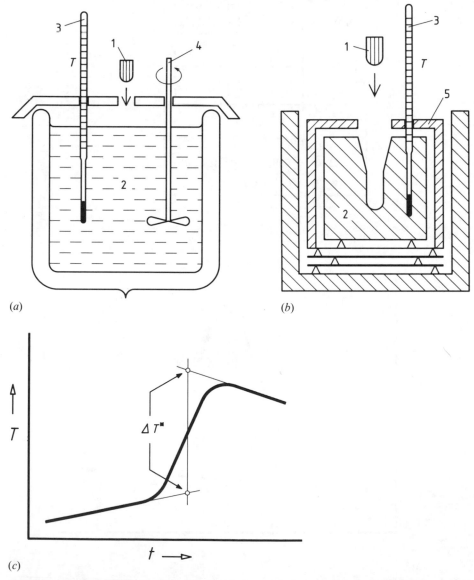

(a)

(b)

(c)

Fig. 2.5. (*a*) Drop calorimeter with liquid; (*b*) drop calorimeter with aneroid: (1) sample, (2) calorimeter substance (liquid or solid), (3) thermometer, (4) stirrer, (5) radiation shields; (*c*) measured curve: $\Delta T^* =$ calculated temperature change; $Q = K \, \Delta T^*$

24

with which the heat is exchanged. The general dependence is $Q = K \Delta T$, where K is the calibration factor and is determined preferably by means of electric calibration.

2.2.2.1 The drop calorimeter (Fig. 2.5)

The sample heat is exchanged with the solid or liquid calorimeter substance which is placed in surroundings of constant temperature (isoperibolic operation). The temperature change of the substance is measured during the 'preperiod' and 'postperiod' in order to calculate the 'true' temperature change according to defined rules. The proportionality factor K (calibration factor) comprises the heat capacity of the calorimeter substance plus that of other calorimeter components which cannot be marked off exactly. K is determined by calibration with Joule's heat.

2.2.2.2 The adiabatic bomb calorimeter (Fig. 2.6)

Here, a combustion bomb filled with oxygen at a high pressure is immersed in water (calorimeter substance). The water temperature is continuously measured before and after the electric ignition. The temperature of the surroundings is continuously adapted (controller) and serves as measured signal. Bomb calorimeters (usually automated) are widely used to measure the calorific value of solids or liquids (fuels) under standardized conditions. There are also 'dry' bomb calorimeters with the 'bomb' itself acting as the calorimeter substance, and bomb calorimeters with isoperibol surroundings.

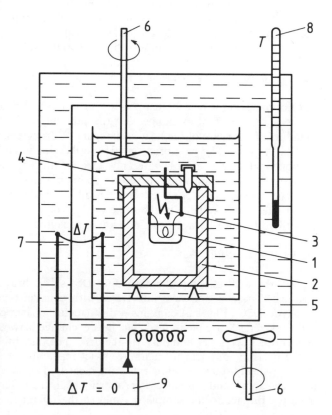

Fig. 2.6. (a) Adiabatic bomb calorimeter: (1) sample with crucible, (2) reaction vessel (bomb), (3) ignition wire, (4) water, (5) adiabatic jacket, (6) stirrer, (7) temperature difference sensor, (8) thermometer, (9) controller

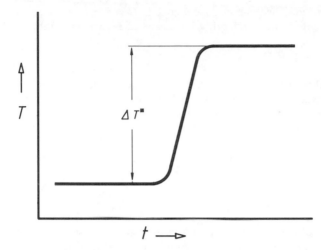

Fig. 2.6. (*b*) Adiabatic bomb calorimeter, measured curve: ΔT^* = temperature change; $Q = K \Delta T^*$

2.2.2.3 The adiabatic calorimeter (Fig. 2.7)

In the case of adiabatic calorimeters, it is not an unknown heat that is determined: a well-known, electrically generated heat W_{el} serves to change the sample temperature by ΔT^*. The temperature of the surroundings is adapted to the measurement temperature (adiabatic jacket). Calorimeters of this type are used to determine the heat capacity $C(T)$ (according to NERNST [6]) with the greatest possible accuracy: $C = W_{el}/\Delta T^*$. Latent heat can also be measured (with $\Delta T^* = 0$ during transition).

2.2.3 Heat-exchanging calorimeters

In calorimeters in which the temperature changes of the calorimeter substance are to be measured, the heat exchange with the surroundings is kept low to maximize the signal ΔT. In the case of the heat-exchanging calorimeters, a well-defined exchange of heat takes place. The measured signal describes the intensity of the exchange and is proportional to a heat-flow rate (not to a heat). This allows observation of time dependences of a transition (the power-compensation DSC also offers this possibility). The twin design of these calorimeters allows disturbances from the surroundings, which affect both systems in the same way, to be eliminated by taking into account only the difference between the individual measured signals.

2.2.3.1 The heat-flux DSC

There are two basic types of heat-flux DSCs [7]: heat-flux DSCs with disk-type measuring systems and heat-flux DSCs with cylinder-type measuring systems. The disk-type DSC (Fig. 2.8) contains a disk (metal, quartz glass) on which the sample and the reference are positioned. The heat exchange between furnace ('surroundings') and samples takes place mainly via the disk. The disk must be a 'heavy' component of the measuring system to guarantee strict repeatability of the heat exchange.

For steady-state heat flow rates, the measured signal ΔT (temperature difference between sample and reference) is proportional to the difference of the heat flow rates from the furnace to the sample ϕ_{FS} and to the reference ϕ_{FR}:

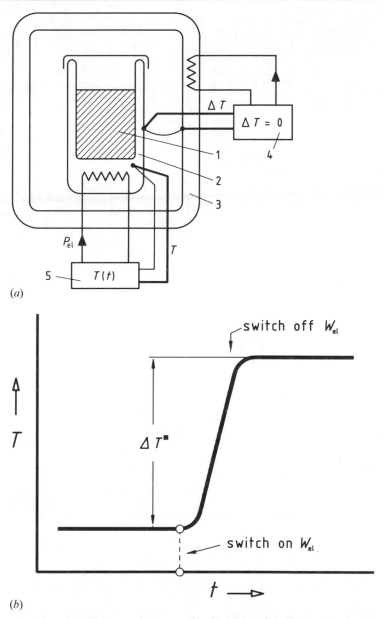

(a)

(b)

Fig. 2.7. (a) Adiabatic calorimeter for measuring heat capacity: (1) sample, (2) sample container with heater, (3) adiabatic jacket, (4) temperature difference controller, (5) controller and programmer; *(b)* measured curve: $C = W_{el}/\Delta T^*$

$$\Delta\phi = \phi_{FS} - \phi_{FR} = -K\,\Delta T$$

(An endothermic effect in the sample, i.e. heat consumption, creates a negative ΔT and a positive $\Delta\phi$, i.e. a heat flow into the sample; the proportionality or calibration factor K must be determined by calibration.)

27

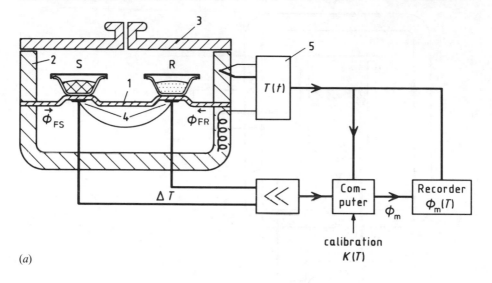

(a)

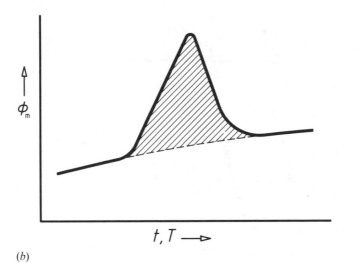

(b)

Fig. 2.8. (*a*) Heat-flux DSC (disk-type): S = sample, R = reference, (1) disk, (2) furnace, (3) lid, (4) temperature sensors, (5) controller and programmer; (*b*) measured curve: $\phi_m = -K\,\Delta T$

In the cylinder-type DSC (Fig. 2.9), the two 'heavy' cylindrical containers for sample and reference are connected with the furnace by one thermopile each. The thermocouple wires are at the same time heat conduction paths and temperature difference sensors. When a differential connection is provided between the outputs of both thermopiles, the difference signal ΔT is again proportional to the differential heat-flow rate from the furnace to the samples. A more detailed presentation of the measuring systems of the widely used heat-flux DSCs and the power-compensation DSCs is given in the section 2.3.

28

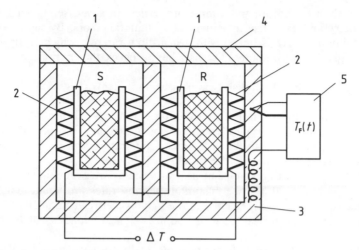

Fig. 2.9. Heat-flux DSC (cylinder-type, according to CALVET [8]): S = sample, R = reference, (1) container (vessel), (2) thermopile, (3) furnace (F), (4) lid, (5) controller and programmer

2.3 DSC measuring systems

In contrast to most of the DTA instruments, DSCs can be calibrated for heat measurements. Two types of DSC are widely used: heat-flux DSCs and power-compensation DSCs. Although in both types the measured signal represents the temperature difference generated by the instruments, the differential heat-flow rate flowing from the furnace to sample and reference is measured in completely different ways. The heat-flux measuring system is a passive system, whereas the power-compensation system is active. The heat-flux DSC registers the full temperature difference between sample and reference, which is built up during sample reaction, while the power-compensation DSC 'suppresses' this difference down to a fraction which is used as the measured signal.

2.3.1 Heat-flux DSCs

The measuring system, i.e. the disk or the sample containers with the temperature sensors, determines the temperature field inside the system. In view of the dominance of the measuring system the sample reaction can then be regarded as a small disturbance of the steady-state temperature field established during heating. As this disturbance must be small, the measured signal itself will also be small, i.e. the noise of the system must be low to allow sensitive measurements.

When the measuring system including the samples is ideally symmetrical, equally high heat-flow rates flow into the samples; the differential temperature signal is then zero. If this steady-state equilibrium is disturbed by a sample transition, a differential signal is generated which is proportional to the difference between the heat-flow rates flowing to the sample and to the reference:

$$\Delta\phi = \phi_{FS} - \phi_{FR} = -K\,\Delta T$$

Ideal thermal symmetry of the measuring system cannot be attained under all operating conditions, i.e. even outside the transition interval a signal ΔT occurs which

29

depends on the temperature and sample properties. In any case, a signal ΔT stands for asymmetry and gives rise to sources of systematic errors. By means of factory-installed calibration, ΔT is linked with a heat-flow rate ϕ_m which is displayed as a measured signal via a computer. The user has to check the calibration.

2.3.1.1 Heat-flux DSCs with disk-type measuring systems (Fig. 2.8)

Here the main heat flow from the furnace to the samples flows through a disk of good thermal conduction. The samples (pans) are positioned on this disk and the temperature sensors are integrated in the disk. Each temperature sensor covers roughly the area of support of one pan in order to allow calibration.

Heat-flux DSCs with disk-type measuring systems are available for temperatures between -190 and $1500\,^\circ$C. The maximum heating rates are about $100\,^\circ$C/min. A typical time constant is 15 s. The noise of the signal is between 5 and 50 μW. The total uncertainty of the heat measurement is about 5 % (heat capacities at least 10 %).

2.3.1.2 Heat-flux DSCs with cylinder-type measuring system (Fig. 2.9)

The cylindrical big sample cavities are connected with the furnace and/or with one another via numerous thermocouples (thermopiles). In the original cylinder-type system (according to CALVET [8]), the outer surfaces of the containers are in contact with a large number of thermocouples connected in series. Both cavities are thermally decoupled; heat exchange takes place only with the furnace. The measured signal is generated by differential connection of both thermopiles. For this type of calorimeter a detailed theory has been developed which quantitatively describes the functional correlation between the instantaneous measured signal and the original heat-flow rate exchanged inside the calorimeter (deconvolution procedure). In modified cylinder-type systems (Fig. 2.10) only a small fraction of the heat flow flows from the furnace to the samples via thermocouples (see e.g. ref. 9). Thermopiles are preferably used to measure directly the temperature difference between sample and reference cavities. The two cavities are no longer thermally decoupled, which makes theory more difficult and calibration and measurements more susceptible to errors. In all cylinder-

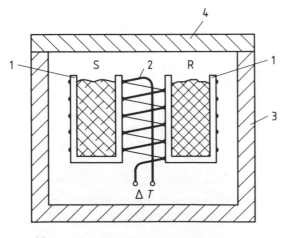

Fig. 2.10. Heat-flux DSC (modified cylinder-type, according to PETIT *et al.* [9]): S = sample, R = reference, (1) container (vessel), (2) thermopile, (3) furnace, (4) lid

type systems, the calibration factor depends on the position of the sample in the cavity.

Compared with disk-type systems, cylinder-type systems offer the advantage of a larger useful volume. However, greater time constants (from about 10 s up to several minutes) are the price to be paid for this. The large cavities allow special inserts (e.g. for electric calibration) and interventions (electrical, mechanical) as well as direct observations (optical, acoustical).

Heat-flux DSCs with cylinder-type measuring systems are available for a temperature range of -190 to $1500\,°C$. The maximum heating rate is about $30\,°C/min$. The instruments used in biology, for example, show a detection limit of about $1\,\mu W$ in isoperibolic operation.

2.3.2 Power-compensation DSCs

Commercial power-compensation DSCs have an isoperibolic mode of operation. The measuring system consists of two microfurnaces, which contain one temperature sensor each, and a heater (Fig. 2.11(a)). Both furnaces are positioned in an aluminum block of constant temperature; they are thermally decoupled. The temperature range extends from -175 to $725\,°C$; the time constant is about 2 s, the maximum scanning rate is up to $500\,°C/min$.

A control circuit (Fig. 2.11(b)) supplies the same heating power to both furnaces in order to change their mean temperature according to the preset temperature–time program [5]. In the case of ideal thermal symmetry, the temperature of both furnaces is the same. When a sample reaction occurs (Fig. 2.11(c)), the symmetry is disturbed and a temperature difference between the furnaces results. This temperature difference is at the same time the measured signal and the input signal of a second proportional control circuit, which tries to compensate the reaction heat-flow rate by increasing or decreasing an additional heating power ΔP. The compensating heating power is proportional to the measured temperature difference ΔT. Again, a heat-flow rate ϕ_m is assigned to the measured temperature difference as a result of a factory-installed calibration which has to be rechecked. The following are then valid:

$$\Delta P = K_1\,\Delta T$$

$$\phi_m = K_2\,\Delta T$$

where K_1 is a fixed quantity of the proportional controller and K_2 must be adjusted via the software (calibration).

Thermal asymmetries of the measuring system can be electronically compensated. In this way it is also possible to 'straighten' the zero line or incline it as desired. At higher temperatures, the heat-flow rates exchanged with the surroundings are large compared with the measured thermal effect. This means that uniformity of the heat exchange (radiation, conduction, convection) between the microfurnaces and the surroundings must be guaranteed. Moreover, the shares of the various heat exchange mechanisms and their contributions in absolute terms must depend only on the temperature, and strict repeatability must be ensured. The furnaces must be covered with caps of the same kind in order to 'harmonize' such differences between sample and reference as might influence the heat exchange (e.g. emissivity) and thus create an additional heat flow that has nothing to do with the sample properties under investigation.

31

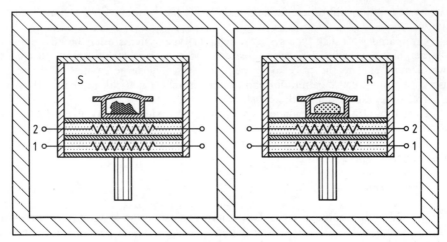

Fig. 2.11. (a) Power-compensation DSC: measuring system (S = sample, R = reference, micro-furnace with heater (1) and temperature sensor (2))

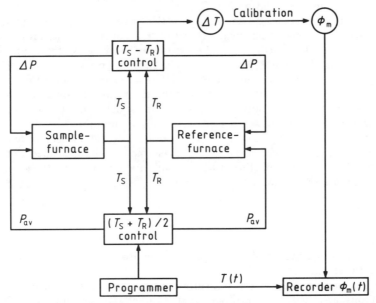

Fig. 2.11. (b) Power-compensation DSC: measuring principle (cf. WATSON et al. [5]) (ΔT = amplified ($T_S - T_R$) signal, ΔP = electric heating power to compensate the difference between T_S and T_R as far as possible, P_{av} = average heating power to maintain the heating rate (based on ($T_S + T_R$)/2), ϕ_m = calibrated ΔT signal (measured heat flow))

Compared with heat-flux DSCs, power-compensation DSCs offer the following advantages.

– The short heat conduction path (sample heater) and the small mass of the microfurnace allow an almost instantaneous response to a sample reaction. Due to the small time constant, deconvolution is required only in special cases.

– Reaction heat flows are rapidly and largely compensated. In other words, only

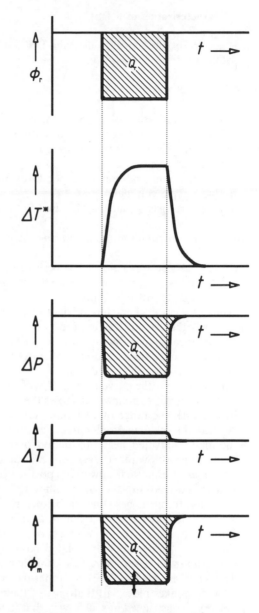

Fig. 2.11. (c) Power-compensation DSC: quantities involved in measuring ϕ_m (ϕ_r = idealized sample reaction heat-flow rate, ΔT^* = temperature difference $(T_S - T_R)$ due to the sample reaction without compensation (cf. heat-flux DSC), ΔP = compensated power, ΔT = real temperature difference in the measuring system, ϕ_m = heat-flow rate signal linked with ΔT by means of calibration)

small temperature differences occur between the two microfurnaces. This means that the calibration factor is practically independent of the intensity and kinetics of the sample reaction (in contrast to the heat-flux DSC).
- The total compensation energy is equal to the reaction heat.
- The temperature dependence of the electric circuit properties is known and repeatable. It can be taken into account by the software. In principle, a single heat calibration is then sufficient to determine the calibration factor (but there are systematic deviations between heat-flow rate calibration and peak-area calibration; see chapter 3).

2.4 Characterization of a DSC

It is important and useful to describe DSCs in terms that clearly show their practical capabilities.

2.4.1 Main characteristics of a DSC

A first characterization of a DSC, which should always be included in the manufacturer's description, should give information on the
- measuring principle (heat-flux or power-compensation DSC)
- temperature range
- potential heating rates and temperature–time programs
- usable sample volume
- atmosphere (gases that may be used, vacuum).

2.4.2 Characterization of the measuring system and test procedures

The material whose reaction must be investigated is one thing, the DSC is another. The DSC gives information on temperature and heat-flow rate. Whether or not it is good enough to solve the problem at hand depends on the performance of the instrument, i.e. the characteristic data of the DSC must be known in order to get an idea of whether or not it is appropriate.

Numerical values that serve to describe the efficiency of the measuring system can be taken from the measured curve. In the following, the terms used to describe a measured curve are described first.
- The *zero line* is the curve measured with the DSC empty, i.e. without samples and without sample containers. It shows the thermal behavior of the measuring system. The smaller the range of variation (see below), the better is the instrument.
- A *peak* in the measured curve occurs when the steady state is disturbed by thermally activated heat production or consumption in the sample (Fig. 2.12). Peaks in heat-flow rate curves (not ΔT curves!), which are assigned to endothermic processes, are plotted 'upwards' (positive range, because in thermodynamics heat that is delivered into the sample shows a positive sign). A peak begins at T_i (first deviation from the base line, cf. below), rises/falls to the peak maximum/minimum T_p and joins the base line again at T_f. The most important characteristic temperature is the *extrapolated peak onset temperature* T_e (intersecting point of tangent at the point of inflection/extrapolated base line). Other definitions are used for the *glass-transition temperature* (see chapter 6).
- The *base line* (see Fig. 2.12) is (i) that part of the measured curve where no sample transition takes place; (ii) that part of the curve which in the range of a peak (see above) is constructed in such a way that it connects the measured curve before and behind the peak as if no reaction heat had been exchanged, i.e. as if no peak had developed.

The following terms characterize the measuring system of a DSC and describe the efficiency of the system:
- the zero-line repeatability
- the base-line repeatability
- the scanning and isothermal noise
- the linearity
- the time constant.

34

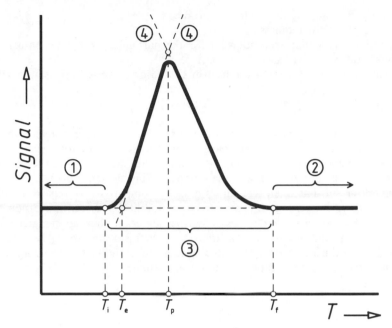

Fig. 2.12. DSC peak (heat-flow rate due to an endothermic first order reaction): T_i, T_e, T_p, and T_f explained in text, ① initial base line, ② final base line, ③ interpolated base line, ④ auxiliary lines

In this context the term 'repeatability' must be explained. The *repeatability* indicates the agreement between the results of a great number of measurements of the same kind carried out with one specific instrument. For DSCs, this term can be used to characterize the measuring system in several ways (cf. sections 2.4.2.1 and 2.4.2.2).

The *repeatability error* (absolute or percentage) indicates the scatter of the results of a series of measurements carried out on samples of the same kind. Example: ten measurements of the heat of fusion of tin, carried out under the same conditions, show a repeatability error (of the peak areas) of $\pm 3\%$ in relation to the mean value. (It must be stated whether the standard deviation, the maximum deviation or another measure serves as 'scatter'.)

The range of variation (fluctuation) of the zero line or base line (absolute or percentage) indicates, for example, the temperature-dependent range of deviations from the mean value of many zero lines or base lines recorded under the same conditions. It is caused by random variations and uncertainties. The smaller the range of variation over the whole temperature range, the better is the DSC.

The *reproducibility* describes the agreement between the results of measurements carried out on a sample using different instruments in different laboratories (round robin). The absolute or percentage deviation of the mean value of the results of an instrument from the total mean value is a measure of the reproducibility. Low reproducibility may point to systematic errors of measurement (see 2.4.3) of certain instruments.

Why is it important to test the DSC and not simply trust the manufacturer's statements?
– Because even the manufacturer may not be aware of important data.

– Because some data are of special importance to the user.
    Testing of a DSC requires
– preparation (data that are important for the particular problem, comparison of various models on the basis of prospectus, etc.)
– time (at least a day for disk-type heat-flux DSC or power-compensation DSCs).

2.4.2.1 The zero-line repeatability (Fig. 2.13)

The zero line represents the behavior of the pure DSC measuring system without the influence of the samples and sample pans. Run at least three or four zero lines over the whole temperature range at medium scanning rate. Superimpose the curves and check the $\pm$ range of variation and the differences in the signal drift.

It is important that when two curves must be compared with regard to absolute heat-flow rate values, the $\pm$ range of the repeatability gives an impression of the uncertainty. Furthermore, the curves show the degree of thermal symmetry of the DSC and the repeatability of the heat transfer mechanism involved. Also, increasing of amplification during scanning at some temperatures allows the scanning noise of the empty DSC at a given scanning rate to be evaluated (see 2.4.2.3).

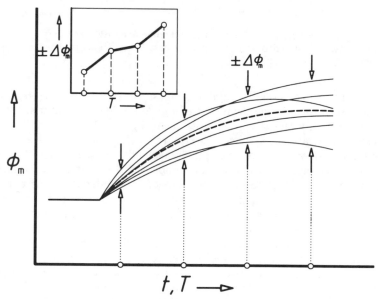

Fig. 2.13. Zero-line repeatability: $\Delta\phi_m$ = maximum deviation of ϕ_m from the average zero-line ($- - - -$), $\pm\Delta\phi_m$ = zero-line repeatability

2.4.2.2 The base-line repeatability (Fig. 2.14)

The base-line repeatability gives an idea of what the scatter is like when the DSC is operated with empty crucibles or with samples. Run the DSC three or four times over the entire temperature range with for example an indium sample and a reference (Pt, Al_2O_3). Superimpose the curves (Fig. 2.14(a)) to obtain the base-line repeatabil-

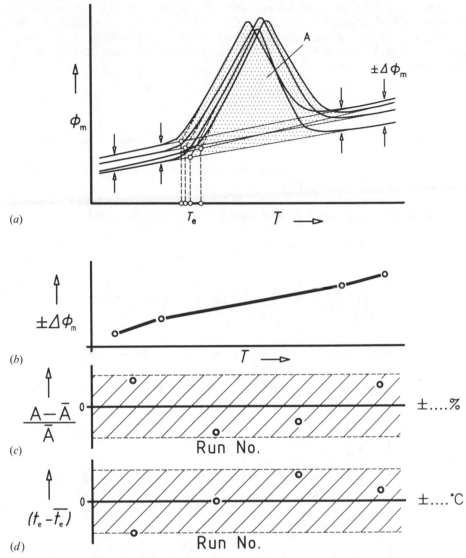

Fig. 2.14. Repeatability: (*a*) superimposed measured curves; (*b*) repeatability of the base line; (*c*) repeatability of the peak area A; (*d*) repeatability of the extrapolated peak-onset temperature t_e (*T*, *t* = temperature, $\Delta\phi_m$ = maximum deviation of ϕ_m from the average base line, $\pm\Delta\phi_m$ = base-line repeatability, $\bar{A}$ = average peak area, $(A - \bar{A})/\bar{A}$ = repeatability of the peak area (in %), $\bar{t}_e$ = average extrapolated peak onset temperature, $(t_e - \bar{t}_e)$ = repeatability of the extrapolated peak onset temperature (in °C))

ity (Fig. 2.14(b)) and the repeatability of the extrapolated onset temperature T_e and the peak area (Figs. 2.14(c) and (d)). If possible, increase the amplification at some temperatures during scanning to get the scanning noise at a given scanning rate, which gives an impression of the signal-to-noise ratio.

The base-line repeatability helps to assess the uncertainty of the shape of the base line during the peak and, as a result, the uncertainty for peak areas.

37

The repeatability of the peak area (taking the base-line uncertainty into account) gives the calibration uncertainty if the DSC is calibrated by means of heats of fusion (in the case of heat-flow rate calibration the base-line repeatability gives the calibration uncertainty). The calibration uncertainty is the smallest uncertainty for heat (heat-flow rate) measurement.

2.4.2.3 The scanning and isothermal noise

The *noise* (Fig. 2.15) of the measured signal (in µW) is indicated in different ways.
– Peak-to-peak noise (pp): maximum fluctuation of the measured signal around the mean signal value.
– Peak noise (p): maximum deviation of the measured signal from the mean signal value; p = 0.5 pp is usually valid.
– Root-mean-square noise (RMS): root of the mean value of the squared instantaneous deviations of the measured signal from the mean signal value. For a

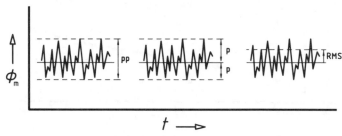

Fig. 2.15. Noise: pp = peak-to-peak noise, p = peak noise, RMS = root-mean-square noise

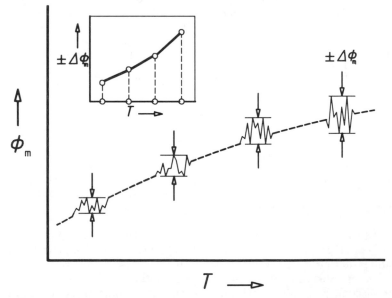

Fig. 2.16. Scanning noise: $\pm \Delta\phi_m$ peak-to-peak scanning noise (pp)

38

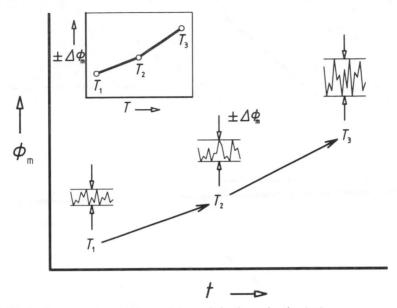

Fig. 2.17. Isothermal noise: $\pm \Delta \phi_m$ peak-to-peak isothermal noise (pp)

sine-shaped measured signal, RMS noise $= 0.35$ pp noise; for a statistical random signal, RMS noise ≈ 0.25 pp noise is valid.

The noise of the DSC depends on the heating rate, the temperature and other parameters (e.g. atmosphere). The signal-to-noise ratio is decisive for the smallest heat-flow rate detectable ('heat-flow rate resolution' of the DSC). This heat-flow rate amounts to about two to five times the noise.

It is often important to know the sample volume- or sample mass-related noise (in $\mu W/cm^3$ or $\mu W/g$) which allows the smallest detectable heat-flow rate to be estimated for a given sample.

To measure the noise, run a curve with high amplification to obtain the scanning noise (Fig. 2.16), e.g. during the zero-line or base-line tests. Then measure the noise at a few selected temperatures in thermal equilibrium, starting with the lowest and ending with the highest temperature stated by the manufacturer in the technical data (Fig. 2.17).

The isothermal noise should be the smallest noise possible (compare the scanning noise of the zero line). It gives an impression of the disturbances from the environment to which the sensors are subject (fluctuation of heat transfer, of furnace temperature) and of the noise of the electronics. It determines the maximum possible signal-to-noise ratio.

2.4.2.4 The linearity

The *linearity* (Figs. 2.18(a)–(c)) characterizes the functional relation between the measured signal ϕ_m and the true reaction heat flux ϕ_r determined on the basis of the calibration tests. If this relation is linear ($\phi_r = K\phi_m$), the behavior of the measuring system is 'ideally linear' (Fig. 2.18(a)). If the proportionality factor K depends—as usual—on parameters, the ratio ϕ_r/ϕ_m is preferably represented graphically as a

39

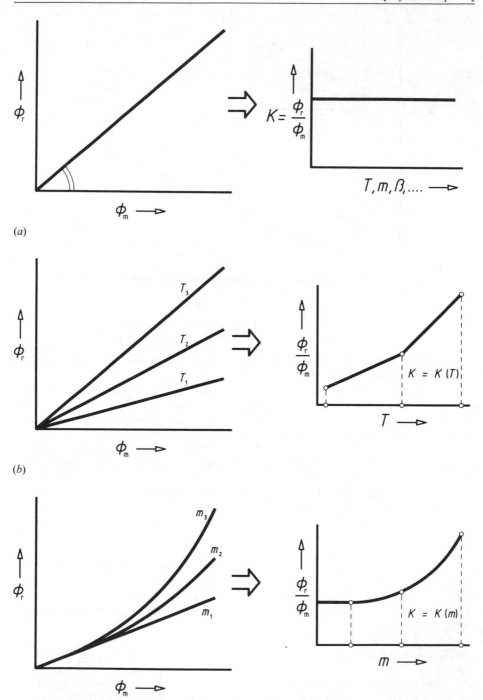

Fig. 2.18. Linearity: (*a*) ideal linearity; (*b*) temperature dependence of K; (*c*) mass dependence of K (β = heating rate)

function of the parameter of interest (temperature, mass, etc.) in order to detect the deviations from linearity and to take them into account in the evaluation of the measurement results (Figs. 2.18(b) and (c)). In some DSCs, the temperature dependence of K is 'linearized' electronically or by means of the software so that $K = $ constant can apparently always be expected. This forced linearity includes the uncertainties of the calibrations used as a basis and the uncertainties of interpolation. It should therefore always be critically examined.

To check the linearity with respect to mass (and/or heat), run indium samples with decreasing mass m (and the appropriate reference) under the same conditions (Fig. 2.19). Evaluate the peak areas A. The scatter of the extrapolated onset temperature T_e can also be checked. With decreasing mass, the minimum heat of fusion that can be detected with reasonable accuracy can be assessed from the noise and the base-line uncertainty. The linearity can be tested with regard to the scanning rate, the sample position, the surface-to-volume ratio, etc.

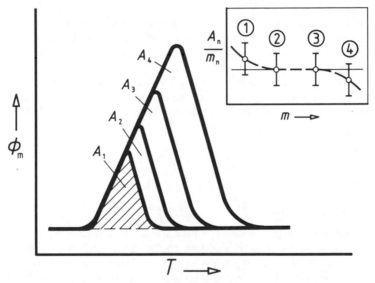

Fig. 2.19. Linearity with regard to mass: A_1 to $A_4 = $ peak areas belonging to sample masses m_1 to m_4 ($m_1 < m_2 < m_3 < m_4$); A_n/m_n with uncertainty range for area evaluation, – – – deviation from ideal linearity

The linearity of ϕ_r or Q_r (true heat-flow rate or heat) with respect to ϕ_m or A (measured heat-flow rate or peak area representing a measured heat) is tested by means of the calibration procedure (cf. chapter 3) which yields the calibration function $K_\phi(T)$ with the parameter T (temperature):

$$\phi_r = K_\phi(T)\phi_m \qquad (\text{or } Q_r = K_Q(T)A_{\text{peak}})$$

2.4.2.5 The time constant

The *time constant* τ of the measuring system (Fig. 2.20) is the time interval in which—after an abrupt change of the measurand (heat-flow rate)—the measured signal ϕ_0 drops down to the eth part (26.8 %) of the original value. At a specific

41

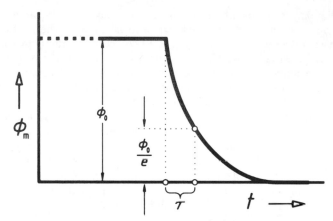

Fig. 2.20. Time constant τ (electric heating): ϕ_0 = steady-state signal

heating rate, the time course of two successive events (peaks) are recorded in a clearly separate way or in the form of a 'summation curve': resolution with time. Knowledge of the time constant is a prerequisite for the desmearing or deconvolution of the measured curve.

How can the time constant be measured? When an electric calibration heater is installed, a constant heat-flow rate ϕ_{el} is switched on until a constant measured signal ϕ_0 is obtained; after heater switch-off, the descending curve is evaluated (Fig. 2.21(a)). The following relations are valid:

$$\phi_m(t) = \phi_0 \exp(-t/\tau)$$

$$d\phi_m/dt = \phi_m/\tau$$

hence

$$\tau = -\phi_m(t^*)/(d\phi_m(t^*)/dt)$$

There is not just one time constant ruling the slope of the curve, in fact there are many, but it is good enough to evaluate the greatest and most important time constant τ_{max}, e.g. by constructing a few tangents at the descending curve (this also works with the ascending part). When the interval Δt (between t^* and the tangent–abscissa intersection) becomes constant, τ_{max} is obtained, which is the most important factor governing the signal transmittance from the sample to the sensor in the most influential way.

When no calibration heater can be installed, the descending section of the curve following the peak maximum when a pure substance is melted is evaluated by the 'tangent method' according to the procedure described above (Fig. 2.21(b)). Towards the end of the descending curve, a constant value results: the (greatest) time constant τ_{max} which gives a good approximation of the thermal inertia of the measuring system.

2.4.3 *The quality of a measurement*

The measurement results are taken from the measured curves. Their quality must be evaluated before they are interpreted and published. This evaluation is made on the basis of the performance data of the measuring system. The following terms are used to describe the quality of measurement results.

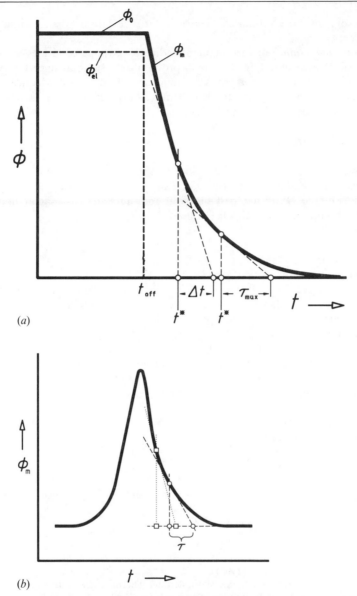

(a)

(b)

Fig. 2.21. Time constant τ: (a) Electric heating; (b) heat-of-fusion peak (t_{off} = switch off electric heating, t^* = time belonging to the tangent, Δt = time interval between t^* and the intersection of tangent and abscissa, ϕ_{el} = electric power (Joule's heat), ϕ_0 = constant measured signal, τ_{max} = maximum Δt interval)

The *accuracy* describes how well the result of a measurement approximates the 'true' value. The accuracy is determined by random and systematic errors of measurement. With the conditions of measurement unchanged, systematic errors are constant; they may comprise known and unknown fractions.

Known systematic errors (deviations) are allowed for in the measurement result as *corrections*. The uncertainty of these corrections, unknown systematic errors and random errors must be estimated. The standard deviation is an estimate of the random errors.

The *total uncertainty* of measurement indicates a range of values in which the true value of the measurand lies with high probability. The total uncertainty is often referred to as accuracy (more precisely, inaccuracy); it is not identical with the repeatability or the reproducibility (which may, however, serve as a basis for the estimation of unknown systematic errors).

2.4.4 Checklist for DSCs

In the characterization of a DSC, the following checklist (according to ref. 1) may be of assistance. The manufacturer of a DSC can provide you with the essential data. The selection of the DSC that best fits your requirements (a difficult undertaking indeed) may become easier if this list is used.

- Manufacturer:
- Type:
- Sample volume: mm^3
- Atmosphere (vacuum?, which gases?):
- Temperature range: from ... to ... °C (or K, here and hereinafter)
- Scanning rates: from ... to ... °C/min; steps: ... °C/min
- Zero-line repeatability: from ± ... mW (at ... °C) to ± ... mW (at ... °C)
- Peak-area repeatability: ± ... % (at ... °C)
- Total uncertainty for heat: ± ... % (at ... °C)
- Extrapolated peak-onset temperature repeatability: ± ... °C (at ... °C)
- Total uncertainty for temperature: ± ... °C (at ... °C)
- Scanning noise (pp) at ... °C/min: from ± ... mW (at ... °C) to ± ... mW (at ... °C)
- Isothermal noise (pp):   from ± ... mW (at ... °C) to ± ... mW (at ... °C)
- Time constant with sample: ... s

References

1 *Hemminger, W., Höhne, G.:* Calorimetry—Fundamentals and Practice. Verlag Chemie, Weinheim, Deerfield Beach (FL), Basel, 1984
2 *Bunsen, R. W.:* Ann. Phys. 141, p. 1 (1870)
3 *Ginnings, D. C., Ball, A. F., Douglas, T. B.:* J. Res. Natl. Bur. Stand. 45, p. 23 (1950)
4 *Brönsted, J. N.:* Z. Phys. Chem. 56, p. 645 (1906)
5 *Watson, E. S., O'Neill, M. J., Justin, J., Brenner, N.:* Anal. Chem. 36, p. 1233 (1964)
6 *Eucken, A.:* Phys. Z. 10, p. 586 (1909)
7 *Hemminger, W. F., Cammenga, H. K.:* Methoden der Thermischen Analyse. Springer-Verlag, Berlin, Heidelberg, New York, London, Paris, Tokyo, 1989
8 *Calvet, E.:* C.R. Hebd. Seances Acad. Sci. 226, p. 1702 (1948)
9 *Petit, J. L., Sicard, L., Eyraud, L.:* C.R. Hebd. Seances Acad. Sci. 252, p. 1740 (1961)

Nomenclature

A	peak area
C	heat capacity
K	calibration factor
m	mass
P	electric power
Q	heat
t	time
T	temperature
W	electrically generated heat
Δ	difference
τ	time constant
φ	heat-flow rate

Subscripts

bl	base line
e	extrapolated
el	electric
f	final
F	furnace
i	initial
m	measured
max	maximum
p	peak
Q	due to heat
r	real, true
R	reference sample
S	Sample
φ	due to heat flow
0	constant, steady state

Chapter 3

Fundamentals of Differential Scanning Calorimetry and Differential Thermal Analysis

Günther W. H. Höhne, University of Ulm, Department of Calorimetry, P.O. Box 4066, D-89069 Ulm, Germany

Contents

3.1 Theoretical fundamentals of DTA and heat-flux DSC

In differential thermal analysis (DTA) and differential scanning calorimetry (DSC) the occurrence of a temperature difference ΔT between the sample and the reference is the primary effect due to a thermal event Φ_r within the sample. In the case of DTA this ΔT is detected with a differential thermocouple and the resulting electrical signal is plotted as a function of time. For most heat-flux DSCs the primary temperature difference likewise leads to an electrical signal which is internally converted, being proportional to a heat-flow rate signal Φ_m which is plotted against temperature.

In every case a ΔT signal is the physically relevant result of a thermal event inside the sample. Therefore, in the following ΔT is always assumed to be the measured signal, i.e. we seek the relation between the real heat-flow rate Φ_r of the sample event and the signal ΔT of the apparatus. This is important for the

- time-related assignment of Φ_r to ΔT (investigation into the kinetics of a reaction)
- determination of partial heats of reaction
- evaluation and assessment of the influence of operating parameters and properties of the measuring system on this relation
- the estimation of the overall uncertainty of measurement.

The relation between Φ_r and ΔT (or Φ_m) can be derived in varying degrees of approximation to real DSCs. Analytical solutions are possible only for simple boundary and initial conditions and for quasi-steady states. Numerical procedures and solutions can approximate the actual conditions more exactly, though less clearly.

The analytical description of the heat transport phenomena in question is presented below in three steps. The results of a numerical simulation are then discussed.

The relevant models are always plotted both in reality and as an equivalent electric circuit. The latter diagram allows better understanding of the interrelations. From a physical point of view, charge transport and heat transport are equivalent processes, and many people find it easier to read electric circuit diagrams than to visualize heat fluxes in real equipment. In the appendix to this chapter the analogies in question are given in the form of a translation table.

The DTA apparatus resembles the heat-flux DSC in some basic principles of heat transport (for classification see chapter 2). Thus the theory is equivalent, and the conclusions drawn below for DSC are also valid for DTA.

3.1.1 The zeroth approximation

Heat-flux DSC (and DTA) is represented by a rather simple linear model (Fig. 3.1) and its equivalent electric circuit. The following simplifications have been made in the model of the zeroth approximation:
- steady-state (constant) heat flow
- only the thermal resistance between furnace and sample is taken into account (i.e. no thermal interaction between sample and reference sample)
- except for the sample and reference, no heat capacities are taken into account
- no difference between sample and probe temperatures
- no heat exchange with the surroundings other than conduction through the support.

Biot–Fourier's law of heat conduction is

48

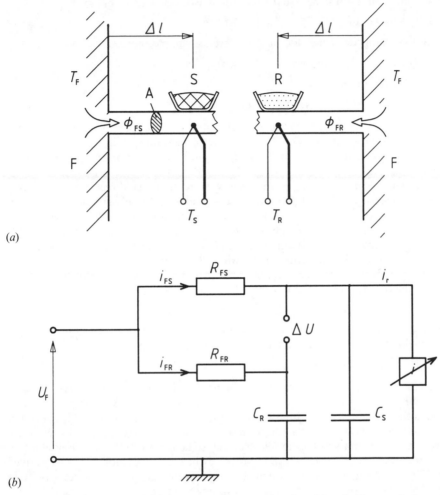

Fig. 3.1. Model of a heat-flux DSC in a zeroth approximation: (*a*) real; (*b*) as equivalent electric circuit

$$\Phi/A = -\lambda \operatorname{grad} T \qquad \text{(or in absolute values, } |\Phi|/A = \lambda|\operatorname{grad} T|) \tag{3.1}$$

The magnitude of the heat flux Φ/A through a certain cross-section A is proportional to the gradient of the temperature there; the thermal conductivity λ is the proportionality factor.

For steady-state conditions in the one-dimensional model referred to above, this equation is reduced as follows for the left-hand and right-hand sub-system $(T_F > T_S, T_R)$:

$$\Phi_{FS}/A = \lambda(T_F - T_S)/\Delta l \quad \text{and} \quad \Phi_{FR}/A = \lambda(T_F - T_R)/\Delta l \tag{3.2}$$

In case of thermal symmetry, $\Phi_{FS} = \Phi_{FR}$ and $T_S = T_R$. When a constant (exothermic) heat-flow rate Φ_r is produced in the sample, T_S increases by ΔT_S, and the temperature difference $T_F - T_S$ and thus the heat-flow rate Φ_{FS} decreases. When the steady state is

49

reached again, $\Delta\Phi_{FS}$ must be equal to Φ_r for reasons of energy balance:

$$\Delta\Phi_{FS} = \Phi_r = -A\lambda\,\Delta T_S/\Delta l$$

Nothing has changed on the side of the reference sample, hence:

$$\Delta T_S = \Delta T_{SR} = T_S - T_R \quad \text{and} \quad \Phi_r = \Delta\Phi_{SR} = \Phi_{FS} - \Phi_{FR}$$

Consequently (from Eq. 3.2):

$$\Phi_r = -A\lambda(T_S - T_R)/\Delta l = -A\lambda\,\Delta T_{SR}/\Delta l = -K\,\Delta T \tag{3.3}$$

$\Delta T = \Delta T_{SR}$ is the measured signal; by definition it reads positive for exothermic effects. The heat-flow rate Φ_r is defined positive for heat flowing into a system; this is the case for endothermic events in the sample. Thus we have opposite signs between Φ_r and ΔT signals.

Equation 3.3 means that in the steady state there is direct proportionality between the measured Φ_r and the measured signal ΔT. In this simple model, K is given completely by the properties of the heat conduction path between the furnace and the samples. A similar case can be achieved when in scanning operation sample and reference show different temperature-independent 'heat capacities'. A greater heat flow will always flow into the sample whose heat capacity is higher, in order that the steady-state heating rate be maintained. In Eq. 3.3 the reaction heat-flow rate Φ_r must be replaced by the differential heat-flow rate

$$\Phi_S - \Phi_R = \Delta\Phi_{SR} = -K\,\Delta T_{SR}$$

For steady-state heat fluxes into sample and reference $\Phi_S = \beta C_S$ and $\Phi_R = \beta C_R$ is valid, β being the heating rate and C the heat capacity, hence

$$\Delta\Phi_{SR} = \beta(C_S - C_R) \quad \text{and thus} \quad \beta(C_S - C_R) = -K\,\Delta T \tag{3.4}$$

As can be seen, a positive difference of heat capacities $C_S - C_R$ gives rise to a negative ΔT signal, i.e. the curve measured with empty crucibles ΔT_0 (empty pan line) lies above the curve plotted with a sample in the sample pan (Fig. 3.2). This is rather unusual for an experimenter thinking in heat-flow terms, who expects a lower signal if there is a smaller mass in the sample crucible. (The former ICTA recommendation to plot positive signals for exothermic processes in heat-flux calorimeters is derived from DTA tradition, but seems not to be very ingenious for calorimeters built to measure heat-flow rates. Nevertheless, it is widely used.)

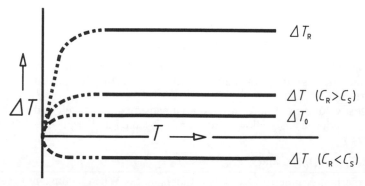

Fig. 3.2. DSC curves from measurements necessary to determine an unknown heat capacity (see text for explanation)

Equation 3.4 describes the shape of the base line (i.e. the steady-state heat-flow curve without thermal effects in the sample), and is the basic equation for determining the heat capacity C_S. For the empty reference crucible ($C_R = 0$), we get

$$C_S = -K\,\Delta T/\beta$$

or with a known heat capacity C_R:

$$C_S = C_R - K\,\Delta T/\beta$$

In practice three measurements are necessary to determine an unknown heat capacity (Fig. 3.2).
 (i) Measurement I: empty crucibles, leading to empty pan line ΔT_0 (equal to zero in case of total symmetry).
(ii) Measurement II: reference substance of known heat capacity C_R in reference crucible, other crucible empty, leading to the measured curve ΔT_R. This is a calibration measurement since the calibration factor can be calculated with the aid of Eq. 3.4:

$$K = C_R\,\beta/(\Delta T_R - \Delta T_0)$$

(iii) Measurement III: sample in sample crucible, reference in reference crucible as before, leading to the measured curve ΔT. From Eq. 3.4:

$$C_S = C_R - K(\Delta T - \Delta T_0)/\beta$$

With K from calibration measurement II, the unknown heat capacity can be calculated:

$$C_S = C_R - C_R(\Delta T - \Delta T_0)/(\Delta T_R - \Delta T_0)$$

From Eq. 3.3 it follows that the calibration factor essentially depends on the thermal conductivity of the path from the furnace to the sample. For a well-defined calibration factor K it is necessary that the sample itself does not affect this thermal conductivity. Already in this rough approximation we can draw the conclusion that any equipment with temperature probes inside the sample (DTA) is appropriate for quantitative heat-flow measurements.

3.1.2 The first approximation

In the first approximation, non-steady-state processes in the sample are also permitted, which are manifested as 'peaks' of the measured curve. This means that ΔT is not constant. For the rest, the same simplified arrangement as for the zeroth approximation is used here (Fig. 3.1).

Φ_{FS} is the heat-flow rate from the furnace to the sample; $\Phi_r(t)$ is the heat-flow rate additionally released/adsorbed (from reaction or transition; exothermic: Φ_r negative) by the sample. The following balance for the heat-flow rates is then valid for the sample with the heat capacity C_S:

$$C_S(\mathrm{d}T_S/\mathrm{d}t) = \Phi_{FS} - \Phi_r$$

With $\Delta T = T_S - T_R$ we get

$$C_S(\mathrm{d}T_R/\mathrm{d}t) + C_S(\mathrm{d}\Delta T/\mathrm{d}t) = \Phi_{FS} - \Phi_r$$

Accordingly, for the reference sample ($\Phi_r = 0$)

$$C_R(\mathrm{d}T_R/\mathrm{d}t) = \Phi_{FR}$$

51

By subtraction between these two balance equations

$$\Phi_{FS} - \Phi_{FR} = (C_S - C_R)(dT_R/dt) + C_S(d\Delta T/dt) + \Phi_r$$

If the reaction heat-flow rate does not change the quasi-steady state, then for the heat-flow rates Φ_{FS} and Φ_{FR}

$$\Phi_{FS} = (T_F - T_S)/R_{FS} \quad \text{and} \quad \Phi_{FR} = (T_F - T_R)/R_{FR}$$

where R_{FS} and R_{FR} are global heat resistances between the furnace and the samples. In the case of thermal symmetry, $R_{FS} = R_{FR} = R$, thus

$$\Phi_{FS} - \Phi_{FR} = -(T_S - T_R)/R = -\Delta T/R$$

$$\Phi_r = -\Delta T/R - (C_S - C_R)(dT_R/dt) - C_S(d\Delta T/dt)$$

This equation links the reaction heat-flow rate Φ_r sought with the measured signal ΔT. The second term takes the asymmetry of the measuring system into account as regards the heat capacities of the sample and reference sample. The third term considers the contribution of the thermal inactivity of the system when a measured signal $\Delta T(t)$ appears. Analogous to the charging or discharging of a capacitor of capacity C, a time constant τ can also be defined for the heat flows and heat capacities: $\tau = C_S R$.

When ΔT is changed, R is the effective thermal resistance to the 'charging or discharging' of the 'capacity' C_S. With this resistance, with $dT_R/dt = \beta$ (heating rate of reference sample) and with the time constant τ defined above:

$$\Phi_r(t) = -\Delta T/R - \beta(C_S - C_R) - (\tau/R)(d\Delta T/dt) \tag{3.5}$$

The measured signal ΔT is not proportional to the heat-flow rate Φ_r at a given moment, but is delayed with time and thus also distorted ('smeared'). Furthermore, when $C_S \neq C_R$ is valid, the measured signal ΔT is not equal to zero—even if Φ_r is equal to zero and steady-state conditions prevail—but has the value $R\beta(C_R - C_S)$ which is the initial deviation after the quasi-steady state has been reached in scanning operation. This contribution is the base line (before/behind a peak) which is parallel to the abscissa if R, ΔC and β are constant (Fig. 3.3). In reality, the term $R\beta(C_R - C_S)$ is not constant but reflects the temperature dependence of R (in general,

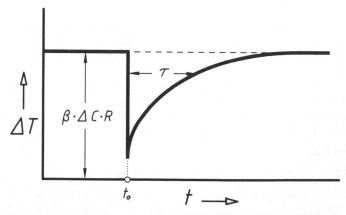

Fig. 3.3. DSC curve (first approximation) for a sudden endothermic event at the moment t_0: signal deflection returns to the base line with a time constant τ

of the heat transition conditions), of ΔC and of β, leading to a temperature dependence of the base line (base-line slope).

If the reaction heat-flow rate Φ_r at a given moment has to be deduced from the measured signal ΔT, the third term in Eq. 3.5 must be taken into account ('desmearing'). For this purpose both R and τ must be obtained from calibration measurements.

On the other hand, the signal of the calorimeter ΔT can be calculated for a given Φ_r from Eq. 3.5. The solution of this differential equation for a heat pulse Φ_r at moment t_0 has the form

$$\Delta T = k_1 \exp(-t/\tau) + K_2$$

Up to the moment that $t = t_0$, the solution is the steady-state base line:

$$\Delta T = \beta(C_R - C_S)R = \beta\,\Delta CR$$

For $t \to \infty$, the function $\Delta T(t)$ returns to this base line (Fig. 3.3).

To get the whole heat of reaction or transition Q_r developed/consumed in the sample, the balance

$$Q_r = \int_{t_1}^{t_2} \Phi_r(t)\,dt$$

can be used, where t_1 and t_2 are the beginning and end, respectively, of the reaction/transition.

From substitution of Eq. 3.5:

$$Q_r = -\frac{1}{R}\left[\int_{t_1}^{t_2} \Delta T(t)\,dt - \int_{t_1}^{t_2} R\,\Delta C\beta\,dt\right] - \frac{1}{R}\int_{t_1}^{t_2} \tau\,\frac{d\Delta T}{dt}\,dt$$

The square brackets correspond to area 1, i.e. the so-called 'peak area' between the measured curve and the (interpolated) base line in Fig. 3.4. As can be seen, the reaction heat Q_r is not proportional to this area, which leads to the following remarks.

– When ΔC and R are not temperature- (or time-) dependent, the base line is parallel to the abscissa. In this case, the slope $d\Delta T/dt$ is zero before and behind the peak. The contribution of the third term vanishes when integration is carried out over the whole peak.

– In the real case, the second term of Eq. 3.5 is not constant, but depends on temperature, i.e. the base line is not parallel to the abscissa. In this case, the third term will not vanish but represents a correction of the values obtained by peak integration.

– For the partial integration (Fig. 3.5) of the peak between t_1 and t^*, the contribution of the third term at the point t^* must be taken into account:

$$Q_r(t^*) = -\frac{1}{R}\left[\underbrace{\int_{t_1}^{t^*} \Delta T(t)\,dt}_{} - \underbrace{\int_{t_1}^{t^*} R\,\Delta C\beta\,dt}_{}\right] - \underbrace{\frac{1}{R}\int_{t_1}^{t^*} \tau\,\frac{d\Delta T}{dt}\,dt}_{}$$

$$\underbrace{① + ② - \qquad ②}_{}\qquad \text{correction}$$
$$\underbrace{\qquad\qquad\qquad}_{}\qquad\qquad \text{term}$$
$$①\quad \text{partial peak area}$$

Partial integrations of the peak are required for kinetic investigations and for determination of the purity.

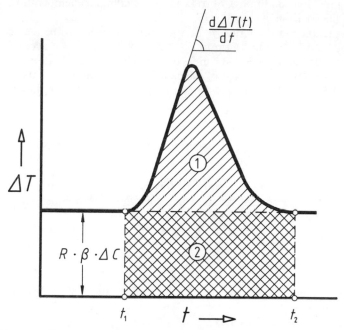

Fig. 3.4. DSC curve of an exothermic reaction: determination of peak area and onset slope (see text for explanation)

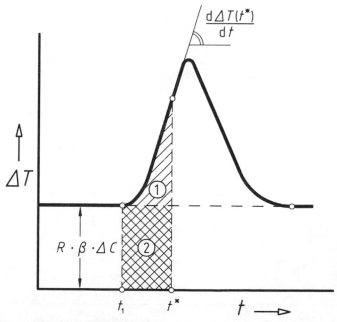

Fig. 3.5. DSC curve of an exothermic reaction: evaluation of partial peak areas (see text for explanation)

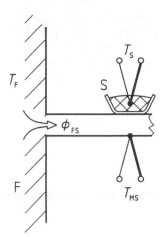

Fig. 3.6. Model of the 'thermometer problem': sample temperature and probe temperature are separated by a thermal resistance

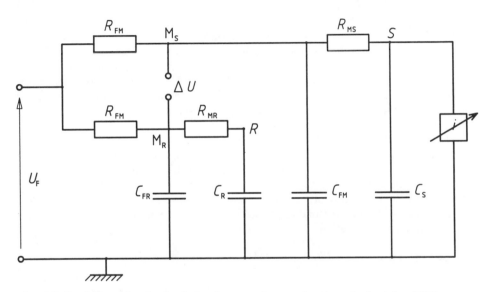

Fig. 3.7. Equivalent electric circuit for the second approximation of a heat-flux DSC

3.1.3 Higher-order approximations

In reality, temperatures of the sample and reference sample (assumed here to be homogeneous) are not measured directly. There is a thermal resistance between the temperature measurement points and the sample or reference sample (Fig. 3.6). Depending on the design of the measuring system, the resistance is made up of several parts, differing in magnitude and originating in the transition layer between sample, bottom of the crucible and support: further resistances exist between support and temperature sensor. When the sample temperature changes, the temperature measurement point reacts after a delay. The analog electric circuit diagram for this second approximation is shown in Fig. 3.7. For this so-called 'thermometer problem':

$$T_{MS} = T_S - \tau_1 \, dT_{MS}/dt$$

where T_{MS} is the temperature of the measurement point (e.g. junction of the sample thermocouple), T_S is the (homogeneous) sample temperature and τ_1 is the characteristic time constant for the temperature balance between sample and temperature measurement point.

By analogy, for the reference side:

$$T_{MR} = T_R - \tau_1 \, dT_{MR}/dt$$

(τ_1 is assumed to be equal for both sides). From these equations, the following results for the difference:

$$\Delta T_{SR} = T_S - T_R = T_{MS} - T_{MR} + \tau_1(dT_{MS}/dt - dT_{MR}/dt)$$
$$= \Delta T_M + \tau_1 \, d\Delta T_M/dt$$

When ΔT_{SR} is used instead of ΔT, the following relation results, analogously to the mathematical procedure leading to Eq. 3.5 of the first approximation:

$$\Phi_r(t) = -\frac{1}{R}\left[\Delta T_M - R(C_R - C_S)\beta + \tau\frac{d\Delta T_M}{dt} + \tau\tau_1\frac{d^2\Delta T_M}{dt^2}\right] \tag{3.6}$$

where $\Delta T_M = T_{MS} - T_{MR}$, the measured temperature difference.

In addition to the first derivative (slope) of the measured curve $\Delta T_M(t)$, the second derivative (curvature) must be used in the second approximation for the reconstruction of the heat flow converted in the sample. Two time constants occur, which must be determined. The second time constant is determined by the thermal resistance and the 'effective heat capacity' between sample and temperature measurement point.

The above-described approximation can be refined as desired. In order that the temperature gradient inside the sample and its influence on the peak shape can be determined, the sample can be considered as having been split into different layers which are connected with one another by heat-conducting boundary layers. An additional differential quotient in the differential equation and another time constant result for each additional thermal resistance in connection with the heat capacity, and we get a differential equation of nth order:

$$\Phi_r(t) = -K\left[\Delta T + K_0 + K_1\frac{d\Delta T}{dt} + K_2\frac{d^2\Delta T}{dt^2} + \cdots + K_n\frac{d^n\Delta T}{dt^n}\right]$$

In reality, the constants K_i are terms into which the thermal resistances and capacities (and thus the time constants) of the arrangement enter. Calculation of the reaction heat-flow rate Φ_r assumes that all K_i and the measured signal ΔT and its time derivatives are known. It can be shown [1] that for practical applications, the second order differential equation is a sufficiently good approximation to calculate the true (desmeared) reaction heat-flow rate; the time constants τ_1 and τ_2 must be known for this purpose.

So far, we have considered only the simplified case where it is assumed that no heat exchange takes place between sample and reference sample. This simplification is certainly not permissible for disk-type measuring systems. Figure 3.8(a) is a more realistic representation of the measuring system of such a calorimeter. When the differential equation is to be set up for this system, it is convenient to use the electric circuit (represented in Fig. 3.8(b)) for the second approximation. In order to form the desired differential equation, according to Kirchhoff's laws, the voltage and current balance are made up for each loop and each node of the electric circuit. For

56

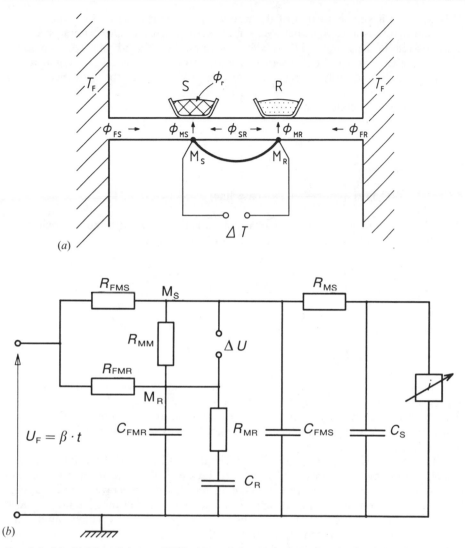

Fig. 3.8. Model of a disk-type DSC: (*a*) and the equivalent electric circuit; (*b*) in a second approximation

Fig. 3.8(b), five equations each for voltages and currents are then obtained. On the basis of the laws of the electric circuits, these ten equations are combined to form one differential equation and retranslated into the language of heat transport. The result for a symmetrical twin design ($R_{FMS} = R_{FMR} = R$, $C_{FMS} = C_{FMR} = C$) is as follows [2]:

$$\Phi_r = -\left(\frac{1}{R} + \frac{2}{R_{MM}}\right)\Delta T - \left\{C + C_S\left[1 + R_{MS}\left(\frac{1}{R} + \frac{2}{R_{MM}}\right)\right]\right\}\frac{d\Delta T}{dt} - R_{MS}CC_S\frac{d^2\Delta T}{dt^2}$$

$$+ (C_R - C_S)\frac{dT_R}{dt} + C_S C_R (R_{MS} - R_{MR})\frac{d^2 T_R}{dt^2} \tag{3.7}$$

57

. This equation is similar to that of the second approximation (Eq. 3.6), but as there is the thermal resistance R_{MM} between sample system and reference sample system, a thermal event in the sample will also affect the reference side and thus T_R. Only in the steady-state case at a sufficiently great distance from peaks is dT_R/dt equal to the heating rate β, and only there is the second derivative of the reference sample temperature equal to zero.

For the steady-state case, with

$$\Phi_r = 0, \quad \frac{d\Delta T}{dt} = 0, \quad \frac{d^2\Delta T}{dt^2} = 0$$

the following is then valid:

$$\Delta T_{st} = (C_R - C_S)\beta \left/ \left(\frac{1}{R} + \frac{2}{R_{MM}}\right)\right. \qquad \text{(base line)}$$

When Eq. 3.7 is integrated over the area of a transition peak, with the approximation $dT_R/dt \approx \beta$, and Q_S is the heat of transition of the sample

$$\int_{t_1}^{t_2} \Phi(t)_r \, dt = Q_S = -\left(\frac{1}{R} + \frac{2}{R_{MM}}\right)\left(\int_{t_1}^{t_2} \Delta T \, dt - \int_{t_1}^{t_2} \frac{C_R - C_S}{\frac{1}{R} + \frac{2}{R_{MM}}} \beta \, dt\right)$$

$$= -\left(\frac{1}{R} + \frac{2}{R_{MM}}\right)\int_{t_1}^{t_2} (\Delta T - \Delta T_{st}) \, dt \qquad (3.8)$$

This integral describes the peak area between measured curve and base line. The approximation is better, as the last two additive terms in Eq. 3.7 are smaller. From this, the rule is derived that when heats of transition are determined with heat-flux DSCs, the sample and the reference sample should be as similar as possible ($C_R \approx C_S$, $R_{MR} \approx R_{MS}$).

The factor $(1/R + 2/R_{MM})$ is decisive for the sensitivity of the calorimeter; the greater the thermal resistance of the disk, the greater is the peak for a given heat of transition. However, as a result, the time constant τ increases as well, i.e. the system becomes more inert. From this factor, it can also be concluded that the ratio of R to R_{MM} plays an important role for the sensitivity. Depending on whether sample and reference sample are arranged on the disk, R (thermal resistance between furnace and sample location) and R_{MM} (thermal resistance between sample and reference sample) change: this means that high reproducibility of the location of the sample and reference sample is of great importance for the calibratibility of a DSC.

In the approximation $C_S \approx C_R$ and $R_{MS} \approx R_{MR}$, Eq. 3.7 changes into the second order differential equation

$$\Phi_r(t) = K \Delta T + K_1 \frac{d\Delta T}{dt} + K_2 \frac{d^2\Delta T}{dt^2}$$

The solution of this equation (i.e. the curve $\Delta T(t)$ recorded by the calorimeter), for a heat pulse Φ_r at $t = 0$ is the superposition of two exponential functions:

$$\Delta T(t) = k_1 \exp(-t/\tau_1) + k_2 \exp(-t/\tau_2)$$

with the time constants τ_1 and τ_2 depending in a complicated way on the coefficients K, K_1 and K_2. Here, τ_1 is determined in approximation by $C_S R_{MS}$ and τ_2 by CR. Such

58

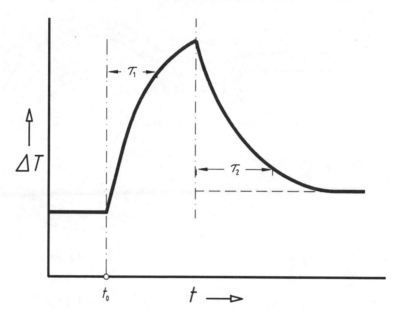

Fig. 3.9. Measured curve after a sudden exothermic event at moment t_0 in a heat-flux DSC (τ_1 and τ_2 are different time constants)

a peak, produced by a heat pulse, is shown in Fig. 3.9. The time constant τ_1 essentially determines the ascending branch, and τ_2 the descending branch.

3.1.4 Numerical simulations

For calculations that also allow for heat exchange by convection and radiation and include non-linearities due to temperature-dependent thermal resistances and heat capacities, it is not advisable to set up differential equations. In this case, numerical simulation by the finite-element method should be applied. Appropriate computer programs are commercially available for personal computers. With the required time and effort, the temperature and heat flow fields and the measured $\Delta T(t)$ curves can be simulated for any complicated arrangement and any thermal process inside the sample. The finite-element method consists of splitting the whole arrangement into sufficiently small 'cells' for which the heat flows and the material properties are defined. On the basis of considerations with regard to the balance, a system of equations is obtained which is solved by conventional methods. In the following, this method is applied to a very simple model of the disk-type measuring system. For the purpose of this model consideration, the disk-type measuring system is divided into a number of concentric rings (Fig. 3.10). For reasons of symmetry, the heat flow from the furnace to the interior will be strictly radial. In first approximation, a double sector with sample and reference sample can be considered as a linear chain of cells of different size, heat conduction between which is ensured by the common boundary. The heat flow will have non-radial components only in the vicinity of reference sample and sample, which, however, will be neglected in this approximation.

The heat capacity of a cell is:

$$C_i = V_i \varrho_i c_i$$

59

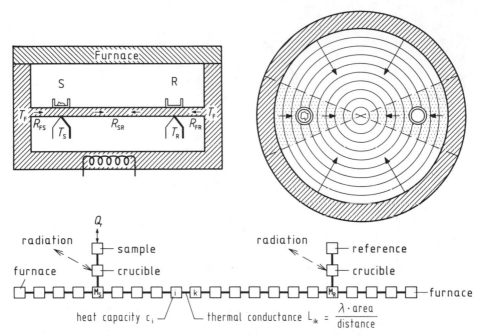

Fig. 3.10. Model of a disk-type DSC and a quasi-linear approximation for finite element calculations: heat flow is marked by arrows

where V_i = volume, ϱ_i = density and c_i = specific heat capacity of cell i. The reciprocal thermal resistance (thermal conduction coefficient) $L = \Phi/\Delta T$ between two cells is given by

$$L_{ik} = A_{ik}\lambda_{ik}/l_{ik}$$

where A_{ik} is the common boundary area, λ_{ik} is the thermal conductivity and l_{ik} is the distance between cells i and k.

The energy balance for the cells yields

$$\sum_k L_{ik}(T_i - T_k)\,\Delta t = C_i(T_i' - T_i) + Q_{\text{reaction}}^{\Delta t} + Q_{\text{radiation}}^{\Delta t} + Q_{\text{convection}}^{\Delta t} \tag{3.9}$$

where $i = 1, 2, 3, \ldots, n$; k = all neighbors of i; n is the number of cells and $Q^{\Delta t}$ is the heat exchanged per time step Δt.

This is a system of recursive equations that allows the temperatures T_i' of all cells at the moment $t + \Delta t$ to be calculated from the temperatures T_i at the moment t, and which allows the inclusion of the heats of reaction Q_r and the heat exchange due to convection and radiation. In this way, the measured signal $\Delta T(t) = T_{\text{MS}} - T_{\text{MR}}$ can be simulated for any Q_r inside the sample. The results for this simple model of a disk-type measuring system and an endothermic phase transition are shown in Fig. 3.11. As can be seen, the shape of the peak depends strongly on the test parameters. The position of the peak maximum, for example, changes with the heating rate, the thermal conduction coefficient of the sample and the mass (or heat of transitions Q_r) of the sample. The slope of the ascending branch is determined by the heating rate and the thermal conduction coefficient of the sample. Only the extrapolated onset temperature T_e of the peak (cf. section 3.3.1.1) is relatively independent of test parameters.

60

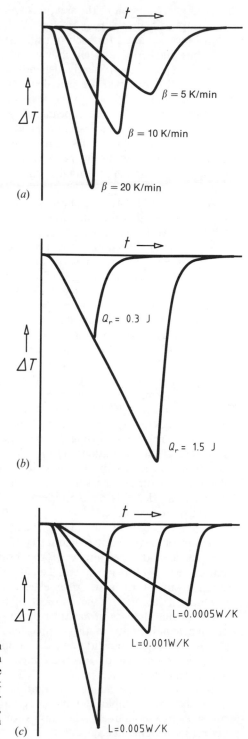

Fig. 3.11. Results of numerical calculation simulating an endothermic phase transition in a disk-type DSC: (*a*) peak shapes of the same sample at different heating rates β; (*b*) peak shapes of two samples with different transition heats Q_r; (*c*) peak shapes of the sample, with different heat conductivity L between sample and crucible

This is why, in addition to the area, this temperature is preferred for characterization of the peak. In contrast, peak maximum T_p and peak width are not suitable for use as characteristic values for such transitions.

It has furthermore been shown that for this model the calibration factor $K_Q = Q_r/A$ (peak area) clearly depends on test parameters (Q_r, λ_{sample}, β) when radiation and convection are included (Table 3.1). This is of great importance to practical scanning calorimetry, and *must* be taken into consideration during calibration. The clear temperature dependence of the calibration factor is also a distinct feature, although the temperature dependence of the heat capacities and heat resistances of the substances has not been included in these model calculations. The reason is to be found in the extensive radiative interchange with the surroundings (heat leak!), which is strongly temperature dependent and not linear to the ΔT signal.

As far as steady-state reaction heat-flow rates are concerned, the above calculations have shown that the calibration factor $K_\Phi = \Phi_r/\Delta T_{st}$ also depends on Φ_r (or the difference between heat capacities C_S and C_R) when radiative interchange is included. In addition, the calibration factors for a peak integration (determination of Q_r) and for steady-state heat-flow rates (determination of C_S) are generally not equal. The calibration factor is also strongly influenced by the surface quality (emissivity) of the containers of the sample and reference sample.

When the temperature dependence of the thermal conductivity and of the heat capacity is included in the calculation, the temperature dependence of the calibration factor is increased even more. Moreover, peak integration may lead to incorrect heats of reaction being obtained, as the calibration factor can no longer be regarded as a constant, even in the case of 'narrow' peaks. A particularly adverse effect is due to the changes of the heat capacity C_S and of the heat transition between sample and measuring system, which take place during transition. In these cases, a quantitative peak evaluation may lead to results affected by systematic errors, as the calibration factor also undergoes substantial changes.

In summary, the following can be stated for heat-flux DSCs.

- For steady-state conditions $\Delta T_{st} = (C_R - C_S)\beta/K_\Phi$ (base line) is valid. With this function, for a known heating rate β, the unknown heat capacity C_S can be determined from the measured ΔT_{st}. It must be borne in mind that the calibration constant K_Φ depends on the difference of the heat capacities ($C_R - C_S$), and of course on temperature.
- For non-steady-state conditions (peak), the shape of the peak changes as a function of heating rate, thermal conductivity, heat capacity and shape of the sample, and the magnitude of the heat of transition. As a result, the peak parameters (width, maximum, height) also change; only the extrapolated peak onset temperature is to some extent independent of sample parameters. The calibration factor K_Q (the quotient of heat of transition and peak area) is not constant but changes upwards or downwards with temperature, heating rate and thermal conductivity of the sample, as well as with the heat of transition, the surface quality (emissivity) and the sample location.
- The calibration factor for peak areas differs from the calibration factor for steady-state heat flows.

It is basically recommended that in DSC measurements the best symmetry possible between sample and reference sample be ensured ($C_R \approx C_S$, $R_{MR} \approx R_{MS}$, identical sample containers). The results of the model calculations are confirmed by practical measurements [3].

62

Table 3.1. Dependence of the calibration factor of a fictional and a real disk-type DSC on different parameters: results of numerical calculations with the aid of the finite-element method in a quasi-linear approximation

Temperature	Changed parameter	Calculated calibration factor K_Q	Relative change of K_Q with relevant parameter	Remarks
I: Disk-type DSC model (steel)				
i: without radiation heat transfer				
300 K	$\lambda_{sample} = 0.01$ W/K $= 0.005$ $= 0.0015$	$4.318 \ 10^{-2}$ J/(Ks) 4.311 4.316	0 %	No change within the numerical precision
ii: with radiation heat transfer				
300 K	$Q_r = 0.3$ J $= 3.0$	$4.423 \ 10^{-2}$ J/(Ks) 4.464	1 %	K_Q increases with Q_r
800 K	$Q_r = 0.3$ J $= 3.0$	$5.327 \ 10^{-2}$ J/(Ks) 5.411	1.5 %	K_Q increases with T
800 K	$\lambda_{sample} = 0.005$ W/K $= 0.001$ $= 0.0005$	$5.427 \ 10^{-2}$ J/(Ks) 5.460 5.490	1.2 %	K_Q decreases with λ_{sample}
800 K	$\beta = \ 2$ K/min $= \ 5$ $= 10$ $= 20$	$5.445 \ 10^{-2}$ J/(Ks) 5.456 5.468 5.507	1.2 %	K_Q increases with β
II: Mettler TA 2000 with radiation heat transfer				
430 K	$Q_r = 0.3$ J $= 3.0$	$5.540 \ 10^{-3}$ J/(Ks) 5.642	1.8 %	K_Q increases with T and Q_r
800 K	$Q_r = 0.3$ J $= 3.0$	$9.819 \ 10^{-3}$ J/(Ks) 9.935	1.2 %	

3.2 Theoretical fundamentals of power-compensation DSCs

In ideal power-compensation DSCs each ΔT signal appearing between sample and reference sample would be immediately compensated by a corresponding change of the heating rate. The differential heating power required for this purpose would be equal to the differential heat-flow rate as the electric energy is completely converted into heat. If the differential heating power were measured and output, this signal would stand directly for the heat-flow rate into the sample, which is the quantity sought. The calibration factor is identical to unity, and the measured signal reflects exactly the thermal effect, i.e. $\Phi_r = \Phi$. In real power-compensation DSCs, the sample is always put in a container and then placed in the heater. At least one heat conduction path and, as a result, at least one time constant τ between the controlled heater and the sample location must therefore be considered. This leads to a measured signal Φ that is 'smeared' in comparison with the processes inside the sample, and a differential equation of at least the first order is therefore obtained:

$$\Phi_r = \Phi + \tau \cdot \frac{\mathrm{d}\Phi}{\mathrm{d}t} \qquad (\text{with } \tau = C_S R_{MS})$$

Such an ideal calorimeter does not exist. Without exception, the power-compensation DSCs commercially available are instruments in which a temperature difference occurs between sample and reference sample (as is the case in heat-flux DSCs). On the one hand, this temperature difference is the measured signal; on the other it is used to compensate electrically the measurement effect by means of a proportional controller. However, a proportional controller can never completely compensate the measurement effect, so even in these calorimeters a ΔT_{SR} remains between the microfurnaces of sample and reference sample, in the steady state (base line) as well as during a peak. The following is then valid:

$$\Phi = K_{prop.} \, \Delta T_{SR}$$

with $K_{prop.}$ set by the proportional controller circuit. As different temperatures also result in a different exchange of radiation and convection with the surroundings, conclusions similar to those for heat-flux DSCs would have to be drawn. This means that, in principle, what has been stated in the previous section is also valid for power-compensation DSCs. However, it is valid to a lesser extent as, due to the compensation, the temperature differences are smaller than in heat-flux DSCs. For the power-compensation DSCs that are commercially available at present it is therefore basically to be expected that
– the shape of the peak depends strongly on sample parameters, the time constant(s) being, however, smaller than those of heat-flux DSCs.
– the calibration factor is not exactly equal to 1 (i.e. calibration is necessary)
– the calibration factors for steady-state determination of the heat capacity and for peak evaluation are not the same
– the calibration factor depends on the temperature, the heating rate, the thermal conductivity of the sample, the heat of transition and the sample location.
Practical measurements have confirmed these assumptions [3]. As expected, the effects are, however, substantially smaller than with heat-flux DSCs. Nevertheless, when power-compensation DSCs are calibrated this dependence on parameters must be taken into consideration. For further details of the theory of power-compensation DSCs see [18, 19, 20].

3.3 Calibration of DSCs

Calibration means the definite coordination of the magnitude of a measured quantity and the 'true one'. For calorimeters the quantities in question are the temperature and the heat or heat-flow rate. As DTA apparatuses are not suitable for quantitative heat measurements, they need only temperature calibration, a procedure similar to that for DSCs. Thus we can proceed without specifying DTA separately.

For temperature and heat calibration of DSCs there are recommended procedures published and in preparation [14, 15, 17] with a solid theoretical and experimental backing. The results of investigations carried out in this respect must be taken into account in the relevant calibration regulations, and they also show the basic limits to the reliability of measurements. To what extent DSCs are capable of being calibrated depends on the quality of the DSC measuring system and on operational parameters, as well as on the availability of precisely measured reference materials.

The fundamentals of the apparative side are described in chapter 2, but it should be mentioned that no complete theory of DSC exists, and for a particular apparatus there may be unknown influences of experimental parameters on the results. For the other side of the problem the situation is quite unsatisfactory. Either certified materials are not available or the certified data of substances available are not thermodynamic equilibrium values but mean values weighted in some way. Finally, one should bear in mind that the overall uncertainty of the calibration yields the minimum systematic uncertainty of the measurements.

3.3.1 Temperature calibration

Temperature calibration means the unequivocal assignment of the temperature 'indicated' by the instrument to the 'true' temperature. The 'true' temperature is defined by fixed points with the aid of reference substances. It is reasonable to use as reference substances, if possible, the substances that serve the realization of the fixed points of the International Temperature Scale (ITS). The temperature 'indicated' by the instrument must be derived from the measured curves, which usually requires extrapolation to zero heating rate to minimize influences of the apparative parameters.

The 'indicated' (or measured) temperature must be taken from the output of the instrument, which is usually a graph or plot. Thus we have to define a method to find the 'measured temperature' in a clear-cut way. It is highly advisable not to trust temperatures automatically calculated with built-in computer software. Interactive construction of auxiliary lines on the terminal screen should at least be possible to influence the calculations.

After calibration has been completed, the potentiometer provided for this purpose will be adjusted until the temperature indicated corresponds to the true one, or adaptation will be ensured via the internal computer program (in a way usually not apparent to the user), or a graph or table will be established showing the relation between the indicated and the true temperatures. In each case, a table should be drawn up-which indicates the variation of the indicated temperature at different heating rates.

A calibration already carried out by the manufacturer must be checked. Regular calibrations provide important information about the repeatability error and any long-term systematic variations (drift). (It should be noted that repeatability error is not the same as uncertainty of measurement.)

3.3.1.1 Necessary definitions

In the case of an endothermic transition, the DSC records the heat-flow rate signal shown in Fig. 3.12. The base line is interpolated by various methods between the peak onset temperature T_i and the peak offset temperature T_f. The linear extrapolation of the first part of the base line into the peak area suffices to fix the characteristic temperature T_e, which is important here.

The auxiliary lines are constructed either as tangents at the point of inflection or as fitted lines through the (almost linear) section of the peak flanks. For temperature calibration, the difference is of no significance within the scope of the repeatability error.

At T_i, the peak onset temperature, the measured curve starts to deviate from the extrapolated base line.

At T_e, the extrapolated peak onset temperature, the auxiliary line intersects the extrapolated base line.

65

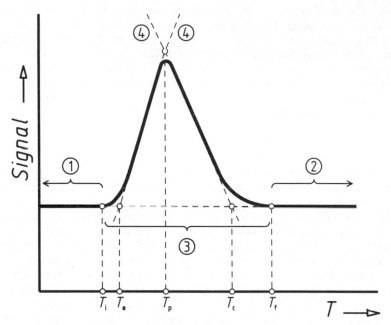

Fig. 3.12. Definitions of characteristic peak quantities: (1), (2) base line; (3) interpolated base line; (4) auxiliary lines; $T_i - T_f$ = characteristic peak temperatures (see text)

At T_p, the peak maximum temperature, the maximum deviation of the measured curve from the interpolated base line is reached.

At T_c, the extrapolated peak offset temperature, the auxiliary line intersects the extrapolated base line.

At T_f, the peak offset temperature, the measured curve again joins the extrapolated base line.

3.3.1.2 Calibration procedure

– Selection of at least three reference substances that cover the desired temperature range as uniformly as possible.
– At least two calibration samples of each substance are prepared. The sample mass should correspond to that commonly used in routine measurements.
– The transition is to be measured with each calibration sample at at least five different heating rates. The second calibration sample of the same substance is also measured at different heating rates.
– Checking of whether or not there is a significant difference between the characteristic temperatures obtained at identical heating rates for the first and the second calibration sample of the same substance. If necessary, checking of whether or not the temperatures depend on other parameters (mass, location of the sample in the crucible).
– If the result of this checking is negative, T_e is represented as a function of the heating rate and the extrapolated value $T_e(\beta = 0)$ is determined for the zero heating rate (Fig. 3.13).

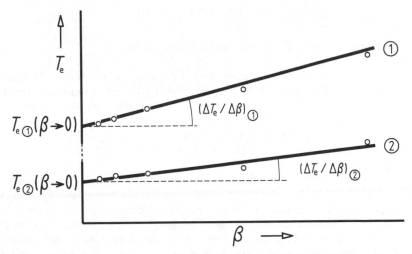

Fig. 3.13. Extrapolated peak onset temperatures T_e as a function of heating rate β for two different substances: in this case the function is linear and the slope is constant but different

- The difference $\Delta T_{corr}(\beta = 0)$ between the value $T_e(\beta = 0)$ obtained in this way and the corresponding fixed-point value T_{fix} or the value T_{lit} taken from the literature [4–6] is either used to change the instrument calibration according to the manufacturer's instructions or enters a calibration table or curve (Fig. 3.14).
- Should T_e depend not only on the heating rate but also on other parameters, these dependencies are to be represented accordingly (location of the sample in the container, position of the sample container in the measuring system, open/closed sample container, sample mass, sample shape (foil, bulk), atmosphere, material of sample container) (cf. [7]).

In each case, a table (or graph) should be compiled that shows the variation of the indicated temperature (or of that read from the measured curve) in relation to the true temperature at different heating rates. The formulas are

$$\Delta T_{corr} = T_{fix} - T_e$$

$$\Delta T_{corr} = \Delta T_{corr}(\beta = 0) + \Delta T_{corr}(\beta)$$

$$\Delta T_{corr}(\beta = 0) = T_{fix} - T_e(\beta = 0)$$

Temperature calibration has then been completed, and the true temperature of a measured transition can be calculated:

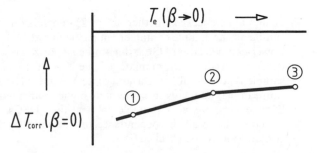

Fig. 3.14. The difference ΔT_{corr} between the true temperature (T_{fix} or T_{lit}) and the quasi-static measured temperature $T_e(\beta = 0)$ plotted against temperature

67

$$T_{\text{true}} = T_{\text{e}} + \Delta T_{\text{corr}} \quad \text{or} \quad T_{\text{true}} = T_{\text{e}}(\beta = 0) + \Delta T_{\text{corr}}(\beta = 0)$$

3.3.1.3 Arguments in support of this calibration procedure

The older recommendations concerning the temperature calibration of DSCs (e.g. ASTM E 967-83, ASTM E 794-85) and the specifications of most manufacturers take a standard heating rate (e.g. 10 K/min) as a basis. This method is not to be recommended for the following reasons.
– Different heating rates require different corrections; shifting of T_{e} is not always linear to the heating rate and different materials have different slopes of the correction curve (cf. Fig. 3.13).
– Only at zero heating rate is the sample temperature equal to the measured temperature.
– The temperature fixed points of the International Temperature Scale are defined for the reference substances in phase equilibrium (i.e. static, zero heating rate).
– As we do not at present have reference substances for cooling experiments (because of the supercooling effects during freezing of most pure materials), a cooling temperature calibration is not possible. A temperature calibration extrapolation to $\beta = 0$ as described above also holds for cooling experiments. Because of symmetry in heat transport phenomena the correction term $\Delta T_{\text{corr}}(\beta = 0)$ should be the same, and the correction due to the heating rate $\Delta T(\beta)$ should in most cases have the same magnitude but the opposite sign (for details see [16]).
Only T_{e} can be used as a characteristic temperature of a peak to define the temperature scale, for the following reasons.
– T_{i} cannot be determined with the required reliability (because of the noise); the same applies to T_{f} (repeatability error between 2 and 15 °C, depending on substance, kind of transition etc.).
– T_{p} and T_{c} depend strongly on the thermal conductivity, mass and layer thickness (volume) of the sample substance, and on the heating rate and the heat transfer from sample to sample container, which may change due to melting (cf. Fig. 3.11).
– T_{e} is the least dependent on heating rate and sample parameters (substance, thermal conductivity, mass, layer thickness), but any potential effect of fusion (heat transfer) should also be checked by carrying out various similar experiments with the same calibration sample.
The temperature calibration method described above is rather time-consuming, but it has a lot of advantages, and results in a higher accuracy of the measured temperatures.

3.3.1.4 Notes

(1) It should be checked whether or not T_{e} depends on the location of the calibration sample in the sample container (in particular at high heating rates). In the case of power-compensation DSCs, due to the isoperibol operation, a temperature profile develops in the bottom plate of the microfurnace which—depending on the sample type—results in earlier (center) or later (boundary) melting of a calibration sample, and thus in a lower or higher indicated temperature T_{e} (Fig. 3.15). Such effects are also possible in heat-flux DSCs. In cylinder-type measuring systems, above all the dependence on the level of the sample in the cylindrical container must be checked.

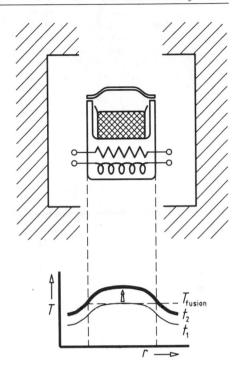

Fig. 3.15. Model of a microfurnace in a power-compensation DSC and the temperature profile in its bottom plate: the temperature at the boundary increases with a certain delay due to the center

(2) In the case of exothermic reactions, the sample temperature is higher than the measured temperature. As no reference materials exist for exothermic processes, the magnitude of this deviation cannot be precisely determined. The assignment of a temperature is therefore useful only at the beginning of the reaction.

(3) In measurements at a negative heating rate (cooling), the sample temperature is higher than the indicated temperature. As a result, the correction $\Delta T_{corr}(\beta)$ from the calibration table or curve must be applied with the sign reversed as compared with heating. This symmetry may be checked [16]. Special cooling calibration is not possible, as the pure calibration substances supercool in an undefined way. In such cases, the equilibrium solidification temperature cannot, of course, be determined, not even by means of the correction ΔT_{corr}, but only the solidification temperature after supercooling and thus the amount of supercooling. Generally, the assignment of the characteristic temperatures for cooling curves is affected by greater uncertainties than in the case of heating.

3.3.1.5 Reference materials

The ITS is defined on the one hand by 'fixed points', i.e. temperatures of two-phase equilibrium of pure substances, and on the other by a table or formula of resistances of the platinum thermometer. As the fixed points are subject to permanent international control, the transition temperatures of these materials are very suitable for temperature calibration, and only a high purity need be certified.

Substances from the ITS 90 fixed point table that can be used for temperature calibration of DSCs are listed in Table 3.2. As can be seen, almost the whole interesting temperature region is covered. We highly recommend that these sub-

Table 3.2. Substances from the ITS-90 fixed point table and some additional substances suitable for DSC calibration (temperatures adjusted to the valid ITS-90 scale) [4, 5, 15]: T = triple point; S = solid–solid transition point; M, F = melting and freezing points respectively at a pressure of 101325 Pa

Substance	State	Temperature		Uncertainty mK	Ref.
		K	°C		
Cyclopentane	S	122.38	−150.77	50	[15]
Cyclopentane	S	138.06	−135.09	50	[15]
Cyclopentane	F	179.72	−93.43	50	[15]
Mercury	T	234.3156	−38.8344	fixed	ITS-90
Water	F	273.15	0	10	IPTS-68
Water	T	273.16	0.01	fixed	ITS-90
Gallium	M	302.9146	29.7646	fixed	ITS-90
Indium	F	429.7485	156.5985	fixed	ITS-90
Tin	F	505.078	231.928	fixed	ITS-90
Lead	F	600.61	327.46	10	IPTS-68
Zinc	F	692.677	419.527	fixed	ITS-90
Lithium sulphate	S	851.43	578.28	250	[15]
Aluminum	F	933.473	660.323	fixed	ITS-90
Silver	F	1234.93	961.78	fixed	ITS-90
Gold	F	1337.33	1064.18	fixed	ITS-90
Copper	F	1357.77	1084.62	fixed	ITS-90
Nickel	F	1728	1455	unknown	IPTS-68
Cobalt	F	1768	1495	unknown	IPTS-68
Palladium	F	1827	1554	unknown	IPTS-68
Platinum	F	2041	1768	unknown	IPTS-68
Rhodium	F	2235	1962	unknown	IPTS-68
Iridium	F	2719	2446	unknown	IPTS-68

stances be used for calibration, and that the use of other materials, however certified, should cease. Nevertheless, substances should be as pure as possible ($\geq 99.999\%$).

In every case metallic material (one grain only!) is to be preferred for calibration, as for powders (non-metallic materials) the transition temperature depends on grain size.

3.3.2 Heat calibration

By means of heat calibration, the proportionality factor (the calibration factor) between the measured heat-flow rate into the sample Φ_m and the real heat-flow rate Φ_r, and between the measured exchanged heat Q_m and the heat really transformed Q_r, is to be determined:

$$\Phi_r = K_\Phi \Phi_m \quad \text{and} \quad Q_r = K_Q Q_m \tag{3.10}$$

This calibration is carried out either directly as 'heat-flow calibration' in the (quasi-) steady state by
– electric heating
– known heat capacity of the calibration sample
or as 'peak area calibration' by integration over a peak which represents a known heat $Q_r = \int \Phi_r \, dt = K_Q \int (\Phi_m - \Phi_{bl}) \, dt$, by means of
– electric heating
– phase transition (melting) of a pure substance.

70

As $Q_r = \int \Phi_r \, dt$ and $Q_m = \int (\Phi_m - \Phi_{bl}) \, dt$, K_Φ should be equal to K_Q. But in reality this is not the case, as K_Φ depends (i) on temperature and (ii) on heat-flow rate (cf. section 3.1.4). Consequently Eq. 3.10 may be integrated, but $K_\Phi(T, \Phi)$ cannot be put outside the integral

$$Q_r = \int \Phi_r(t) \, dt = \int K_\Phi[T(t), \Phi(t)](\Phi_m - \Phi_{bl}) \, dt \neq K_\Phi \int (\Phi_m - \Phi_{bl}) \, dt \qquad (3.11)$$

On the other hand, from the right-hand half of Eq. 3.10 we can derive

$$Q_r = K_Q Q_m = K_Q \int (\Phi_m - \Phi_{bl}) \, dt \qquad (3.12)$$

Comparison of Eqs. 3.11 and 3.12 shows that K_Q is not equal to K_Φ, but to a kind of integral average of all K_Φ values due to the $\Phi(t)$ function of the peak. In practice, the difference between the two calibration factors is not very large (1 %–10 %, depending on the apparatus), but nevertheless separate calibrations should be done.

Another consequence of the theoretical discussions in section 3.1 is that the thermophysical behavior of the calibration sample and the sample to be measured must be as similar as possible. As this can be achieved only approximately, systematic deviations exist, which must be estimated and included in the overall uncertainty of measurement.

3.3.2.1 Heat-flow calibration

In almost all DSCs commercially available, a heat-flow rate signal Φ is internally assigned to the actual measured signal ΔT. (When the measured signal is read as a voltage the following applies analogously, Φ being replaced by ΔU). The heat-flow calibration defines the functional relation between the heat-flow rate into the sample Φ_m and the heat-flow rate Φ_r absorbed or emitted by the sample:

$$\Phi_r = K_\Phi(X)\Phi_m$$

(steady state, Φ_m with subtracted empty pan line, X is any parameter).

The proportionality factor (calibration factor, calibration function) K_Φ usually depends on parameters such as temperature, and of course also on the heat flow. In some DSCs, K_Φ is set to unity by electronic or software means. In these cases, too, the relation between Φ_m and Φ_r must be carefully checked.

3.3.2.2 Electrical calibration

The installation of an electric calibration heater instead of or in the sample offers the following advantages.
- The electric power (heating power) can be measured easily and with high accuracy ($\Phi_r = P_{el} = u \cdot i$).
- Heat flows of differing intensity can be generated without a change of calibration set-up.
- Adjustment of the steady state can be maintained for any period of time desired; the resulting conditions are very similar to those of a C_p measurement.
- The heater can be switched on or off at any temperature desired so that the position of the base line can also be checked in between.
- By appropriate presetting of the development of the heating rate with time,

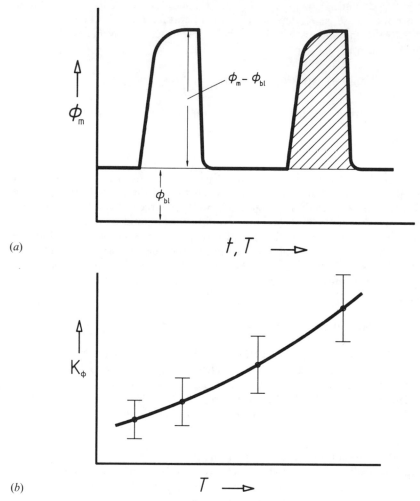

(a)

(b)

Fig. 3.16. Electrical heat flow calibration: (*a*) measured curve (with zero line subtracted), due to electrical power $\Phi_r = u \cdot i$; (*b*) resulting calibration factor. (The hatched area may be evaluated for simultaneous peak area calibration)

measurement effects (peaks) can be 'repeated' (simulated) so that heat flows leading to such a peak can be assigned without deconvolution procedures.
– The time constant of the measuring system can easily be determined.
The disadvantages of the electric calibration heater are as follows.
– The heater cannot be installed in the small crucibles of a disk-type measuring system.
– Heaters permanently installed in measuring systems are not situated at the sample location; this leads to systematic errors.
– Heat flows in the wires (heat leak) lead to systematic uncertainties (ensure thermal symmetry of sample and reference side).
– Wires have a non-zero resistance and thus an additional heat production $i^2 R$.

In practice the calibration heater should have a large resistance compared with that of the wires, which should be as thin as possible to avoid additional heat flows.

Figure 3.16(a) shows the electrical calibration in a DSC with a cylinder-type measuring system. The calibration curve $K_\phi(T) = \Phi_r/(\Phi_m - \Phi_{bl})$ is represented in Fig. 3.16(b). (Peak-area calibration may be carried out simultaneously with the aid of the peak areas.)

3.3.2.3 C_p calibration

Heat-flow calibration can also be carried out with a sample of known heat capacity (Fig. 3.17). For the heat-flow rate absorbed by the sample in (quasi-)steady state:

$\Phi_r = C_p\beta$ (without reference sample)

so that

$K_\phi(T) = C_p(T)\beta/\Phi_m(T)$

The advantages of this calibration method are that
– it is applicable in all DSCs
– calibration is at the sample location
– no wires are required, and thus there is no heat leak
– reference materials with certified C_p values are available.
The disadvantages are
– that calibration cannot be switched off in between, i.e. checking of the empty pan line is not possible during the run (this leads to uncertainties in Φ_m)
– the temperature profile inside the sample (yields 'mean C_p').

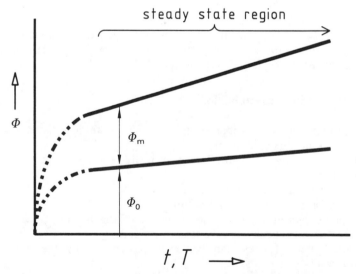

Fig. 3.17. Measured curves due to a C_p calibration of a DSC; Φ_m: heat-flow rate into the sample, Φ_0: empty pan curve

3.3.2.4 Peak-area calibration

In peak-area calibration a known heat Q_r (exothermic or endothermic) is compared with the area of the resultant peak (Fig. 3.18):

73

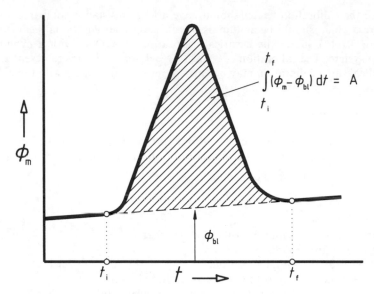

Fig. 3.18. Peak-area calibration method

$$Q_r = K_Q(T) \int [\Phi_m(t) - \Phi_{bl}(t)] \, dt = K_Q A$$

The integral A is the peak area. Integration must be carried out over the whole peak to ensure that contributions of terms with the first or higher derivatives of the measured signal can be neglected (cf. section 3.1.2).

3.3.2.5 Electrical peak-area calibration

If electrical calibration is possible (Fig. 3.19), the avantages listed above apply analogously to peak-area calibration:
- easy and accurate measurement of the electrically generated heat
  $Q_r = \int P_{el} \, dt = W_{el} = u \cdot i \cdot \Delta t$
- peaks of differing size can be produced
- calibration is possible at any temperature
- by appropriate presetting of the development of the heating power with time, measurement effects (peaks) can be 'traced' (simulated)
- measurement effects (peaks) can be 'encompassed' during the run by similar calibration peaks.

The disadvantages are the same as those stated above:
- the electrical heater cannot be installed in disk-type measuring systems
- systematic errors result if heaters are permanently installed (different locations of sample and heater)
- wires give rise to systematic uncertainties (heat leaks).

It must be ensured that electrical calibration is carried out over only a small temperature interval.

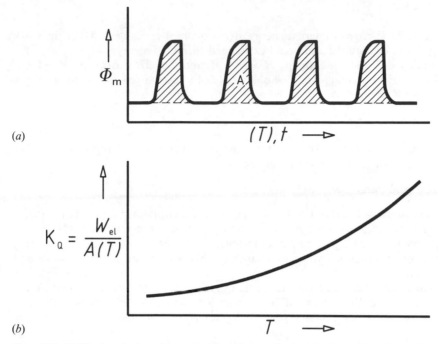

(a)

(b)

Fig. 3.19. DSC electrical peak-area calibration method: (a) measured curve; (b) resulting calibration factor

3.3.2.6 Peak-area calibration with enthalpy references

Peak-area calibration of DSCs is usually carried out with the aid of the heats of fusion of pure substances. The advantages of this method are as follows:
- applicable in all DSCs
- calibration at the sample location
- no wires required
- simultaneous temperature and heat calibration is possible with some reference materials.

The disadvantages are as follows:
- calibration practicable only at discrete temperatures
- no adaptation to a certain peak profile is possible
- systematic uncertainties due to the special shape of the sample temperature curve $T_s(t)$ during the calibration procedure
- uncertainties resulting from the determination of the area (definition of the integration limits and shape of the base line).

    A very important feature of the relation between enthalpy difference ΔH and ΔQ measured must be taken into consideration. The enthalpy difference of a phase transition of a substance is a difference of two functions of state, and is thus well-defined. But in a calorimeter the heat exchanged (during the transition) is actually measured, which is not a function of state and thus may depend on the path of the reaction. The thermodynamic connection of these quantities is as follows. The first law is

75

$$dU = dQ + dW + \sum_i dE_i$$

where dU is the change of internal energy, dQ is the heat exchanged, dW is the work exchanged and dE_i are the additional exchanged forms of energy.

From the definition of enthalpy $H = U + pV$ and thus $dH = dU + p\,dV + V\,dp$, and the restriction that work should only be exchanged as volume work ($dW = -p\,dV$) we get

$$dH = dQ + V\,dp + \sum_i dE_i$$

If we set pressure $p =$ a constant and assume that energy is exchanged in no other form (e.g. as surface energy, electrical energy) we get

$$dH = dQ \quad \text{and thus} \quad \Delta H = \Delta Q$$

That is, the measured heat of transition is equal to the enthalpy difference only under such conditions that the pressure is constant (no vapor pressure change!) and no other energy forms change during phase transition, i.e. no change of surface energy, no change of deformation energy, no change of cohesion energy between sample and crucible and so on.

Furthermore, it should be borne in mind that the enthalpy function is defined for infinite phases and the process of phase transition must be done quasi-static (reversibly). As a result, the demands on enthalpy references for calibration are that they be
– very pure substances with well-known ΔH of transition
– large grains (single crystals are the best) that do nothing other than phase transition.

Calibration should be conducted at low heating rates. As these demands cannot be met in practice, the calibration may include uncertainties. The magnitude of these uncertainties cannot be specified because it has not been properly investigated. From round robin results, an estimate of the uncertainty of about 1 % seems to be realistic.

The calibration procedure with the aid of the phase transition heat is as follows.
– Selection of calibration substances that cover the desired temperature range and whose thermophysical characteristic data are similar to those of the sample.
– Weighing-in of such masses as approximately generate the heat effect found in normal measurements.
– Adjustment of customary heating rates. (Be careful in the case of reference substances that melt close to the starting temperature: the quasi-steady state must have been reached, the effects due to the start must have faded; apply a lower heating rate if necessary.)
– Evaluation of peak (area and extrapolated onset temperature T_e).
– Determination of $K_Q(T_e)$, drawing of a calibration curve or compilation of a table, or input of measured data into the calorimeter according to the manufacturer's specifications (potentiometer or in software).
– Estimation of the uncertainty of the calibration (uncertainty of the weighing, of the base line, of the integration limits, of the values for the heat of fusion taken from the literature; the estimated uncertainty from theoretical considerations must be taken into account).
– Measurement of the repeatability errors of the calibration factors (or calibration curve). This repeatability error must be clearly smaller than the estimated overall

uncertainty of the calibration (see above). The repeatability error is the minimum possible uncertainty of measurement of caloric measurements.

The same calibration procedure is applied to power-compensation DSCs, but in this case calibration with one reference material (e.g. indium) is in general sufficient, as the calibration factor depends on temperature to only a minor extent.

3.3.2.7 Remarks

It must be taken into account that calibration factors for heat flow and heat (peak area) measurements are different and must therefore be measured separately.

It is to be expected that the heat-flow calibration factor depends on heat flow itself. This should be checked with the aid of two samples that have a large (but well-known) difference in heat capacity, or by a built-in electric heater.

The peak-area calibration factor may depend on sample mass, heating rate, peak shape and size, temperature and other parameters. These dependencies should be tested during the first calibration, and the results of these checks must be taken into account for evaluation of the uncertainty of any measurement.

In principle, the heat calibration run should be as similar to the measuring run as possible. Ideally, the measured peak should be simulated by suitable variation of the current into a built-in electrical heater.

Because of the systematic error sources due to the construction principle, heat-flux DSC yields results in heat measurements with an uncertainty of about 5 %, which may be decreased by very careful special calibration.

For power-compensation DSC the uncertainty of heat measurements is about 1 %–2 %.

Heat flow or heat capacity measurement in heat-flux DSC is always problematical; there is usually an uncertainty of 10 %–20 %. Higher accuracy requires a great deal of calibration.

Power-compensation DSCs can normally measure heat capacities with an uncertainty of 3 %–5 %, which may be decreased to 1 %.

It can therefore be stated that heat-flux DSC needs much more effort at calibration in order to give results with the same precision as obtained from power-compensation DSC. For high accuracy measurements, power-compensation DSC is preferable. On the other hand, the less expensive heat-flux DSC is often sufficient for routine measurements.

3.3.2.8 Reference materials

For heat-flow calibration, air-inert substances with a well-known specific heat capacity should be selected. An α-alumina single crystal (Al_2O_3, sapphire) is usually employed, but platinum and copper (in an inert atmosphere) are also suitable (Table 3.3) [17].

For peak-area calibration, pure substances with a well-known transition heat should be chosen. For precise measurement of the heat of transition adiabatic calorimeters are needed. Such measurement results are available for only a few substances. When several independent measurements have been carried out on one substance, the ranges of uncertainty stated often do not overlap, so that the estimate of the best value and of the overall uncertainty is problematic. Nowadays the heats of fusion of pure metals seem to be the most reliable (Table 3.4) [17].

77

Table 3.3. Specific heat capacity of substances suitable for heat-flow calibration [6, 10, 17]

Substance	Temperature K	Specific heat capacity	
		J/g per K	J/mol per K
α-Alumina (sapphire) [6, 17]	75	0.0558	5.685
	100	0.1261	12.855
	150	0.3133	31.95
	200	0.5014	51.12
	250	0.6579	67.08
	300	0.7788	79.41
	350	0.8713	88.84
	400	0.9423	96.08
	450	0.9975	101.71
	500	1.0409	106.13
	550	1.0756	109.67
	600	1.1039	112.55
	650	1.1273	114.94
	700	1.1466	116.92
	750	1.1632	118.59
	800	1.1783	120.14
	850	1.1908	121.41
	900	1.2030	122.66
	950	1.2137	123.75
	1000	1.2237	124.77
	1100	1.2418	126.61
	1200	1.2578	128.25
	1400	1.2856	131.08
	1600	1.3080	133.36
	1800	1.3253	135.13
	2000	1.3387	136.50
	2200	1.3508	137.73
	2250	1.3540	138.06
Platinum [6]	298.15	0.1326	25.87
	400	0.1357	26.47
	500	0.1383	26.98
	600	0.1411	27.53
	700	0.1437	28.03
	800	0.1464	28.57
	900	0.1490	29.06
	1000	0.1514	29.54
	1100	0.1542	30.09
	1200	0.1568	30.58
	1300	0.1597	31.16
	1400	0.1625	31.71
	1500	0.1651	32.21
Copper [17, 10]	100	0.2519	16.01
	150	0.3223	20.48
	200	0.3549	22.56
	250	0.3729	23.70
	300	0.3850	14.46
	298.15	0.0919	5.840
	400	0.0950	6.036

Table 3.3. (continued)

Substance	Temperature K	Specific heat capacity	
		J/g per K	J/mol per K
Copper [17, 10]	500	0.0974	6.192
	600	0.0996	6.328
	700	0.1015	6.451
	800	0.1034	6.570
	900	0.1054	6.703
	1000	0.1078	6.850
	1100	0.1109	7.046
	1200	0.1148	7.296
	1300	0.1210	7.687

Table 3.4. Substances with well-known heat of fusion suitable for peak-area calibration [8, 9, 17] (temperatures adjusted to valid ITS-90 scale)

Substance	Fusion temperature °C	Specific heat of fusion J/g	
Cyclopentane	− 150.77	69.6 ± 0.3	(0.5 %)
Cyclopentane	− 135.09	4.91 ± 0.02	(0.5 %)
Cyclopentane	− 93.43	8.63 ± 0.04	(0.5 %)
Gallium	29.7646	79.88 ± 0.7	(0.9 %)
Indium	156.5985	28.62 ± 0.11	(0.4 %)
Tin	231.928	60.40 ± 0.3	(0.5 %)
Aluminum	660.323	398 ± 6	(1.4 %)

The stated uncertainties of the heat of fusion are based on measurements of the Physikalisch-Technische Bundesanstalt Braunschweig, and on an evaluation of data found in the literature [8, 9, 17].

On the basis of an analysis of the relatively large number of measured values of In and Sn obtained by different calorimetric methods, we assume a minimum uncertainty of ∼ ±0.5 % of all values stated in the literature [8, 17]. When only a few measured values have been taken as a basis, these can be affected by substantially greater actual uncertainties, as evidenced by, for example, the maximum deviations and standard deviations of the values available for In and Sn.

3.4 Evaluation of results from DSC curves

The investigator using DTA methods is interested in the temperature and in the character of an event (transition or reaction) in the sample. The user of DSC techniques needs more quantitative additional information on heat flow and heat involved in the sample processes. This information has to be evaluated from the measured curves.

Usually the DTA apparatus delivers a differential temperature ΔT_{SR} as a function of time. As a second curve, the measured temperature is plotted.

Present-day DSC in most cases plots a heat-flow rate Φ as a function of temperature T. The heat-flow rate is internally converted from the measured differential

79

temperature signal: $\Phi = K(T)\, \Delta T$; and the abscissa is usually the measured reference temperature or the set value of the heating controller, both of which are strictly proportional to time, with the heating rate β selected being the proportionality factor: $T = T_0 + \beta t$.

The information the experimenter is interested in is the temperature of the sample at a certain moment on the one hand, and the reaction heat-flow rate in the sample Φ_r as a function of sample temperature T_s on the other. The interdependence of this function and the measured one has in part been illustrated theoretically in section 3.1. Below we explain the practice of evaluation of the interesting data from the apparatus output.

3.4.1 Evaluation of the true sample temperature

During heating, the sample temperature is lower than the probe temperature, and during cooling it is higher, thus a correction has to be carried out in both cases. The magnitude of this correction depends on the thermal resistance (R in Fig. 3.20) between the sample and the place where the temperature is measured, and on the heat flux in the sample. In terms of section 3.1.3:

$$\Delta T_{MS} = T_{MS} - T_S = R_{MS}\Phi_{MS}$$

Substituting $\Phi_m = \Delta\Phi_{SR} = \Phi_{MS} - \Phi_{MR}$ and differentiating this equation yields

$$\frac{dT_{MS}}{dt} - \frac{dT_S}{dt} = R_{MS}\left(\frac{d\Phi_m}{dt} + \frac{d\Phi_{MR}}{dt}\right)$$

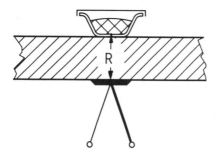

Fig. 3.20. Section of a disk-type DSC with thermal resistance causing the delay between sample and measured temperature

If we consider a phase transition of a pure substance, we get as a first approximation $T_S = $ constant, $\Phi_{MR} = $ constant and $dT_{MS}/dt \approx \beta$, thus:

$$\beta = R_{MS}\frac{d\Phi_m}{dt} \quad \text{or} \quad \frac{d\Phi_m}{dt} = \beta/R_{MS} = \text{constant}$$

As a result the thermal resistance and the temperature delay between sample and temperature probe can be determined from the constant slope of a melting peak of a pure substance. For this reason the characteristic temperature of an event is evaluated from the measured curve as follows.

Determine the onset slope $\Delta\Phi/\Delta T$ of the melting peak at the proper heating rate (Fig. 3.21(a)) and draw auxiliary lines with the same slope from characteristic points of the measured curve to the interpolated base line to get the correct temperatures in question (Fig. 3.21(b)). This slope read-off is different from vertical read-off in the

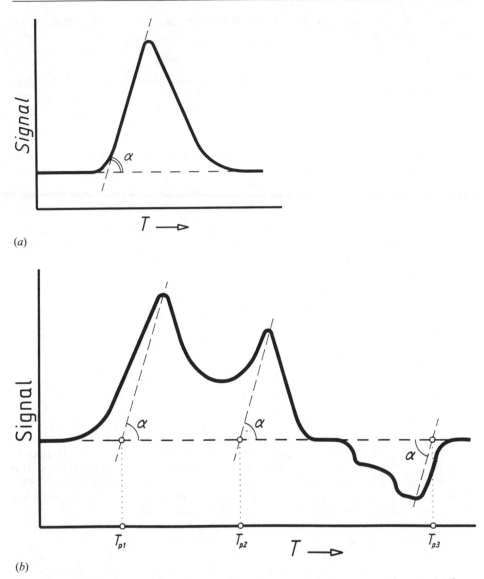

(a)

(b)

Fig. 3.21. Construction of the correct reaction temperatures from a measured curve: (*a*) for a melting peak of a pure substance; (*b*) for a complex reaction

case of rather narrow and high peaks; for broad and low reaction heat-flow rate curves it may not be necessary.

Note that, for exothermic reactions, only the beginning of a peak can correctly be determined by this method. As the reaction proceeds the sample is heated more and more, and its temperature may thus rise almost above the temperature indicated by the probe. This temperature increase depends on the reaction in question in a non-defined way.

The temperatures (T_{p1}, T_{p2}, T_{p3}: Fig. 3.21(b)) are the 'measured temperatures' (as are the extrapolated onset temperatures T_e of phase transitions of pure substances), and have to be corrected in a subsequent step due to the results from calibration (cf. section 3.3.1):

$$T_{true} = T_{meas} + \Delta T_{corr}$$

There are two ways of carrying out the correction: when high accuracy is not required, the correction term is calculated from the correction at zero heating rate $\Delta T_{corr}(\beta = 0)$ and the average slope of the $T_e(\beta)$ curves (cf. Fig. 3.13) as follows:

$$\Delta T_{corr} = \Delta T_{corr}(\beta = 0) + \beta \langle \Delta T_e / \Delta \beta \rangle$$

For higher accuracy requirements the substance in question is measured at various heating rates and the desired characteristic temperature is determined by extrapolation to zero heating rate, to give the quasi-static temperature:

$$T_{true} = T_{meas}(\beta = 0) + \Delta T_{corr}(\beta = 0)$$

The second method is practicable only if the reaction to be investigated does not change when the heating rate is changed. This is the case only for time-independent phase transitions, not for chemical and other kinetic reactions.

3.4.2 Evaluation of the reaction heat

The peak area must be determined in order to calculate heats of transition or reaction. As shown in section 3.1, the peak area is defined between the measured curve Φ and the (invisible) base line Φ_{bl} in the region of the peak (cf. Fig. 3.4). Thus the base line has to be interpolated in the region of the peak.

This is no problem if the base line is a linear function of temperature and does not change during the peak; interpolation can be done with the aid of a ruler. But how can the base line be constructed in other cases?

For irreversible transitions, which occur only once, without a change of heat capacity, the base line can be found by repeating the DSC run with the same sample.

In all other cases the base line has to be constructed in such a way that it changes from the base line before to the base line behind the peak along a 'smooth' path. The method most commonly used [21] is based on the assumption that the base line changes proportionally to the turnover of the reaction, which in turn is proportional to the partial area of the peak up to the moment considered.

The construction is explained in Figs. 3.22 and 3.23. Starting in Fig. 3.22(a) with a zeroth approximation of the base line as a linear extrapolation, the first approximation of the turnover is determined (Fig. 3.22(b)). From this, the first approximation of the base line can be found (Fig. 3.22(c)). This base line is the starting condition for the next approximation with the same method (Fig. 3.23).

The construction may be continued, but the second approximation is usually sufficient, in particular if one bears in mind that base-line changes depend not only on c_p changes but also on changes of heat transfer conditions. As the effects of these changes on the base-line change are unknown, there is always an uncertainty in getting the 'right' base line, and every sophisticated method is as uncertain as a simple interpolation by eye.

After the interpolation of the base line, the peak area can be determined by integration (cf. Figs. 3.18 and 3.24):

$$A = \int [\Phi(t) - \Phi_{bl}(t)] \, dt$$

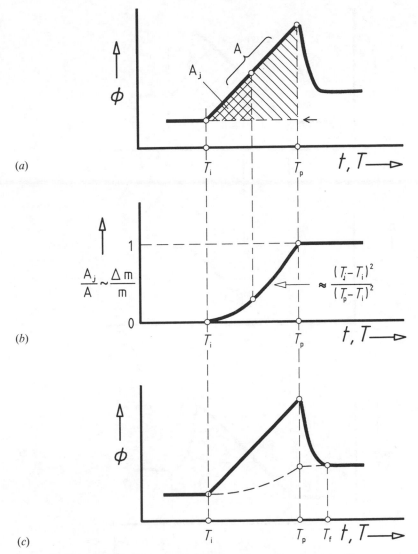

(a)

(b)

(c)

Fig. 3.22. Construction of the base line: (a) zeroth approximation; (b) turnover of the reaction from partial areas of (a); (c) resulting first approximation

Depending on the abscissa of the measured curve, the integration must be conducted with respect to time or with respect to program temperature. The variables may be transformed with the aid of the equation:

$$T = T_0 + \beta t \qquad (\beta = \text{heating rate})$$

This integration is usually performed by the computer. From the peak area, the reaction heat can be calculated by multiplication by the proper calibration factor (cf. section 3.3.2):

83

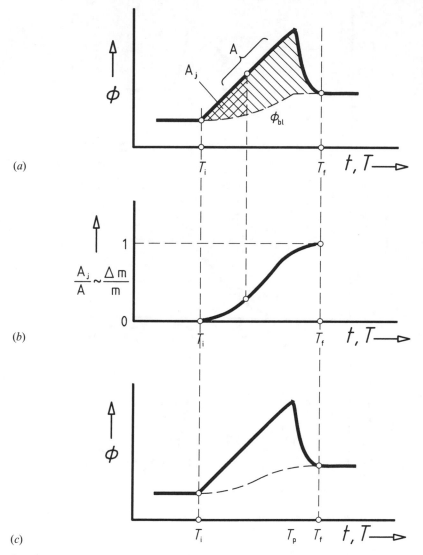

Fig. 3.23. Construction of the base line (continued): (*a*) first approximation from Fig. 22(a); (*b*) turnover of reaction from partial areas of (*a*); (*c*) resulting second approximation

$$Q_r = K_Q A$$

Finally, the uncertainty should be evaluated:

$$\delta Q_r = \pm (A \, \delta K_Q + K_Q \, \delta A)$$

The peak-area uncertainty δA should include both the standard deviation of several measurements and the systematic uncertainty of the base line. Usually, reaction and transition heats determined in a heat-flux DSC have an uncertainty of about 5 %. Power-compensation DSC yields somewhat better results.

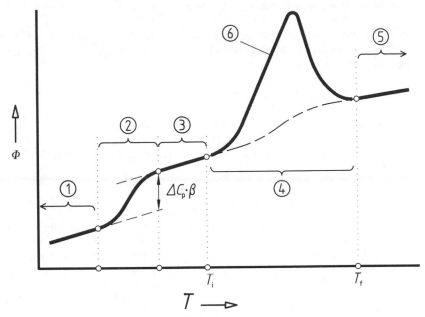

Fig. 3.24. DSC curve of a polymer sample: (1), (3), (5) base line; (2) glass transition; (4) interpolated base line; (6) transition peak

3.4.3 Evaluation of heat capacities

This procedure is explained in section 3.1.1 (cf. Fig. 3.2). For satisfactory results, the DSC must be calibrated very carefully (cf. section 3.3.2) and the apparatus should have a very high reproducibility of the empty pan line. On these premises, heat capacities are measured as follows.

The substance in question (sealed in a crucible) is put in the sample container and an empty crucible is put in the reference container. Choose a medium heating rate (10 K/min) and measure the heat-flow rate curve Φ. Replace the sample crucible with an empty one (of the same mass!) and execute a second measurement, yielding the empty pan line Φ_0. The unknown heat capacity can then be calculated (cf. Fig. 3.17):

$$C_p = K_\Phi(\Phi - \Phi_0)/\beta = K_\Phi \cdot \Phi_m/\beta$$

In the case of measurement of a step change in heat capacity ΔC_p (e.g. glass transition, Fig. 3.24) the formula is

$$\Delta C_p = K_\Phi \, \Delta\Phi/\beta$$

where $\Delta\Phi$ is the step change of the heat-flow rate curve.

Usually several measurements have to be taken for both sample and empty pan line runs, and the average results should be taken for calculation of C_p to reduce uncertainty.

Finally, in this case also the uncertainty δ must be evaluated carefully:

$$\delta C_p = \pm[\delta K_\Phi(\Phi - \Phi_0) + K_\Phi(\delta\Phi + \delta\Phi_0)]/\beta$$

3.4.4 Evaluation of the true reaction heat-flow rate (desmearing)

Up to now, the measured signal Φ (or ΔT) has not been changed at all. But for some investigations it is necessary to calculate the true heat-flow rate Φ_r into the sample from the measured curve Φ.

3.4.4.1 Desmearing by solution of differential equations

The mathematical interrelation is presented (approximately) in section 3.1. Thus we can calculate the desired function $\Phi_r(t)$ from the measured function Φ (or $\Delta T(t)$, which is connected to Φ by the equation $\Phi = -K \, \Delta T$) with the aid of Eq. 3.6 (section 3.1.3) and the relation $\Phi_m = \Phi - \Phi_0$:

$$\Phi_r(t) - \Phi_m + a_1 \frac{d\Phi_m}{dt} + a_2 \frac{d^2\Phi_m}{dt^2} \tag{3.13}$$

The sought function is a simple sum of terms which include the measured function and its derivatives. The coefficients a_1 and a_2 contain the time constants τ and τ_1 of the apparatus, and thus the thermal resistances and capacities of the equipment. Equation 3.13 can easily be realized by an electronic circuit with operational amplifiers. Thus we can get the 'desmeared' signal 'online' (simultaneously) from the apparatus if the empty pan line Φ_0 is neglectible. Mostly one has to calculate the signal from the (stored) function $\Phi_m(t)$ after the measurement has finished.

Although the solution of the problem is easy to realize (mathematically or by electronics), the determination of the proper coefficients a_1 and a_2 (Eq. 3.13) is not so easy in practice. The time constants τ and τ_1 (Eq. 3.6) can be found from the measured function of a pulse-like event (cf. Fig. 3.9), but the real connection to a_1 and a_2 may be more complicated than the approximative calculations in section 3.1.3 predict. There is no other way than to adjust these two coefficients until the results of the desmearing procedure (online or calculated) coincide with the original heat-flow rate Φ_r inside the sample. The best method is to switch on a constant heat-flow rate (for a certain time) with the aid of a built-in electric heater.

3.4.4.2 Deconvolution

Another method has its roots in the theory of linear response [18]. It is valid for all apparatuses that operate linearly, which means that the measured signal for two distinguished pulse-like events in the sample is the sum (superposition) of the single function from each event alone (Fig. 3.25). Another condition is that all the measured functions of different pulse-like events should have the same shape, in other words, all these measured functions divided by their peak areas (normalizing) must yield the same function, the so-called apparatus function $a(t)$.

If these conditions are fulfilled, then

$$\Phi_m(t) = \int \Phi_r(t')a(t - t') \, dt' \equiv \Phi_r(t) * a(t) \qquad \text{(convolution)} \tag{3.14}$$

This equation holds for all DSCs that work in the linear manner described above. The only knowledge necessary is the 'apparatus function' $a(t)$, which can easily be found from a pulse-like event (usually produced with a built-in electric heater).

The unpleasant side of this desmearing method is the rather ambitious mathematics required to solve the integral equation (Eq. 3.14) for the function of interest,

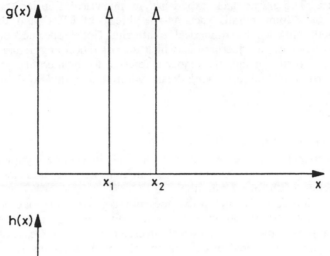

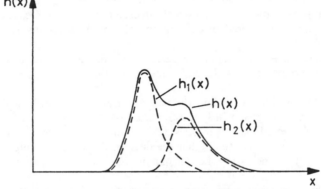

Fig. 3.25. Explanation of linear response: for two pulse-like events $g_1(x_1)$ and $g_2(x_2)$, the apparatus responds with the sum of the two single responses ($h(x) = h_1(x) + h_2(x)$)

$\Phi_r(t)$. Essentially, there are two methods for this solution. Both require numerical calculations with a computer.

The Fourier transform represents an integral operation:

$$\mathscr{F}(f(x)) \equiv 1/\sqrt{2\pi} \int_{-\infty}^{\infty} f(y) \exp(ixy)\ \mathrm{d}y$$

Applied to the convolution integral it yields (see mathematics textbooks):

$$\mathscr{F}(\Phi_m(t)) = \mathscr{F}(\Phi_r(t))\mathscr{F}(a(t)) \qquad \text{(convolution theorem)}$$

Thus the convolution product turns into an ordinary product which can be solved for

$$\mathscr{F}(\Phi_r(t)) = \mathscr{F}(\Phi_m(t))/\mathscr{F}(a(t))$$

The sought function is obtained via an inverse Fourier transform:

$$\Phi_r(t) = \mathscr{F}^{-1}(\mathscr{F}(\Phi_r(t))) = \mathscr{F}^{-1}(\mathscr{F}(\Phi_m(t))/\mathscr{F}(a(t)))$$

This method can be applied in all cases. Nowadays the Fourier transform is included in the program library of every computer. The drawbacks of this procedure lie in its laborious course and abstract nature, since the calculations are performed in

87

Fourier space. Those who lack experience in numerical Fourier transforms are advised to study some pitfalls such as the 'break-off effect' and the 'sampling theorem', both obtained by numerical treatment. Under specific conditions this simulates periodicities and fluctuations which do not reflect any actual processes in the sample. For further details the reader is referred to the literature [11, 23].

The recursion method for solving the convolution integral (Eq. 3.14) starts from the recursion formula:

$$\Phi_{ro}(t) = \Phi_m(t)$$
$$\Phi_{rn}(t) = \Phi_{r(n-1)}(t) + [\Phi_m(t) - a(t) * \Phi_{r(n-1)}(t)]$$

(3.15)

The deviation between the 'reconvoluted', still inaccurate synthetic function $\Phi_{mn}(t) = a(t) * \Phi_{r(n-1)}(t)$ and the measured function $\Phi_m(t)$ is used additively for a simple correction of the approximation.

The recursion formula (Eq. 3.15) does not converge for all event functions. Abrupt changes and steps (on-off effects and similar phenomena) generate oscillations of the approximation function which diverge rapidly. In the case of the smooth curves commonly encountered in calorimetry, the procedure converges quickly and without problems.

The desmeared curve $\Phi_r(t)$ is the basis for all further kinetic evaluation procedures, since these procedures depend heavily on the exact shape of the reaction turnover.

The greatest problem is to find the true apparatus function in question. Several methods have been suggested in literature [22, 24].

It should be noted that every desmearing procedure increases the noise. The better the resolution in time, the higher the noise will be. Desmearing should thus be carried out only if it is necessary, i.e. for narrow peaks due to the time constant of the DSC.

So far, the desmearing procedures have not changed the variable (time or probe temperature) of the transformed functions. But for special purposes it may be necessary to calculate the true reaction heat flux Φ_r as a function of sample temperature instead of time (or probe temperature). As this is a rather difficult method it is only mentioned here, and the interested reader is referred to specialist literature [12, 13, 18].

Appendix: equivalences between heat and electrical transport phenomena

Translation table

Heat transport	Charge transport
C: Heat capacity Φ: Heat flux Q: Heat R: Thermal resistance ΔT: Temperature difference	C: Capacitance i: Current q: Charge R: Resistance Δu: Potential difference (voltage)
$$Q = C\,\Delta T$$ $$\Phi = \frac{\mathrm{d}Q}{\mathrm{d}t} = C\,\frac{\mathrm{d}T}{\mathrm{d}t}$$	$$q = C\,\Delta u$$ $$i = \frac{\mathrm{d}q}{\mathrm{d}t} = C\,\frac{\mathrm{d}u}{\mathrm{d}t}$$
Biot–Fourier's law (one dimension): $$\Phi = \frac{A\lambda}{l}\,\Delta T = (1/R)\,\Delta T$$ (A = area, l = distance, λ = conductivity)	Ohm's law: $$i = \frac{A}{l\cdot\varrho}\,\Delta u = (1/R)\,\Delta u$$ (A = area, l = distance, ϱ = specific resistance)

Literature

Hemminger, W., Höhne, G.: Calorimetry, Fundamentals and Practice. Verlag Chemie, Weinheim, Deerfield Beach (FL), Basel, 1984

References

1   *Loeblich, K.-R.:* Thermochim. Acta 83, p. 99 (1985); Thermochim. Acta 85, p. 263 (1985)
2   *Höhne, G. W. H.:* Thermochim. Acta 69, p. 175 (1983)
3   *Höhne, G. W. H., Glöggler, E.:* Thermochim. Acta 151, p. 295 (1989)
4   *IPTS-68:* Metrologia 12, p. 7 (1976)
5   *ITS-90:* Metrologia 27, p. 3 (1990)
6   *Marsh, K. N. (Ed.):* Recommended Reference Materials for the Realization of Physicochemical Properties. Blackwell Scientific Publications, Oxford, London, Edinburgh, Boston, Palo Alto, Melbourne, 1987
7   *Callanan, J. E., Sullivan, S. A., Vecchia, D. F.:* Standards Development for Differential Scanning Calorimetry. National Bureau of Standards, Boulder, CO, SP 260-99
8   *Hemminger, W., Raetz, K.:* PTB-Mitteilungen 99-2, p. 83 (1989)
9   *Raetz, K.:* Thermochim. Acta 151, p. 323 (1989)
10   *Hultgren et al.:* Selected Values of Thermodynamic Properties of the Elements and Alloys. American Society for Metals, Metals Park, OH, 1973
11   *Bracewell, R. M.:* The Fourier Transform and its Applications. McGraw-Hill, New York, 1965
12   *Davies, B.:* Integral Transforms and their Applications. Springer Verlag, New York, Heidelberg, Berlin, 1978
13   *Schawe, J. E. K., Höhne, G. W. H.:* Thermochim. Act, in preparation (1994)
14   *Höhne, G. W. H. et al.:* Thermochim. Acta 160, p. 1 (1990)
15   *Cammenga, H. K. et al.:* Thermochim. Acta 219, p. 333 (1993)
16   *Höhne, G. W. H. et al.:* Thermochim. Acta 221, p. 129 (1993)
17   *Sarge, S. M. et al.:* Thermochim. Acta, in preparation (1994)
18   *Höhne, G. W. H., Schawe, J. E. K., Schick, C.:* Thermochim. Acta 229 or 230, in press (1993)

19 *Tanaka, S.:* Thermochim. Acta 210, p. 67 (1992)
20 *Schawe, J. E. K., Schick, C., Höhne, G. W. H.:* Thermochim. Acta 229 or 230, in press (1993)
21 *Hemminger, W. F., Sarge, S. M.:* J. Thermal Anal. 37, p. 1455 (1991)
22 *Flammersheim et al.:* Thermochim. Acta 187, p. 269 (1991)
23 *Müller, K. H., Plesser, T.:* Thermochim, Acta 119, p. 189 (1987)
24 *Schawe, J. E. K., Schick, C.:* Thermochim, Acta 187, p. 335 (1991)

Nomenclature

a apparatus function
A area
c specific heat capacity
e Energy
F operator of Fourier transformation
H enthalpy
i current
k coefficient
K calibration factor
l distance
L thermal conduction coefficient
p pressure
P power
Q heat
R thermal resistance
t time
T temperature
u voltage
U internal energy
V volume
W work
β heating rate
Δ difference
λ thermal conductivity
ϱ density
τ time constant
Φ heat flow (rate)

Subscripts

bl   baseline
c   extrapolated offset
e   extrapolated onset
f   offset
F   furnace
Φ due to heat flow (rate)
i number i
i   onset
m   measured
M   measuring point
p   peak
P   power
Q   due to heat
r   real (reaction)
R   reference
S   sample
st   steady state
0   empty pan line

90

Chapter 4

DSC on Polymers: Experimental Conditions

Mike J. Richardson, National Physical Laboratory,
Teddington, Middlesex, TW11 OLW, Great Britain

Contents

4.2   *Thermal lag and experimental variables* — 93

4.3   *Experimental design* — 98

4.4   *Calibration* — 100
4.4.1   Temperature — 100
4.4.2   Heat capacity and enthalpy — 101

       Literature — 103

       Nomenclature — 103

4.1 Background requirements

Differential scanning calorimetry (DSC) is one of the easiest of modern analytical techniques to use: minimal sample preparation is required and quantitative information can be obtained from only a few milligrams of material. This chapter considers experimental procedures that are common to DSC work on polymers: a brief theoretical introduction is followed by a discussion of the effect of experimental parameters on the observed signal. Features specific to individual instruments are naturally covered in the various manufacturers' manuals, and no attempt will be made to reproduce this level of detail.

The signal in a DSC experiment is related to the difference between the 'thermal response' of the sample and reference cells as the two are heated or cooled at the same rate (q_+ or q_- respectively; the subscript is omitted if comments are valid for both heating and cooling conditions) or maintained at a constant temperature (this is useful, for example, in some crystallization or polymerization experiments). 'Thermal response' is a deliberately general term which includes contributions from both the sample and the instrument; the latter may involve complex heat transfer functions ΔQ but their form is immaterial provided they are reproducible from one run to another. In its most general form, the signal S from a DSC may be written

$$S \propto (\Delta C + \Delta Q) \tag{4.1}$$

where Δ emphasizes the differential nature of the signal, which may be a differential power or temperature. ΔC is the overall difference in the total heat capacity of the two cells, and can be subdivided into instrumental and sample contributions

$$\Delta C = \Delta C_E + mc_p \tag{4.2}$$

where ΔC_E is the difference between the two cells in the empty state (subscript E) and mc_p is the additional heat capacity due to the presence of a mass m of material of specific heat capacity c_p. Any additional thermal process leading to the evolution or absorption of energy perturbs the normal heat capacity curve, and much of the value of DSC in the study of polymers concerns the wealth of information that can be derived from curves that feature crystallization or melting, annealing, change of crystal structure, the glass transition or a (usually) exothermic polymerization reaction. A full derivation of the heat capacity–temperature curve is not always required—changes with thermal treatment or from one sample to another may provide sufficient information—but the full procedure is discussed here so that potential users will be aware of the significance of the various steps and able to form a judgement of the level of sophistication needed for their particular application.

Equations 4.1 and 4.2 can be combined to give

$$kS = mc_p + (\Delta C_E + \Delta Q) \equiv mc_p + Z \tag{4.3}$$

where k is a conversion factor that relates the signal to the magnitude of the thermal process; the bracketed expression ($\equiv Z$) is difficult to determine and generally no attempt is made to do this—it will be seen below that it is sufficient to ensure that Z is repeatable from one run to another. For quantitative work three runs are required to cover a given range of temperature, and the sample cell will contain (i) empty pan; (ii) pan + sample (subscript S); (iii) pan + calibrant (subscript C). The reference cell remains undisturbed throughout (it normally contains an empty pan to 'back off' the contribution of the empty pan in (i)). The order of runs (i)–(iii) is immaterial—an 'unknown' sample will have to be run first to define the temperature

92

range that is of interest. If several samples are to be run over the same range it is sensible to arrange for the empty and calibration sets to be in the middle of the sequence so that, if there is any instrumental drift during the day, some sort of average characterizes the calibrant data.

The three forms of Eq. 4.3 that correspond to (i)–(iii) above are

$$kS_E = Z \tag{4.4}$$

$$kS_S = m_S c_{pS} + Z \tag{4.5}$$

$$kS_C = m_C c_{pC} + Z \tag{4.6}$$

subtraction of Eq. 4.4 from Eq. 4.5 and from Eq. 4.6 and division gives

$$c_{pS} = \frac{S_S - S_E}{S_C - S_E} \cdot \frac{m_C}{m_S} \cdot c_{pC} = \frac{k(S_S - S_E)}{m_S} \tag{4.7}$$

which is the basic operating equation for quantitative DSC work. The resultant 'heat capacity' curve may contain peaks and troughs due to a variety of physical and/or chemical processes. In some of these, for example melting, a detailed theoretical treatment is much more complex than that given above, and peak shapes may appear grossly distorted. However, if a particular instrument is operating in a truly calorimetric mode, integration will give the overall enthalpy change for a given process—the subsequent interpretation is discussed in many of the following chapters.

The derivation of Eq. 4.7 is valid only if Z in Eqs. 4.4–4.6 is repeatable and care must be taken to ensure that physical conditions remain as constant as possible (mass and location of pans and lids, and thermal contact). Whatever operating conditions are used, it is important to remember that the sample and reference pans are not in an adiabatic environment. There will be internal thermal gradients which are better reproduced if sample pans are lidded to present a consistent surface to their immediate environment; otherwise heat transfer (especially at high temperatures) from, for example, a carbon black filled sample will be very different from one that contains a white pigment. The second form of Eq. 4.7 shows the role of k and of the contribution from the empty pan. k is the ordinate-to-heat capacity conversion factor; it should be independent of temperature for both power compensation and most heat-flux DSC—in the former case this is an inherent instrumental feature and in the latter case electronic compensation generally ensures that this is so. In both cases the repeatability of k, with respect to experimental conditions and from day to day, provides a good test of instrumental performance. The empty pan makes a twofold contribution to the overall signal through its magnitude and its change with temperature—the shape of the empty pan line. By its very nature S_E will usually be appreciably less than S_S (or S_C), although the trend to smaller samples (perhaps microtomed from a molding or a GPC fraction) calls for increased care in the definition of S_E. Empty pan line curvature can conceal or distort a variety of effects—a simple example is provided by the many curves in the literature which appear to show that $\partial c_{pl}/\partial T < 0$ ($1 \equiv$ liquid), whereas most polymers show modest increases in c_{pl} with temperature.

4.2 Thermal lag and experimental variables

Many of the experimental aspects of DSC are conveniently discussed by considering the effects of mass, geometry and q on the effective signal S_S (or S_C) $- S_E$. An idealized experiment, with no thermal lag between sample and sensor or within the

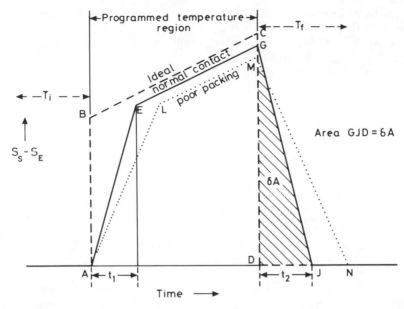

Fig. 4.1. Schematic DSC heating curves showing (greatly exaggerated) the effect of thermal lag and the area (hatched, δA) used for its calculation

sample itself, would give the instantaneous response ABCD (Fig. 4.1). A real instrument requires a finite time t_1 for the sample to achieve programmed temperature conditions and also to equilibrate at the final program temperature T_f (t_2). Both t_1 and t_2 depend on sample mass and geometry, and the progression from a well- to a poorly-packed sample is represented by curves AEGJ and ALMN of Fig. 4.1. In all cases the area enclosed by the curves is proportional to the overall enthalpy (H) change $H(T_f) - H(T_i)$, so that area ABCD = AEGJ = ALMN and the average specific heat capacity $\bar{c}_p$ is given by, say, $F \cdot \text{AEGJ}/(T_f - T_i)m$, where F is the area-to-enthalpy calibration factor (discussed in more detail in section 4.4.2). A major advantage of DSC is that it allows study of a variety of sample geometries—films, powders, granules and molten materials are all acceptable—and the calculation of $\bar{c}_p$ is one way whereby differing thermal lags can be accommodated. There is, however, no need to restrict results to $\bar{c}_p$, because the detailed $c_p - T$ curve may be obtained if a corrected temperature is available. This clearly requires an individual calibration for each sample; this can only be done by using a specific feature from each curve—the 'enthalpy lag' (hatched area, Fig. 4.1) is an obvious candidate. Appropriate procedures are described in more detail in section 4.4.1; it is sufficient to note here that the curves of Fig. 4.1 can be reduced to a common $c_p - T$ curve, the only indication of inefficient packing being a slight reduction in the temperature range covered (t_1 is increased and there is larger thermal lag at the end).

Thermal contact can be improved for irregular chunks (e.g. pellets) by cutting and/or light rubbing on a fine grade of abrasive paper to give a flat base for optimum contact with the DSC pan. This is especially useful if the as-received condition is of interest: without such preparation the sample may collapse (depending, through the viscosity, on the molar mass (MM)) on passing through the glass transition T_g or melting T_m regions to give spurious effects that can be very misleading. If this

94

behavior is suspected it may be confirmed by repeat experiments on fresh samples which will show generally similar effects but no reproducibility on a fine scale. The interfaces between sample and pan and between pan and DSC are major barriers to heat transfer, and better thermal contact can be obtained by bridging the gaps with a drop of silicone oil or other inert liquid. It is difficult to ensure that equal amounts of oil are present in successive experiments corresponding to the conditions of Eqs. 4.4–4.6, so this procedure is best suited to optimization of curves that may show fine structure in peak areas.

Other occasions when heat transfer media (metal powders are another possibility) are useful are the study of fibers and other oriented materials which retract on passing through T_g or T_m—again giving false effects unless precautions are taken. Freeze-dried polymer of high MM can be very difficult to pack into a DSC pan, and it is tempting to compact such material into a disk using an appropriate die and a small hand press. This operation needs great care: it is easy to store energy in a polymer and the subsequent DSC curve may be much distorted when this energy is released—perhaps on passing through the T_g region (see section 6.5.1).

It is clear from Eq. 4.7 that $S_S - S_E$ is directly proportional to m_S, so it might be thought advantageous always to use the maximum amount of material (conventional pans hold $\sim 30 \text{ mm}^3$) to optimize the signal-to-noise ratio. This reasoning has some validity for simple heat capacity work (i.e. $c_p - T$ curves that are free of thermal events), although even here it is important to ensure that the increased thermal lag due to the large sample is correctly calculated. However, most practical applications of DSC to polymers normally feature some kind of thermal 'event', and a large mass can seriously distort the resultant signal because the sensors, which are not in the sample, can only record an averaged response. Figure 4.2 shows melting curves for the *same* high density polyethylene. The peak temperature T_p, conventionally taken as the melting temperature *of a polymer*, is clearly very sensitive to experimental conditions, so that values reported as 'T_m' should be treated with great caution. For materials of low molar mass a more accurate value of T_m is obtained from the point of intersection of the extrapolated 'leading edge' with the baseline (X, Fig. 4.2) and this theoretically sound construction gives more consistent results for the polymer of Fig. 4.2 than does T_p. Polymers, however, melt over a range of temperature so that any single value of T_m must represent some sort of average and T_p is a simple well-defined point which, in spite of the obvious problems suggested by Fig. 4.2, has value for differentiation between samples of *comparable shape and size*.

Although T_p varies for the curves of Fig. 4.2, the overall enthalpy change through the melting region is constant; this emphasizes the *calorimetric* potential of DSC as opposed to its more common use for the measurement of temperature which, as seen above, must be heavily qualified to be of value. Overall enthalpy increases are unaffected by structural changes en route to the molten state—these merely change the shape of the enthalpy–temperature curve. A gross example of this effect is shown schematically in Figs. 4.3(a) and (b), where the broken curve represents the melting of a quenched polymer with a metastable structure α—a high heating rate q_1 has been used to prevent structural changes during the course of the DSC experiment. The full curve shows the corresponding result at q_2 ($\ll q_1$) when the sample can revert to the more stable structure β during the longer experimental timescale; in the latter case any T_m, however defined, is unrelated to that of the *original* material. In contrast, enthalpy increments are obtainable for changes from both the α and β phases to the liquid state. By using the construction shown in Fig. 4.3, heats of fusion

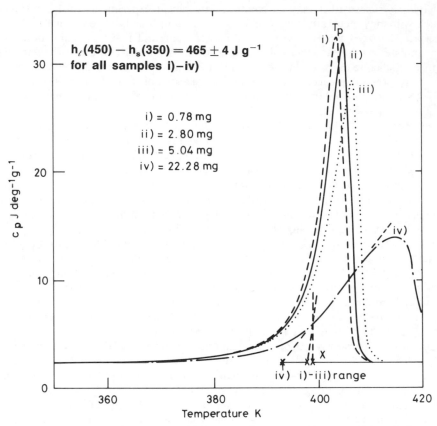

$h_\ell(450) - h_s(350) = 465 \pm 4 \text{ J g}^{-1}$
for all samples i)–iv)

i) = 0.78 mg
ii) = 2.80 mg
iii) = 5.04 mg
iv) = 22.28 mg

iv) i)–iii) range

Fig. 4.2. Melting curves for various thicknesses of constant diameter disks of the same polyethylene

and/or transition can be estimated over a range of temperatures—an important consideration for polymers when it may be necessary to reduce data for several materials to a common temperature.

The heating or cooling rate q does not specifically appear in Eq. 4.7 but it may be introduced by noting that in a power-compensation DSC, for example, a doubling of q will be directly reflected in a doubling of the differential power requirement as given by the signal $S_S - S_E$. In theory, therefore, $k_q = k_{q=1}/q$, but this must be checked experimentally—in our experience there are usually small systematic trends (of the order of 1 % when q is doubled) which are of little practical consequence since both sample and calibrant are equally affected and the ratio (i.e. c_p) remains independent of q. These remarks apply only to simple heat capacities; when a thermal event occurs the usual 'steady rate of temperature change' conditions are perturbed and a melting peak, for example, is spread out, with the peak amplitude increasing by only $\sim(q_2/q_1)^{1/2}$ rather than by the direct ratio of heating rates q_2/q_1. For this reason the melting of materials of low molar mass is much better resolved at low q_+. For polymers, however, there is always a conflict between this instrumental constraint and the need to minimize the opportunity for structural changes in the

96

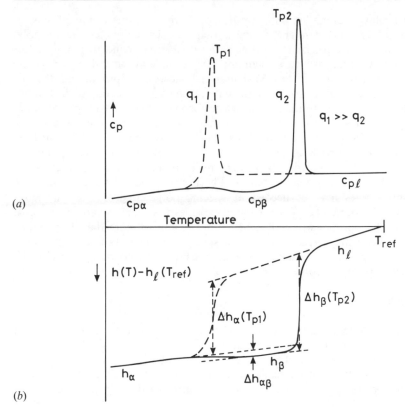

Fig. 4.3. Schematic (*a*) specific heat capacity and (*b*) enthalpy curves for the melting of an unstable quenched structure at high (----) and low (——) heating rates: constructions used to determine $\Delta h(T)$ are shown; $\Delta h_{\alpha\beta}$ is the specific enthalpy of transition for the phase change $\alpha \rightarrow \beta$

calorimeter (high q_+). The importance of the latter can only be decided for individual samples by specific experiments at different heating rates.

The practical implications of the above discussion can be summarized as follows: 10–30 mg samples with $q = 10$–20 K/min give c_p–T curves that are accurate to $\pm 1\%$ (as judged by comparison with results for standard reference materials). Heats of fusion or reaction normally require only a few milligrams of polymer, although if some distortion of peak shape (e.g. curve (iv), Fig. 4.2) can be tolerated, a larger sample may be used to define additionally the low and high temperature c_p–T curves. A note of caution should be sounded here for studies of cure (chapter 7), which often require a careful determination of the temperature at which the reaction rate is at a maximum; it is important to use only small samples in such work, otherwise the required temperature may be perturbed by self-heating.

The heat capacity of a molten isotropic polymer usually increases slowly with temperature; this region is of interest for various predictive schemes for the estimation of c_p. Increase of temperature eventually leads to the onset of thermal degradation, which is reflected in a change of slope in the c_p–T curve and is important for the development of stabilizer systems. In general, however, the interesting DSC features are found well below degradation temperatures, and problems due to loss of mass in a simple physical experiment might not be anticipated. Unfortunately, many polymers contain up to a few per cent of water or residual solvent and this may be lost during a run: it is always sensible to weigh any DSC sample at the end, as well as the start, of an experiment. Loss of low MM material shows as a broad, endothermic event which is, of course, absent on an immediate rerun but, if due to adsorbed water, may reappear after exposure to the laboratory atmosphere. Quantitative performance is naturally degraded when there is a loss of mass: the problem is not the simple change of mass—m(original) and m(final) can both be inserted in Eq. 4.7 to at least define the corresponding parts of the $c_p - T$ curve—but the uncertain correction for the associated heat of vaporization. Although residual solvent and adsorbed water are generally undesirable, and may be removed by pretreatment (dessicants, pumping under vacuum) studies may be needed of specific polymer/solvent or plasticizer systems which will require one of the sealable capsules that are available from most manufacturers. The capsules will withstand from 1–2 atm to more than 100 atm, or the whole DSC head may be controlled at a specific pressure. Whatever system is used, it is always important to check the final composition by reweighing, piercing the capsule and evaporating to constant mass.

4.3 Experimental design

When an appreciable temperature range is to be covered it is often sensible to split this into two or three shorter intervals because potential errors, essentially due to the non-equivalence of Z in Eqs. 4.4–4.6, are at their maximum in the middle of any run. Figure 4.4 shows the effect schematically for both heating and cooling curves—cooling is valuable because, provided curves refer to undisturbed samples, errors are of

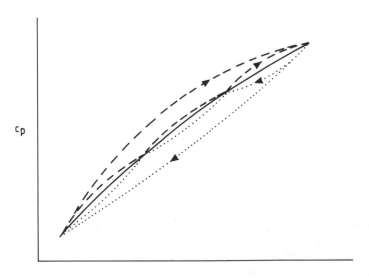

Temperature

Fig. 4.4. Apparent specific heat curves obtained in heating (----) and cooling ($\cdots$) when experimental conditions are not matched throughout the empty/calibrant/sample sequence: solid line shows true curve; errors are reduced when a long run is split into shorter steps

opposite sign to those in the heating mode. As a result, lens-shaped curves similar to those of Fig. 4.4 are common at high temperatures (where emissivity can be difficult to control) and may sometimes be found for some of the modern engineering polymers for which processing temperatures of 600 K or more are needed. Ideally, short runs should connect smoothly, and if this is the case some confidence can be placed in the equivalent single, lengthy run. An alternative procedure is to make one short run in the mid-point region of the lengthy experiment—agreement between the two again gives confidence in the main experiment. Agreement between results in heating and cooling modes is an excellent test of instrumental performance, especially the validity of any thermal lag correction. Polymers can anneal or premelt many degrees away from a melting or crystallization event, and care should always be taken to avoid stopping a scanning experiment in such a region, otherwise the resultant drift will perturb the isothermal base line. Cooling experiments involving phase transitions can show enormous hysteresis effects (many tens of degrees), but this can be turned to advantage (in addition to its value in simulating many practical processing operations) as a check on the thermodynamic capability of an instrument: the overall enthalpy change for the cycle $h_l(T_2) \to h_s(T_1) \to h_l(T_2)$ should be zero (1 is the isotropic liquid and s is the solid state; this is partially crystalline in Figs. 4.5(a) and

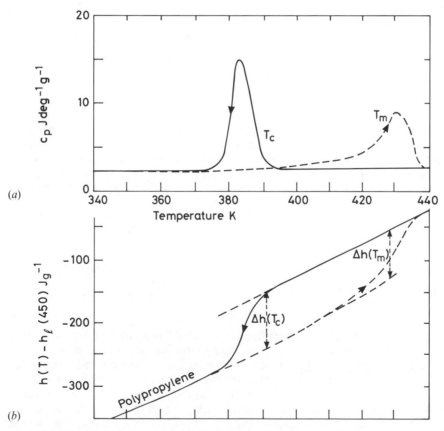

Fig. 4.5. (*a*) Specific heat capacity and (*b*) enthalpy curves in crystallization (——) and melting (---) regions (T_c and T_m respectively)

99

(b)). The relevant conditions for this test must be emphasized—the isotropic liquid state at T_2 (440 K in Fig. 4.5), rather than some ill-defined solid, is chosen to ensure a well-defined cycle, and the temperature T_1 must be sufficiently low (340 K in Fig. 4.5) to prevent structural changes on the DSC timescale (i.e. there are no unrecorded annealing effects in the equilibration period at T_1). One point that emerges from Figs. 4.5(a) and (b) is that the two peaks should *not* be of the same 'size' (however measured) because $\Delta h(T_m) \neq \Delta h(T_c)$ for the example shown. A practical point relevant to cooling experiments is that differential contraction sometimes results in a previously-molten material (not necessarily a polymer) pulling away from its sample pan; this process may be reflected by local noise in the DSC curve.

All manufacturers now offer facilities, of varying sophistication, for the storage of DSC data in digital form for subsequent processing. Although this is a welcome development it is important to understand, in principle if not in detail, what a given operation is doing. This is especially relevant to the data generation itself, since the sampling and averaging procedures may sometimes be manipulated to produce seemingly noise-free curves which may in fact 'average out' a minor event. It is always best to have access to raw data that can be smoothed, if required, at the calculation stage—only operator experience can decide the correct approach.

4.4 Calibration

Calibration procedures are placed at the end of this chapter because they are fully described in manufacturers' manuals. Some alternative approaches that emphasize the calorimetric potential of the DSC technique are discussed below.

4.4.1 Temperature

It is a very simple matter to observe the correctness, or otherwise, of the indicated instrument temperature T_1 under isothermal conditions. A standard reference material of known T_m is taken through the melting region in increments of, say, 0.1 K—the melting region is very obvious under these conditions (Fig. 4.6, see also 7.3.1). The isothermal correction is $\delta T_I = T_m - T_1$ and may be of either sign. The thermal lag

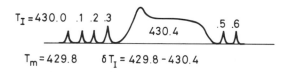

$T_I = 430.0 \quad .1 \quad .2 \quad .3$ 430.4 $.5 \quad .6$

$T_m = 429.8 \qquad \delta T_I = 429.8 - 430.4$

Fig. 4.6. Definition of the isothermal temperature (all in K) by stepwise melting of indium

under dynamic conditions δT_d is obtained from the 'enthalpy lag' (δA, Fig. 4.1) using $mc_p \, \delta T_d = F \, \delta A$ (here it is sufficient to use an approximate value of c_p, obtained from Eq. 4.7 neglecting thermal lag). δT_d is always negative for heating and positive for cooling experiments; it varies slightly with temperature but directly with sample mass and heating rate (Fig. 4.7). The extrapolation to zero mass gives an instrumental constant, the lag at the sample face in contact with the pan, that is equivalent to that obtained by the conventional 'leading edge' procedure. The latter (X, Fig. 4.2) is recommended by most manufacturers and software is generally available to give corrected temperatures. These are based on a comparison of 'observed' (i.e. onset)

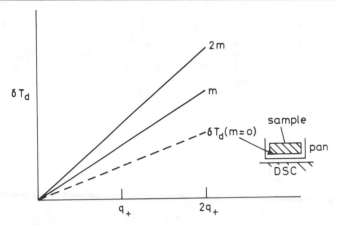

Fig. 4.7. Thermal lag as a function of heating rate q_+ and sample mass

and true transition or melting temperatures for at least two calibrants at the relevant heating rate. The precautions needed and the requirement, for accurate work, of extrapolation to $q = 0$ (equivalent to the stepwise procedure of Fig. 4.6) are emphasized (see literature).

4.4.2 Heat capacity and enthalpy

The ordinate-to-heat capacity conversion factor k (Eq. 4.7) is determined directly by measurements on materials of known c_p. By far the most popular is α-alumina, for which consistent data are available to temperatures much higher than the range of stability of polymers. At temperatures below 270 K, $\partial c_p/\partial T$ becomes large for alumina so that, if there is any uncertainty in T, there is a corresponding problem with c_p and it is better to use benzoic acid, which has a much flatter $c_p - T$ curve in this region. Benzoic acid should not, however, be used much above room temperature because it starts to sublime far below its melting point of 395.5 K. It is mentioned above that k should be independent of temperature for modern calorimeters (both heat-flux and differential-power) and it is useful to be able to display this as a check both on the consistency of absolute values from day to day and on the validity of any instrumental settings that may need to be made to ensure temperature independence. Some deviation from a truly constant value of k is tolerable provided it is reproducible—our own experience with a power-compensation calorimeter has shown that $k_+ \neq k_-$, and at extremes of temperature k may change by a few per cent. The resultant heat capacities, however, appear to be independent of experimental conditions even up to $q = \pm 80$ K/min: a confirmation of the calorimetric value of the technique.

Integration of a heat capacity–temperature curve gives an enthalpy change, and the area so defined can be used to calculate the area-to-enthalpy conversion factor F. Although the principle is simple, the practice is more complicated because it is common to use a heat of fusion rather than an enthalpy change derived only from heat capacity sources. Any DSC measurement of 'heat of fusion' must cover a temperature interval ΔT and the overall signal contains contributions from both ΔH and $C_p \Delta T$ (ΔH and C_p refer to the total, rather than specific, heat of fusion and heat capacity). The distinction between these can be made, but it is not normal to do this;

101

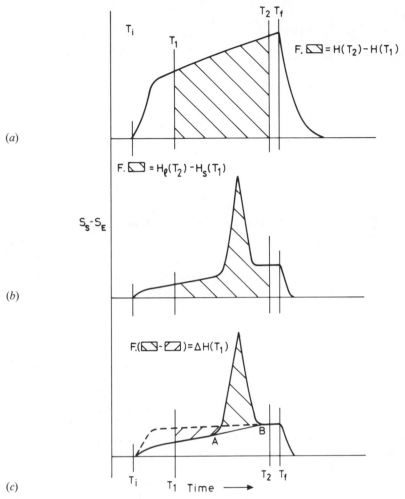

Fig. 4.8. Area-to-enthalpy calibration: (a) heat capacity only; (b) heat capacity and fusion; (c) fusion using the supercooled liquid as an internal calibrant

on the contrary, some arbitrary construction is used to derive ΔH (e.g. AB, Fig. 4.8(c)) and the result is a thermodynamically ill-defined quantity that is a poor foundation for subsequent quantitative work. Several types of enthalpy calibration are shown in Figs. 4.8(a)–(c), where hatched areas correspond to the quantities shown. In Fig. 4.8(b) the change is given by

$$H_1(T_2) - H_s(T_1) = [H_1(T_2) - H_1(T_m)] + \Delta H(T_m) + [H_s(T_m) - H_s(T_1)] \qquad (4.8)$$

$$\text{(i)} \qquad\qquad\qquad \text{(ii)} \qquad\quad \text{(iii)}$$

and quantities (i)–(iii) should be available for a suitable calibrant (recall that $\Delta H(T_m)$ is *defined* as $H_1(T_m) - H_s(T_m)$; here and in Eqs. 4.8 and 4.9, subscript s refers to the normal crystalline state for a calibrant of low molar mass—for a crystallizable polymer, which is not normally used as a calibrant, it refers to the relevant degree of crystallinity). The examples of Figs. 4.8(a) and (b) both show

102

areas that are defined by the empty pan line; if the calibrant can be supercooled (a relatively common property) it is possible to define an area (Fig. 4.8(c)) that is equivalent to

$$[H_1(T_2) - H_s(T_1)] - [H_1(T_2) - H_1(T_1)] = \Delta H(T_1) \tag{4.9}$$

The correction from $\Delta H(T_1) \rightarrow \Delta H(T_m)$ depends on ΔC_p ($= C_{pl} - C_{ps}$) through $\Delta C_p(T_m - T_1)$. All the constructions of Fig. 4.8 give simple well-defined quantities that are free of the uncertainties associated with an arbitrary base line such as AB in Fig. 4.8(c). All the methods of Fig. 4.8 should lead to equivalent results if a particular DSC is behaving as a true calorimeter.

Literature

General descriptions of the procedures for quantitative DSC are:
Richardson, M. J., in: Developments in Polymer Characterisation. *Dawkins, J. V.* (Ed.), Vol. 1, p. 205, Applied Science, London, 1978
Richardson, M. J., in: Compendium of Thermophysical Property Measurement Methods. *Maglic, K. D., Cezairliyan, A., Peletsky, V. E.* (Eds.), Vol. 1, p. 669; Vol. 2, p. 519, Plenum, New York, 1984, 1992

Further discussion of heat capacity relationships:
Mraw, S. C.: Rev. Sci. Instrum. 53, p. 228 (1982)

Analysis of transition regions:
Gray, A. P., in: Analytical Calorimetry. *Porter, R. S., Johnson, J. F.* (Eds.), Vol. 1, p. 209, Plenum, New York, 1968
Baxter, R. A., in: Thermal Analysis. Vol. 1, p. 65, Academic Press, New York, 1969

Comments on temperature calibration:
Höhne, G. W. H., Cammenga, H. K., Eysel, W., Gmelin, E., Hemminger, W.: Thermochim. Acta 160, p. 1, 1990

Appropriate materials for all aspects of calibration:
Marsh, K. N. (Ed.): Recommended Reference Materials for the Realization of Physicochemical Properties. Blackwell, Oxford, 1987

Nomenclature

c_p	specific heat capacity at constant pressure: additional subscripts may refer to sample (S) or calibrant (C), and solid (s) or liquid (l) states
$\bar{c}_p$	average value of c_p over temperature range $T_i \rightarrow T_f$
C_p	total sample heat capacity ($= mc_p$)
F	area-to-enthalpy conversion factor (Fig. 4.8)
$H_x(T), h_x(T)$	total and specific enthalpy ($H = mh$) at temperature T in the solid ($x = s$) or liquid ($x = l$) state
k	ordinate-to-c_p conversion factor (Eq. 4.7): when necessary, subscript $+$ or $-$ refers to heating or cooling experiment respectively
m	mass, subscript may refer to sample (S) or calibrant (C)
MM	molar mass
q	heating or cooling rate (when necessary, the sign is shown as a subscript)
S	DSC signal relative to the isothermal base line: subscripts refer to conditions when the instrument is empty (E) or loaded with sample (S) or calibrant (C)
t_1, t_2	times to achieve steady-state conditions in the programmed and final isothermal regions respectively (Fig. 4.1)
T	temperature
T_c, T_g, T_m	crystallization, glass transition or melting temperature

T_i, T_f	initial and final isothermal temperatures that define the range of a heating or cooling run
T_I	read-out of instrumental temperature under isothermal conditions
T_p	apparent (peak) temperature at which the rate of melting is at a maximum
T_1, T_2	any two temperatures (Fig. 4.8) in regions of 'steady rate of temperature change' (EG, LM, Fig. 4.1)
Z	total instrumental contribution to the signal S (Eq. 4.3)
ΔC	overall difference in heat capacity between DSC sample and reference cells (Eq. 4.1)
ΔC_E	as ΔC but with sample pan empty (Eq. 4.2)
$\Delta h(T_y)$	specific enthalpy of melting ($y = m$) or crystallization ($y = c$) [$\Delta H(T_y) = m \ \Delta h(T_y)$]
ΔQ	contribution of heat transfer terms to the DSC signal S (Eqs. 4.1, 4.3)
δT_d	thermal lag under dynamic (programmed change of temperature) conditions
δT_I	correction to T_I needed to give the true isothermal temperature

Chapter 5

Thermal Characterization of States of Matter

Vincent B. F. Mathot, DSM Research, P.O. Box 18, NL-6160 MD Geleen,
The Netherlands

Contents

5.1 Heat capacity

The heat capacity of a material is an extremely important thermal property [1, 2]. It provides direct experimental information on the possibilities of motion of the molecules and parts thereof; quantities derived from it, e.g. enthalpy, entropy and free enthalpy, can provide important information about the state of the material. From a technological point of view, quantitative C_p and enthalpy data are important. The temperature dependence of the quantities [3–10] just mentioned is particularly important in crystallization and melting processes of semi-crystalline polymers, as these often cover wide temperature ranges, and it is also a factor in the study of glass transition phenomena taking place in more or less amorphous polymers.

Figure 5.1 shows an overall picture of how the heat capacity of a linear polymer changes as a function of temperature, depending on the physical state of the polymer [11]. The states that have been distinguished are: completely (perfect-equilibrium)

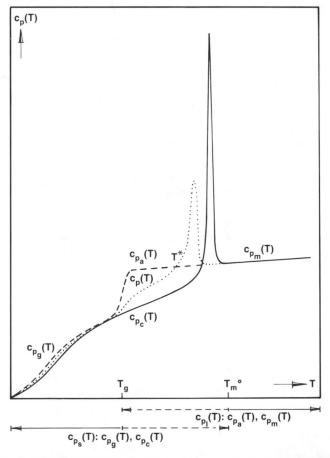

Fig. 5.1. Specific heat capacity curves of a linear polymer for some possible states: completely (perfect-equilibrium) non-crystalline, semi-crystalline and completely (perfect-equilibrium) crystalline (see text for explanation of symbols used)

crystalline/completely (perfect-equilibrium) molten states, with T_m^o as equilibrium crystal–melt transition temperature and $c_{Pc(rystal)}$ below $T_m^o/c_{Pm(elt)}$ above T_m^o; completely (perfect-equilibrium) non-crystalline states, i.e. vitrified/amorphous states with T_g as glass transition temperature and $c_{Pg(lass)}$ below $T_g/c_{Pa(morphous)}$ above T_g, and a semi-crystalline state with c_p. Specific quantities are indicated throughout by lower-case letters, for instance c_p in J/g per K. In addition, the term solid (subscript s) is used to cover glass (subscript g) and crystal (subscript c), and the term liquid (subscript l) to cover amorphous (subscript a) and melt (subscript m). In the literature the term 'melt' is normally used for materials that can occur in a (semi-) crystalline state, while the term 'amorphous' is used when there is no such state. However, many authors use these terms interchangeably.

In practice, a polymer is often either completely non-crystalline (usually referred to as 'amorphous') or partly crystallizable. The 'completely (perfect-equilibrium) crystalline' and 'completely (perfect-equilibrium) non-crystalline' states are called 'reference states' here; in most cases they represent extremes.

5.1.1 Measuring: calibration

With a relative method such as power-compensating DSC, measurement starts with calibration (see chapters 3 and 4), which, of course, is carried out according to the manufacturer's instructions and often with pure substances supplied by the manufacturer. The standard calibration procedure, which serves for temperature calibration as well as enthalpy calibration, is to melt indium, of which $T_m^o = 156.6\ ^\circ C$ and $\Delta h(T_m^o) = 28.5\ J/g$. As a second temperature-calibration material, lead ($T_m^o = 327.4\ ^\circ C$) can be used, for example. If the calibration is performed correctly, the DSC can normally be used immediately for quantitative determinations, in the sense that the heat capacities of polymers above room temperature can be measured with an accuracy of a few per cent [12]. The accuracy of the measurements of transition temperatures and enthalpies is then so high that variations in the material from which the sample was taken, e.g. inhomogeneities or a slightly different thermal history, start to play a role in the variance of the measured results.

It is tempting to perform only this simple calibration and leave it at that, but for several reasons a more extensive calibration procedure is usually necessary. It is advisable, for example, to check the calibration at different temperatures, e.g. with zinc or tin [13, 14], taking care to avoid oxidation and to always use fresh samples, preferably of the same mass and shape. Especially in the case of measurements below room temperature, an additional calibration in the temperature range of the measurement is required [15].

A procedure that is seldom applied but is nevertheless quite desirable, especially in the case of polymers, is to perform an additional calibration for cooling [16]. After all, we can hardly assume a DSC instrument to be so symmetrical that, except for the sign, calibration in cooling is the same as calibration in heating [17]. For this purpose we use 4,4-azoxyanisole, a substance for which the temperature corresponding to the transition between isotropic and anisotropic melt turns out to be virutally independent of whether the transition occurs during cooling or during heating. In many cases, sample measurements are performed at different scanning rates, which necessitates calibration for each rate [13, 16, 18, 19]. If, to improve calorimeter stability, the block surrounding the DSC unit is thermostatted at a specific temperature, the temperature calibration depends on that temperature. In general, regular checks and a suspicious attitude are advisable.

107

For measurements where maximum accuracy of the enthalpy calibration is required, a heat capacity check should be performed in the measuring range concerned by measuring the heat capacity of sapphire (Al_2O_3) and comparing the result with literature data [20–22].

Many publications [19] describe the influence of sample shape, sample mass, scanning rate, etc. on the shape of the DSC curve: see chapters 3 and 4 and their references. The sample mass should be kept as low as possible, to minimize thermal lag [18]. The lowest value attainable should be determined for the combinations of the instrument used and the samples to be measured.

5.1.2 Measuring: methods

The choice of measuring range depends not only on the phenomenon to be studied, but also on whether or not isotherms can be found before and after the occurrence of the phenomenon where the measuring signal is not influenced by the phenomenon.

In the study of crystallization and melting processes, on the low-temperature side of the process the ideal is to take an isotherm below the glass transition, but it is often possible to take an isotherm above the glass transition, provided that it is sufficiently far removed from the peak corresponding to the process on the temperature scale. On the high-temperature side of the process an isotherm in the melt region can usually be taken. If the phenomenon studied extends into the glass transition region, it is definitely necessary to take an isotherm below this transition, a difference of a few tens of degrees being recommended in many cases.

A common measuring procedure is to record a (first) heating curve on as-supplied material, which often provides insight into the thermal history of the sample. Next, from the 'liquid' state (here referred to as l), which can be a molten state (m) or an amorphous state (a), the sample is cooled at a rate of choice, down to the solid state (s), which may be a crystalline state (c) or a glassy state (g) or a combination of the two. A completely crystalline state is hardly ever reached, the final state usually being semi-crystalline. After this (first) cooling curve, a (second) heating curve is recorded, which is usually affected by the thermal history imposed by the experimenter. Standard temperature ranges and scanning rates are usually applied.

In general, DSC and DTA yield better quantitative results in heating than in cooling; the cooling curves often show more artefacts (curvature etc.). However, this is no reason to report heating curves only, as is often done, for cooling curves are equally important, especially in the case of polymers.

The measuring signal is usually expressed in dQ/dt, the heat input or output per unit time in mJ/s at constant pressure. Since the instrument is usually linked to a computer, it is nowadays customary to give the specific signal, in mJ/mg per s, which is the measuring signal divided by the sample mass. It is also possible to normalize the measuring signal to a particular sample mass which is different from the actual sample mass. The danger of a simple manipulation like this is that incomparable measuring signals can be made to seem comparable, for example when the sample masses differ widely, resulting in different thermal lags and differently distorted curves. Further, in linear time–temperature scans the signal can be divided by the scanning rate and expressed in J/g per K. This is already the dimension of the specific heat capacity, although along the ordinate no absolute scale can yet be indicated.

Whatever high-quality measuring head is used, a certain asymmetry of sample holder and reference holder cannot be avoided, which means that there will be a net measuring signal in the absence of sample, which can show a certain degree of curvature across a wide temperature range. First, this curvature should be minimized according to the manufacturer's instructions. Further, it is advisable to regularly store in the computer memory the curve obtained with the two empty holders or, in the case of heat capacity measurements, with the holders holding empty pans plus lids. This curve, which corresponds to what we shall term the empty-pan measurement (and has *nothing* to do with the so-called base line), is then subtracted from each curve taken in the same temperature range, so that the net measuring signal with possible curvature is effectively eliminated. Such a procedure is referred to as a 'pseudo-C_p' measurement (see section 5.2.6). In the case of isothermal measurements, an analogous procedure can be followed, for example to correct for phenomena caused by instrument starting.

This method is essentially already the procedure of heat capacity (C_p) measurements. In actual C_p measurements according to the classical method [23], a complication arises, especially with polymers (see ref. 24 and its references). The usual step width of 25–30 °C is unsuitable if during the isothermal waiting periods processes take place, such as crystallization or melting, so that no stable isotherms are to be expected before or after a temperature increment and truly quantitative measurement is impossible. In such cases, the step width will have to be enlarged to, say, 100–200 °C [25–28], and the stability of the calorimeter will have to be improved. This measuring method is referred to here as the 'continuous' method, as opposed to the 'stepwise' method.

Figure 5.2 shows DSC specific heat capacity [$c_p(T)$] cooling curves (cc) and heating curves (hc) for Al_2O_3 and an amorphous ethylene–propylene copolymer, EJU 193 (44 mol-% ethylene, $M_n = 6$, $M_w = 15$, $M_z = 26$ (kg/mol) according to size

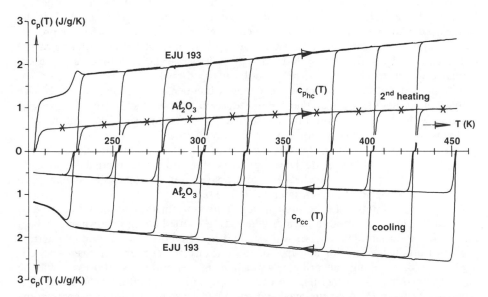

Fig. 5.2. DSC-2 continuous (isothermal stays at 203 K and 453 K) and stepwise (25 K steps) specific heat capacity curves for cooling $c_{p_{cc}}(T)$ and subsequent heating $c_{p_{hc}}(T)$ for sapphire and an ethylene–propylene copolymer: scanning rates $-/+10$ K/min; × indicates literature values for sapphire

exclusion chromatography (SEC)) [22, 29, 30]. It is clear that in heating as well as in cooling, the continuous method (in this case measurement in one step of 250 K) gives results that are in fair agreement with those obtained by the stepwise method, here applied with steps of 25 K. There is also agreement with the literature data for Al_2O_3. At the lowest temperatures, glass transition phenomena are seen for EJU 193, while the temperature dependence of c_p is strikingly linear over more than 200 K. The cooling and heating curves are symmetrical with respect to the temperature axis, which inspires confidence in the measurements from a thermodynamic viewpoint. It should be noted that c_p values measured in cooling are scarce, although such measurements are at least equally important as the usual c_p measurements in heating, as remarked above. Furthermore, combination of the two types of value in principle allows the testing of the degree of thermodynamic irreversibility during the measuring procedure [31].

The conventions currently used for the presentation of endothermic and exothermic processes can cause confusion. For DTA, the standard procedure according to ICTAC [32], which has also been adopted by IUPAC, ISO, ASTM and AFNOR, is that a positive temperature differential ($dT = T_S - T_R > 0$) of sample with respect to reference is plotted as an upward deflection, and a negative differential ($dT < 0$) as a downward deflection. Since an exothermic process causes the sample temperature to rise above the reference temperature ($dT > 0$), the peak has to be plotted upwards. By the same reasoning, an endothermic peak ($dT < 0$) is plotted downwards.

In heat-flux DSC, until recently the DTA convention was customarily applied, i.e. endothermic effects were plotted downwards and exothermic effects upwards, also when the measuring signal has been converted to dQ/dt or dQ/dT.

For power-compensating DSC, no ICTAC standard has yet been established. If, as in the case of DTA, the terms 'positive' and 'negative' refer to the condition of the sample relative to the reference, but now with dQ/dt instead of dT, an exothermic peak should be plotted downwards and an endothermic peak upwards, which is the convention used in practice. After all, at a constant pressure, in an exothermic process the enthalpy of the sample H_S decreases by dQ with respect to the enthalpy of the reference H_R ($dQ = H_S - H_R < 0$; $dQ/dt < 0$), whereas in an endothermic process it increases with respect to the reference ($dQ > 0$; $dQ/dt > 0$).

It should be borne in mind that power-compensating DSCs also measure temperature differences, which they try to reduce to zero as quickly and effectively as possible with a relatively low time constant ($\sim 2\,s$ for the DSC-7) by means of a feedback system, the energy required for these reductions being recorded.

It is of course confusing when DTA signals and heat-flux DSC signals on the one hand, and power-compensating DSC signals on the other, are plotted in opposite directions [33], especially since heat-flux and power-compensating DSC signals have been brought under the same heading (DSC) [34].

A practical solution to the present non-uniformity in presentation would be to maintain the DTA convention in cases where dT is plotted along the ordinate (which means that endothermic processes will usually be plotted downwards and exothermic processes upwards), and for all types of DSC to use the convention given above if dQ/dt is plotted along the ordinate (which means endothermic processes will be plotted upwards and exothermic processes downwards). In that case there will be a difference between DTA on the one hand and DSC (heat-flux and power-compensating) on the other. With regard to this difference, it should be borne in mind that the ordinates (dT and dQ/dt respectively) are also essentially different, so that any attempt to impose complete uniformity would be artificial.

The presentation of the results of C_p measurements has not been completely standardized either. Normally there is no problem, because the measurements reported in literature usually concern a heating run. Again applying the terms 'positive' and 'negative' to the condition of the sample with respect to the reference, and assuming that no extra heat process occurs, a temperature increase $(dT > 0)$ in the case of polymers corresponds to an increased enthalpy of the sample with respect to the reference $(dQ > 0)$ and a positive C_p $(dQ/dT > 0)$. The same holds for an endothermic process, e.g. melting, and the usual procedure is to plot the curve upwards. Also in heating, exothermic processes can occur, e.g. cold crystallization above the glass transition and recrystallization. The C_p will then decrease because it is now the 'normal' C_p $(dQ/dT > 0)$ plus a negative contribution of the C_p corresponding to the exothermic process $(dQ/dT < 0)$. If this contribution is sufficiently large, the measured C_p can become negative [28]. If measurement takes place in a cooling run and no extra thermal processes occur, C_p becomes positive again, of course, since the enthalpy decreases $(dQ < 0)$ but the temperature also decreases $(dT < 0)$. The same holds if an exothermic process occurs, e.g. crystallization.

To preserve the link with the (power-compensating) DSC presentation, in the figures given here the C_p curves are mostly plotted upwards in heating and downwards in cooling, because in heating endothermic processes and in cooling exothermic processes usually occur. Since, as just shown, C_p will usually be positive in both cases, the corresponding ordinates are both taken positive (in the sense that a heating curve is positive when it lies above and negative when it lies below the x-axis, whereas a cooling curve is positive below and negative above the x-axis). As an example, Fig.

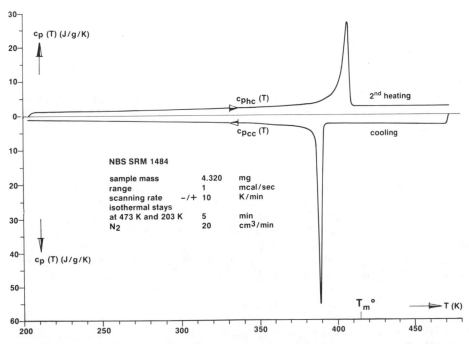

Fig. 5.3. DSC-2 continuous specific heat capacity curves for cooling $c_{p_{cc}}(T)$ and subsequent heating $c_{p_{hc}}(T)$ for an LPE between 203 K and 473 K: peak temperatures 389.4 K and 407.0 K

5.3 gives DSC cooling and heating curves for the linear polyethylene NBS SRM 1484 ($M_n = 91$, $M_w = 121$, $M_z = 154$ (kg/mol) according to SEC), measured by the continuous method [24]. The instrument starting signals are also plotted. In this particular case, the run spanned 270 K. The c_p measurements show crystallization followed by melting.

Detailed information on thermal analysis and calorimetry equipment and methodology can be found in refs. 8, 35–46 and their references.

5.1.3 Measuring: instrumentation

Because measurements had to be performed across broad temperature ranges in single runs, extra attention was paid to the stability of the DSCs. To ensure this stability, the block surrounding the DSC unit was thermostatted at +5 or +10 °C (measurements above ambient temperature) or −120 °C (other measurements) with the aid of liquid nitrogen controlled by a Cryoson TRL 5 unit or a Lauda cryostat RM6. In all these cases, the measurements were performed with Perkin-Elmer calorimeters, types DSC-2 and DSC-7.

The DSC-2 measurements were performed online with either a Tektronix 4052 PC or a Vectra ES 12 PC (with two basic language processors), a Hewlett-Packard 3495A scanner-multiplexer and a Hewlett-Packard 3455A digital voltmeter ($5\frac{1}{2}$–$6\frac{1}{2}$ digit). In measurements across less/more than 200 °C, the temperature and the corresponding measuring value were recorded every 0.1/0.2 °C. Further detailed information can be found in ref. 47.

The DSC-7 was coupled to a Hewlett-Packard A310 computer through a TAC-7 controller, the resolution of which was 1 μW at a dynamic range of ±320 mW, using the RS232-C interface. Measurements were fully computer controlled. Peripheral equipment, such as liquid nitrogen cooling, could be controlled in the same way. The programming language was HP BASIC running under a BASIC (or UNIX) operating system. Temperature and signal were sampled at 1 Hz and stored. The signal was not filtered. At a scanning rate of 5 °C/min the typical noise level was ±1 μW at a dynamic range of ±320 mW.

Very dry nitrogen was used as a purge gas. In case of heat capacity measurements the sample masses were determined to the nearest 1 μg with a Mettler ME 22/36 electronic microbalance.

5.1.4 The liquid state, $T > T_g$; $T > T_c$ or T_m

When a material is in the liquid state (l) (i.e. amorphous (a) above its glass transition temperature T_g, molten (m) above its melting transition temperature T_m or supercooled above its crystallization transition temperature T_c), its heat capacity as a function of temperature $C_{p_l}(T)$ is often fairly accurately approximated by a straight line with a positive temperature coefficient over a reasonably broad temperature range [6, 8, 10] (selenium is an exception [8, 48, 49]). A clear example of this is the linearity across 200 K of the heat capacity of the ethylene–propylene copolymer in Fig. 5.2.

5.1.5 The solid state, $T < T_g$; $T < T_c$ or T_m

When the material is in the glassy state (g), from the glass transition region down to ~150 K, the change of C_p with temperature is often linear or quadratic [1, 28]. The increase in C_{p_s} (the term 'solid' (s) is used for glass (g) as well as crystal (c)) with

112

temperature is always stronger than the increase in C_{p_l} with temperature. Unlike some non-polymeric materials [50], polymeric materials exhibit no appreciable difference between C_{p_g} and C_{p_c} at temperatures below the glass transition temperature down to ~ 60 K. In this region, the heat capacity is practically independent of morphology, crystallinity, chain conformation and configuration, etc. It is even possible to estimate the C_p within 5 % through summation of the characteristic contributions of chemically different groups in the polymer molecule [51, 52], and the theoretical explanation of the temperature dependence of c_{p_s} has been considerably improved [8].

5.1.6 The semi-crystalline state, $T_g < T < T_c$ or T_m, and the two-phase model

Normally, crystallizable polymers are in an amorphous or semi-crystalline state in the temperature range below T_c or T_m, i.e. they have not crystallized or have crystallized only partly. This may be due to the polymerization process, the molecular structure, the thermal history or processing conditions applied, the presence or absence of additives, nucleating agents, etc.

In an approach based on the 'semi-crystalline' concept, the material is regarded as being made up of finite crystalline and amorphous regions [53, 54], with each region additively contributing its (perfect-equilibrium) amorphous or (perfect-equilibrium) crystalline share to the ultimate C_p. If, as a first approximation, we consider these regions to be sharply defined, without transition regions [53, 54], in other words, if we regard the crystal surfaces as two-dimensional defects, we are working according to the two-phase model [55–58]. For such a semi-crystalline state interfacial (free) enthalpies are usually not taken into account explicitly.

In the two-phase model outlined here, the heat capacity $C_p(T)$ will then lie between the extreme values corresponding to the completely crystalline state $C_{p_c}(T)$ and the completely amorphous state $C_{p_a}(T)$, as a weighted average, *provided that no 'excess' heat capacity*, $C_{p_e}(T)$, *occurs* [59, 60] due to crystallization or melting. This heat capacity, which we shall call the 'base-line' heat capacity $C_{p_b}(T)$ can then be written as follows (specific quantities are indicated by lower-case letters):

$$c_{p_b}(T) = w^c(T)c_{p_c}(T) + w^a(T)c_{p_a}(T) \tag{5.1}$$

with

$$w^c(T) + w^a(T) = 1 \tag{5.2}$$

where $w^c(T)$ is the mass fraction crystallinity ($W^c(T)$ is the mass percentage crystallinity).

This means that if $c_{p_c}(T)$, $c_{p_a}(T)$ and $w^c(T)$ are known (the last one may be obtained by other means, such as X-ray or density measurements), $c_{p_b}(T)$ can be calculated or, conversely, the crystallinity $w^c(T)$ can be calculated from a measured value of $c_{p_b}(T)$.

Since Eq. 5.1 holds only for $c_{p_b}(T)$, not for $c_p(T)$, the validity of this expression for $w^c(T)$ is limited. In sections 5.2.4 and 5.2.7 the general definitions, based on $c_p(T)$, are dealt with.

If $c_{p_a}(T)$ cannot be directly measured on an amorphous sample (obtained by, for example, quenching), $c_{p_l}(T)$ or an extrapolation thereof can be used instead (see Fig. 5.1), possibly complemented with values for oligomers with a low T_c.

$c_{p_c}(T)$ is more difficult to obtain. Only in very rare cases (e.g. polyethylene [61, 62], polyoxymethylene [63–65]) is it possible to obtain nearly 100 % crystalline samples, with near-perfect-equilibrium (fully) extended chain crystals whose surface area is small compared to their volume, and to measure them directly. If the fully crystalline state cannot be reached, the value of $c_{p_c}(T)$ just above T_g can be approximated by extrapolation from $T < T_g$.

If it is impossible to obtain a completely amorphous and a completely crystalline state experimentally, while in a certain temperature range there are no excess phenomena involved, there is yet another possibility. The obvious solution is then to combine the $c_{p_i}(T)$ values—and possibly the extrapolated $c_{p_a}(T)$ values—with values for $w^c(T) = 0$ and $w^c(T) = 1$ obtained by extrapolating c_{p_b} with respect to w^c at the desired temperatures with the aid of Eq. 5.1.

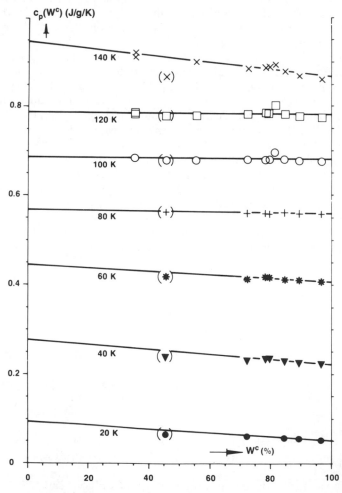

Fig. 5.4. (a) Specific heat capacity $c_p(W^c)$ data from LPEs fitted by means of straight lines at the temperatures 20–140 K: for the fit, the sample with $W^c = 45 \%$ was disregarded.

In order for this procedure to work, samples with maximally different crystallinities must be available, and $c_{p_b}(T)$ measurements must be carried out on them. In Fig. 5.4 this is illustrated for linear polyethylene (LPE). It is clear that in the case of LPE the extrapolation to $c_{p_{W^c=0\%}}(T)$ is still uncertain, and that only further research on low-crystallinity samples can reduce this uncertainty. Up to 290 K, for $c_{p_a}(T)$ the extrapolation based on W^c was applied, and from 290 K various melt data were used. The analysis was focused on LPEs precisely to avoid excess phenomena; BPEs (branched polyethylenes) already start to show excess phenomena at 220 K. For details, see refs. 6, 10, 11, 66, 67.

It is likely that in this way a consistent picture can be built up and that reference values for the specific heat capacity for the extreme states in which the polymer can occur, completely (perfect-equilibrium) crystalline/completely (perfect-equilibrium) non-crystalline, can be established.

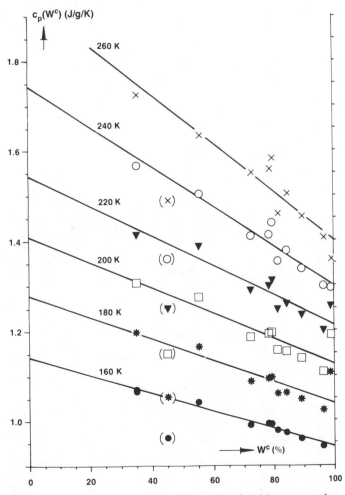

Fig. 5.4. (b) Specific heat capacity c_p (W^c) data from LPEs fitted by means of straight lines at the temperatures 160–260 K: for the fit, the sample with $W^c = 45\%$ was disregarded.

One should be aware that the values for the extreme states thus obtained are first approximations and that all kinds of refinements are conceivable. For example, the enthalpy of a sample consisting almost entirely of crystals that are small and therefore have a relatively large surface area will differ from the enthalpy of a sample that consists almost entirely of extended chain crystals, which have a relatively small surface area. Another factor that can influence the extrapolation is the crystal surface structure, e.g. tight folds, loops, cilia, and chains with end-groups paired. One might try to incorporate these refinements in a two-phase model [59, 68–70] or, if the effects are substantial, in a three-phase model (see for example section 5.3.4).

The ATHAS data bank [1, 4, 10] contains data for the extreme states, obtained by the procedures described above, for many polymers. For the theoretical background of the C_p reference values see ref. 5. Figure 5.5 shows the results of a complete analysis of linear polyethylene [6]. The data-points represent the values

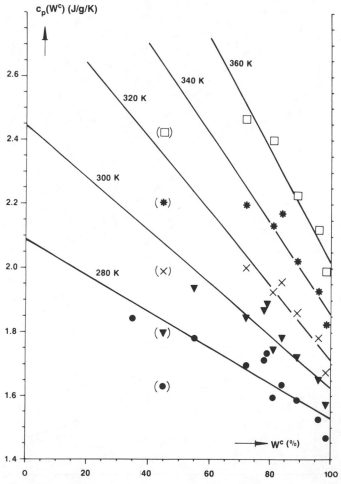

Fig. 5.4. (*c*) Specific heat capacity c_p (W^c) data from LPEs fitted by means of straight lines at the temperatures 280–360 K: for the fit, the sample with $W^c = 45\%$ was disregarded.

116

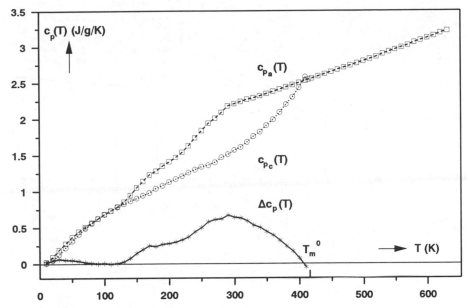

Fig. 5.5. The specific heat capacity reference values $c_{p_c}(T)$, $c_{p_a}(T)$ and $\Delta c_p(T)$ for LPE: $c_{p_c}(T)$ from $c_{p_{Wc}=100\%}(T)$ up to 240 K, supplemented with $c_{p_{Wc}=98.5\%}(T)$ data at 240 K and above; $c_{p_a}(T)$ from $c_{p_{Wc}=0\%}(T)$ up to 290 K, supplemented with values obtained by extrapolation of polyethylene melt data and paraffin melt data at 290 K and above

from Table 4 in ref. 6 (they have been rounded, which explains the slightly irregular pattern in the data, see for instance the high temperature $c_{p_a}(T)$ data).

From 120 K to 290 K there is a gradually increasing difference between the $c_p(T)$ curves for $W^c(T) = 0\%$ and $W^c(T) = 100\%$, possibly due to an extended glass transition range. For the sake of convenience, the curve with $W^c(T) = 0\%$ is referred to as $c_{p_a}(T)$ and all related functions have also been given the subscript a.

Figure 5.6 shows the heat capacity curves for cooling $c_{p_{cc}}(T)$ and subsequent heating $c_{p_{hc}}(T)$, for an LPE, NBS SRM 1484, as well as the c_p reference curves in the same temperature range [24] (see Fig. 5.5). $c_p(T)$ polynomials were used to obtain the analytical expressions for c_p and h reference curves in this chapter and in chapter 9 [24]. Figure 5.6 also shows that the results of c_p measurements performed stepwise in temperature ranges where stable isotherms could be obtained are in agreement with the results of the continuous c_p measurements.

At the beginning of this section it is said that $c_p(T)$ lies between $c_{p_c}(T)$ and $c_{p_a}(T)$ if there is no excess heat capacity, and that then $c_p(T) = c_{p_b}(T)$. In Fig. 5.6 this is the case between 200 K and about 300 K, as in this temperature range the crystallinity is constant. Above 300 K, the experimental heating curve rises faster than the c_p reference curves, and from 350 K upwards it lies above $c_{p_a}(T)$. This excess phenomenon is due to partial melting starting around 300 K. Likewise, crystallization during cooling gives rise to an excess contribution down to about 300 K. This means that above 300 K, $c_p(T)$ contains a contribution not only from $c_{p_b}(T)$ but also from $c_{p_e}(T)$:

$$c_p(T) = c_{p_b}(T) + c_{p_e}(T) \tag{5.3}$$

117

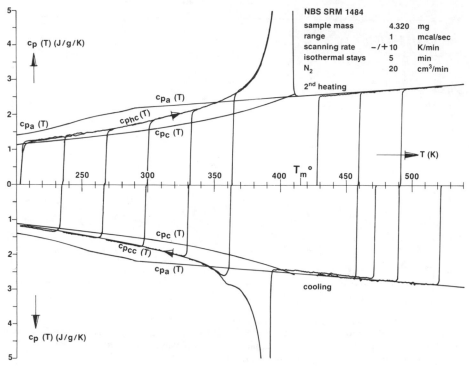

Fig. 5.6. DSC-2 continuous (same measurement as in Fig. 5.3, scale ten times more sensitive) and stepwise (32 K steps) specific heat capacity curves for cooling $c_{p_{cc}}(T)$ and subsequent heating $c_{p_{hc}}(T)$ for an LPE: reference curves $c_{p_c}(T)$ and $c_{p_a}(T)$ as in Fig. 5.5, using polynomials

5.1.7 The change of c_p in the glass transition range

In the glass transition range, amorphous samples, and often also semicrystalline samples, exhibit a change in heat capacity [51, 71–76], $c_p(T > T_g) - c_p(T < T_g)$ (see Fig. 5.1). The higher the crystallinity, the smaller is the change, so the crystallinity can be assessed roughly from the magnitude of the change. This magnitude can be determined, either from a cooling experiment or from a heating experiment, by extrapolating $c_{p_b}(T)$ from $T > T_g$ to the low-temperature side, $c_{p_b}(T_g)$, extrapolating $c_{p_s}(T)$ from $T < T_g$ to the high-temperature side, $c_{p_s}(T_g)$, both up to T_g, and taking

$$c_p(T > T_g) - c_p(T < T_g) = c_{p_b}(T_g) - c_{p_s}(T_g) \qquad (5.4)$$

Peaks caused by enthalpy relaxation (h decreases) and recovery (h increases) in heating have to be corrected for. Moreover, extrapolation is readily feasible only when no (further) crystallization or melting occurs in the same temperature range, in other words, when there is only a 'base-line c_p' and no additional 'excess c_p' (i.e. $c_p(T) = c_{p_b}(T)$) (see Fig. 5.7). The mass fraction of crystallinity defined according to Eq. 5.1 at T_g is

$$w^c(T_g) = \frac{c_{p_a}(T_g) - c_{p_b}(T_g)}{c_{p_a}(T_g) - c_{p_c}(T_g)} \qquad (5.5)$$

118

or, written with the actual heat capacity jump in the numerator (see Fig. 5.7):

$$w^c(T_g) = 1 - \frac{c_{p_b}(T_g) - c_{p_c}(T_g)}{c_{p_a}(T_g) - c_{p_c}(T_g)}$$

Using $c_{p_s}(T_g)$ as an approximation of $c_{p_c}(T_g)$, if the latter is not known, Eq. 5.5 can be written as

$$w^c(T_g) \approx 1 - \frac{c_{p_b}(T_g) - c_{p_s}(T_g)}{c_{p_a}(T_g) - c_{p_s}(T_g)} \tag{5.6}$$

so that only $c_{p_a}(T_g)$ needs to be determined, which is not much of a problem, as indicated above.

Probably the best-known semi-empirical relation for the description of the maximum possible jump in heat capacity for amorphous polymers

$$\Delta c_p(T_g) = c_{p_a}(T_g) - c_{p_s}(T_g) \tag{5.7}$$

which, for crystallizable polymers, can be approximated by

$$\Delta c_p(T_g) \approx c_{p_a}(T_g) - c_{p_c}(T_g)$$

is the rule of constant Δc_p per small mobile group (bead) [51, 52, 77], which leads to an average of 11.5 ± 1.7 J/mol per K for 32 polymers. This could be regarded as an overall average, as has been suggested for the combination of groups with a contribution of 8 J/mol per K, such as $-CH_2-$, and groups with a contribution of 14 J/mol per

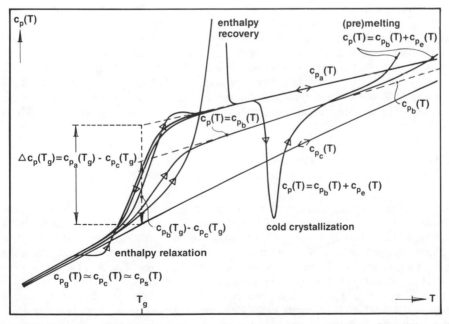

Fig. 5.7. Some examples of jumps in specific heat capacity at T_g and of possible excess phenomena: the actual jump in specific heat capacity $c_{p_b}(T_g) - c_{p_c}(T_g)$ and the maximum difference $c_{p_a}(T_g) - c_{p_c}(T_g)$, which are needed for the crystallinity calculation, are also shown

119

K, such as –CH(CH$_3$)– [6]. For a large number of groups, deviating and usually larger values for the contribution to $\Delta c_p(T_g)$ must be assumed [6, 8].

5.1.8 Determination of the glass transition temperature

Although the choice of T_g [78] has a small influence on the determination of the magnitude of the jump in specific heat capacity at T_g, $c_{p_b}(T_g) - c_{p_s}(T_g)$, a clear definition of T_g is important. At first sight, T_g seems easy to determine, as it is often defined in the dq/dT or $c_p(T)$ plot as the point of inflection, the temperature halfway up the c_p jump, etc. However, this is a purely empirical and arbitrary procedure, which may give not only wrong results but even wrong trends [28]. A procedure on a solid basis, which can best be carried out in a computerized measuring set-up, is to integrate the dq/dT or $c_p(T)$ signal and define T_g in the $\int (dq/dT)\, dT$ plot or the $h(T)$ plot respectively [76, 79–84]. T_g is then defined as the point of intersection of $h_l(T)$ and $h_s(T)$. The corresponding, and accurate, construction in the $c_p(T)$ plot is given in chapter 6.

Definition of T_g in the $h(T)$ plot is also important in the study of the kinetics of the glass transition, in particular enthalpy relaxation (see references in ref. 6; cf. also physical aging [85]), which is the decrease in enthalpy as a function of time, cooling rate, etc. below the (strongly kinetics-dependent) experimental glass transition temperature, to a state of increased thermodynamic stability. Further, heating curves become much more meaningful when the above-mentioned procedure is used, especially with regard to the degree of superheating during the heating cycle in the study of enthalpy recovery. Also, correlations with dilatometric results can then be established.

As well as in the study of enthalpy relaxation and recovery and the kinetics of the glass transition (see chapter 6), DSC is used for determination of the glass transition characteristics in cases where other techniques, e.g. dynamic mechanical analysis, which in themselves are more sensitive than DSC, cannot be used because a high cooling rate is called for [76], or the amount of sample is too small, or the sample has an unsuitable form (powder etc.). DSC is also useful for rapid screening of T_g to support exploratory research on polymerization, in studying the influence of processing conditions, in determining the homogeneity of a blend [86, 87], etc.

5.2 Heat-capacity-related functions
5.2.1 Enthalpy, entropy and free enthalpy reference functions

If a complete heat capacity analysis has been performed for a semi-crystalline polymer and $c_{p_{w c = 0}}(T)$ and $c_{p_{w c = 1}}(T)$ are known, the corresponding enthalpy $h(T)$, entropy $s(T)$ and free enthalpy $g(T)$ functions can be derived in the usual way, using calibration values in view of the need to integrate. These extreme states in which the material can occur are referred to below as reference states. The ATHAS data bank [1, 4, 10] contains such data on a great number of polymers.

As an example, the h, s and g reference states of LPE are shown in Fig. 5.8 (for details see ref. 6). In the case of LPE (see also section 5.1.6), the glass transition range is ill-defined, which means that it is not clear where the material is amorphous and where it is vitrified. For convenience, curves with $w^c = 0$ have been given the subscript a.

120

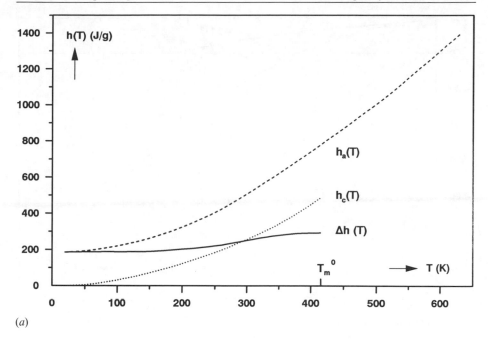

(a)

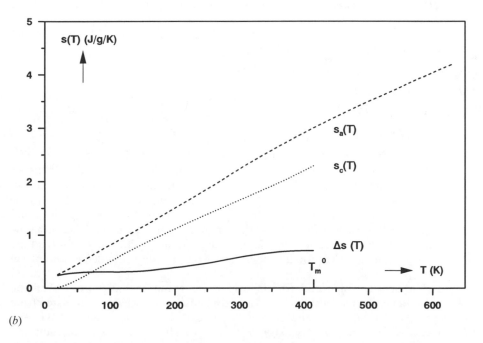

(b)

Fig. 5.8. Reference functions for LPE based on the specific heat capacity values shown in Fig. 5.5: (a) specific enthalpy functions $h_c(T)$, $h_a(T)$ and $\Delta h(T)$; (b) specific entropy functions $s_c(T)$, $s_a(T)$ and $\Delta s(T)$

121

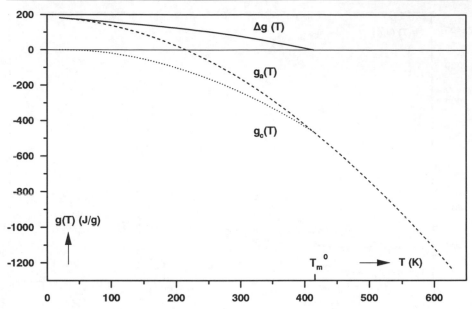

Fig. 5.8. (continued). (c) Specific free-enthalpy functions $g_c(T)$, $g_a(T)$ and $\Delta g(T)$ (T_m^o is the equilibrium crystal–melt transition temperature)

5.2.2 Heat capacity, enthalpy, entropy and free enthalpy reference differential functions

Not only the reference functions as such, but also the differences between them with respect to the extreme states are important [6]. Examples are

$$\Delta c_p(T) \equiv c_{p_a}(T) - c_{p_c}(T) \qquad \text{(see also Eq. 5.5)} \tag{5.8}$$

$$\Delta h(T) \equiv h_a(T) - h_c(T) \tag{5.9}$$

$$\Delta s(T) \equiv s_a(T) - s_c(T) \tag{5.10}$$

$$\Delta g(T) \equiv g_a(T) - g_c(T) \tag{5.11}$$

$\Delta h(T)$ is a frequently-used function (see for example refs. 27, 28, 55, 62, 88–90). In many publications, the 'equilibrium heat of fusion $\Delta h(T_m^o)$'—i.e. the difference between the enthalpies of the reference states at the 'equilibrium crystal–melt transition temperature T_m^o'—is (mis)used in the calculation of the crystallinity. The 'misuse' concerns cases where the temperature dependence of Δh is such that it cannot be neglected in the calculation of T (other than T_m^o). In calculating the enthalpy-based mass crystallinity (see section 5.2.4) at, say, room temperature, one should use the values of h_a, h and Δh at room temperature, which is often 100–200 K lower than T_m^o.

$\Delta s(T)$ plays an important role in discussions about the so-called Kauzmann paradox [91], i.e. the tendency of $\Delta h(T)$ and $\Delta s(T)$ functions when extrapolated from $T > T_g$ to pass through zero and become negative at temperatures between 0 K and T_g. Since it is not clear how the extrapolation should be done, it is not certain that the paradox exists at all. By the same token, the temperature T_2—the lower limit of the range of T_g observed in experiments of infinite time [92]—defined by $\Delta s(T_2) = 0$ (see for example refs. 75, 93) is indeterminable [6].

122

$\Delta g(T)$ is called the specific driving force of melting between T_g and T_m°; $-\Delta g(T)$ is then the specific driving force of crystallization, and has to be dealt with in any theory of crystallization. The use of the term 'driving force' for the $\Delta g(T)$ function refers to the tendency of any system towards thermodynamic equilibrium: below T_m° towards $g_c(T)$ and above T_m° towards $g_m(T)$.

5.2.3 The temperature dependence of the differential functions

We shall now discuss the temperature dependence of the differential functions [6]. For the specific heat capacity differential function:

$$\Delta c_p(T) = -T[d^2\Delta g(T)/dT^2]_p \tag{5.12}$$

$$= [d\Delta h(T)/dT]_p \tag{5.13}$$

$$= T[d\Delta s(T)/dT]_p \tag{5.14}$$

Besides the general boundary conditions

$$\Delta g(T_m^\circ) = 0 \tag{5.15}$$

$$\Delta g(0) = \Delta h(0) \tag{5.16}$$

the following expressions hold:

$$[d\Delta g(T)/dT]_p = -\Delta s(T) \tag{5.17}$$

Since

$$s_c(T) < s_a(T) \tag{5.18}$$

so that always $\Delta s(T) > 0$, it follows that

$$[d\Delta g(T)/dT]_p < 0 \tag{5.19}$$

This means that with increasing supercooling ΔT, defined by

$$\Delta T \equiv T_m^\circ - T \tag{5.20}$$

the driving force $\Delta g(T)$ always increases, regardless of how $\Delta c_p(T)$ progresses.
$\Delta g(T)$ can be calculated from the Δc_p function as follows:

$$\Delta g(T) = \Delta h(T) - T\,\Delta s(T) \tag{5.21}$$

$$= \Delta s(T_m^\circ)\,\Delta T - \int_T^{T_m^\circ} \Delta c_p(T)\,dT + T\int_T^{T_m^\circ} \frac{\Delta c_p(T)}{T}\,d(T) \tag{5.22}$$

with

$$\Delta s(T_m^\circ) = \frac{\Delta h(T_m^\circ)}{T_m^\circ} \tag{5.23}$$

As the first term in Eq. 5.22 represents the tangent to $\Delta g(T)$ in T_m°, this expression is often used as an approximation at low supercoolings. Because the corresponding $\Delta c_p(T)$ function is zero, this approximation is indicated by CP0:

$$\Delta c_{p_{CP0}} = 0 \tag{5.24}$$

$$\Delta h_{CP0} = \Delta h(T_m^\circ) \tag{5.25}$$

123

$$\Delta s_{CP0} = \Delta s(T_m^\circ) \tag{5.26}$$

$$\Delta g(T_m^\circ)_{CP0} = \Delta s(T_m^\circ)\,\Delta T \tag{5.27}$$

For greater degrees of supercooling, however, the temperature dependence of the differential functions has to be taken into account. It is logical to use heat capacity data for this purpose. Therefore, it is very important that the required C_p data be measured in the next few years. Besides, a theoretical explanation of the temperature dependence is not only desirable for a better understanding, but also necessary so that reliable predictions can be made. That such an integrated approach can be fruitful is indicated by the fact that an additivity approach enables calculation of the C_p of a polymer from the contributions of the individual component groups [8].

Section 5.1.7 shows that

$$\Delta c_p(T_g) > 0 \tag{5.28}$$

Sections 5.1.4 and 5.1.5 mention the experimental finding that, almost without exception:

$$dc_{p_s}(T)/dT > dc_{p_l}(T)/dT > 0 \tag{5.29}$$

so that

$$d\Delta c_p(T)/dT < 0 \tag{5.30}$$

This finding, which is increasingly backed by theory [8], means that, just above T_g, $\Delta c_p(T)$ is a positive function which decreases with increasing temperature. At T_m° the function might already be negative. In general, it can be derived [94] that if

if $\Delta c_p(T) > 0$, then	and if $\Delta c_p(T) < 0$, then	
$\Delta h(T) < \Delta h(T_m^\circ) = \Delta h_{CP0}$	$\Delta h(T) > \Delta h(T_m^\circ) = \Delta h_{CP0}$	(5.32)
$\Delta s(T) < \Delta s(T_m^\circ) = \Delta s_{CP0}$	$\Delta s(T) > \Delta s(T_m^\circ) = \Delta s_{CP0}$	(5.33)
$\Delta g(T) < \Delta g(T)_{CP0}$	$\Delta g(T) > \Delta g(T)_{CP0}$	(5.34)

(5.31)

The behavior of the differential functions in the very important (from a practical point of view) temperature range between T_g and T_m° has now been charted.

Figure 5.9 illustrates two possibilities: one that is based on a heat capacity function that is positive in the range concerned (Fig. 5.9(a)) and one that is based on a heat capacity function that, with increasing temperature, becomes negative at a point far before T_m° (Fig. 5.9(b); the distance from T_m° is exaggerated). In the second case, the CP0 approximation for $\Delta g(T)$, which is a tangent to the $\Delta g(T)$ curve at T_m°, can clearly be a reasonable approximation of that curve over a fairly wide temperature range.

If an experimentally derived $\Delta c_p(T)$ function can be described by a polynomial of the nth degree, then the other differential functions can be calculated as follows:

$$\Delta c_p(T) = \sum_{i=1}^{i=n+1} a_i T^{i-1} \tag{5.35}$$

$$\Delta h(T) = \Delta h(T_m^\circ) + \sum_{i=1}^{i=n+1} (a_i/i)(T^i - T_m^{\circ i}) \tag{5.36}$$

$$\Delta s(T) = \Delta s(T_m^\circ) + a_1 \ln(T/T_m^\circ) + \sum_{i=1}^{i=n+1} (a_{i+1}/i)(T^i - T_m^{\circ i}) \tag{5.37}$$

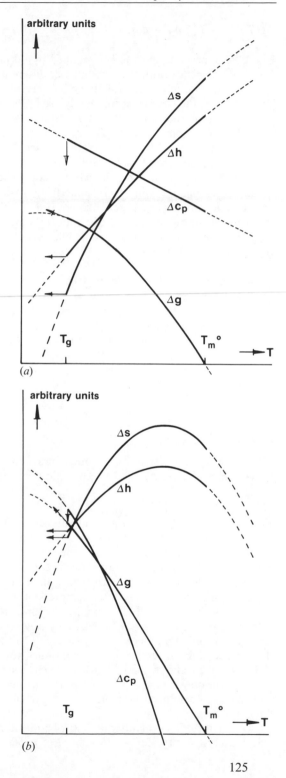

Fig. 5.9. Differential functions $\Delta h(T)$, $\Delta s(T)$ and $\Delta g(T)$ for the same values of T_g, T_m°, $\Delta c_p(T_g)$ and $\Delta h(T_m^\circ)$ but for different $\Delta c_p(T)$: in (*a*) and (*b*) the ordinates are arbitrary but equal; arrows indicate the change in the functions if $\Delta c_p(T) = 0$ for $T < T_g$; interrupted lines indicate extrapolations outside the temperature range (T_g, T_m°)

125

$$\Delta g(T) = \Delta s(T_{\mathrm{m}}^{\circ})\, \Delta T - a_1 T \ln(T/T_{\mathrm{m}}^{\circ}) + \sum_{i=1}^{i=n+1} [(a_i - a_{i+1}T)/i](T^i - T_{\mathrm{m}}^{\circ i}) \qquad (5.38)$$

The CP0 approximation can be obtained by taking all a_is zero.

The temperature dependence of the differential functions was recognized long ago (see for example refs. 91, 95). This has resulted in a number of approximations; Table 5.1 lists the ones most frequently used [96], which are explained in detail in ref. 6. The table shows that the SDM functions [94] are midway between the CP0 and HSC [97] functions. For $\Delta g(T)$, $\Delta h(T)$ and $\Delta s(T)$, the HT_{∞} functions [97] are approximated very well by the HSC functions. The $\mathrm{HT}_{\infty} = 0$ [98] and SDM approximations were developed partly on the basis of the data then known for polyethylene.

Since their introduction, the various approximations have not always been used in strict compliance with their underlying assumptions, which means that their application amounts to a purely numerical use of the differential functions. Of course, this was due to the scarcity of experimental data. This no longer being the case, the C_p-data-based reference functions should be used. The polynomial expressions are of practical use in this respect.

5.2.4 Enthalpy-based crystallinity from heat capacity curves

As discussed in section 5.1.6, a polymer is often regarded as consisting of regions whose thermal properties are those of the reference states; in other words, it is assumed that there are no transition layers between the regions and no contributions from the interfaces. In this simple two-phase model, in which additivity of certain properties is assumed, the following expressions hold between T_g and T_{m}°:

$$h_{\mathrm{c}}(T) \leq h(T) \leq h_{\mathrm{a}}(T) \qquad (5.39)$$

$$s_{\mathrm{c}}(T) \leq s(T) \leq s_{\mathrm{a}}(T) \qquad (5.40)$$

$$g_{\mathrm{c}}(T) \leq g(T) \leq g_{\mathrm{a}}(T) \qquad (5.41)$$

For the heat capacity an analogous expression can be formulated for the specific 'base-line' heat capacity $c_{p_{\mathrm{b}}}(T)$ (see section 5.1.6):

$$c_{p_{\mathrm{c}}}(T) \leq c_{p_{\mathrm{b}}}(T) \leq c_{p_{\mathrm{a}}}(T) \qquad (5.42)$$

For $c_p(T)$, Eq. 5.42 has no general validity because there is usually a contribution from $c_{p_{\mathrm{e}}}(T)$, the excess heat capacity, as well as the $c_{p_{\mathrm{b}}}(T)$ contribution. Figure 5.10 is a good example of this. It shows the specific heat capacity curves for cooling and heating, $c_{p_{\mathrm{cc}}}(T)$ and $c_{p_{\mathrm{hc}}}(T)$, respectively, for a heterogeneous 1-octene VLDPE (very low density polyethylenes are ethylene/1-alkene copolymers with densities lower than $\sim 915\ \mathrm{kg/m^3}$) with a density after compression molding of $899\ \mathrm{kg/m^3}$ and a 1-octene content of $6.6\ \mathrm{mol}$-% [47]. In the melting region, the curves measured are in good agreement with $c_{p_{\mathrm{a}}}(T)$. Virtually nowhere in the extensive measuring range do the curves lie between $c_{p_{\mathrm{c}}}(T)$ and $c_{p_{\mathrm{a}}}(T)$. From this we can conclude that, during a cooling run, crystallization continues down to very low temperatures (at first sight, down to about $-60\,^{\circ}\mathrm{C}$, where the glass transition region is entered), while during a heating run the polymer starts to melt immediately after the instrument starting signal. Both phenomena cause significant excess heat capacities at very low temperatures.

A characteristic quantity in the two-phase model is the enthalpy-based mass fraction crystallinity [5, 24, 26–28, 47, 55, 56, 58, 59, 88–90, 99–101], which is

126

Table 5.1. The various approximations for the driving force $\Delta g(T)$ and the corresponding differential functions $\Delta c_p(T)$, $\Delta h(T)$ and $\Delta s(T)$, and the polynomial expressions: for T_m read T_m°

Code[1]	Δc_p	Δh	Δs	Δg
Polynomial	$\displaystyle\sum_{i=1}^{n} a_i T^{i-1}$	$\displaystyle\Delta h(T_m) + \sum_{i=1}^{n}\frac{a_i}{i}[T^i - T_m^i]$	$\displaystyle\Delta h(T_m)/T_m + a_1\ln[T/T_m] + \sum_{i=1}^{n}\frac{a_{i+1}}{i}[T^i - T_m^i]$	$\displaystyle\Delta h(T_m)\,\Delta T/T_m - a_1 T\ln[T/T_m] + \sum_{i=1}^{n}\frac{a_i - a_{i+1}}{i}[T^i - T_m^i]$
CP0	0	$\Delta h(T_m)$	$\displaystyle\frac{\Delta h(T_m)}{T_m}$	$\displaystyle\frac{\Delta h(T_m)}{T_m}\,\Delta T$
HT$_\infty$	$\displaystyle\frac{\Delta h(T_m)}{T_m}\frac{T_g + T_m}{T_m}$	$\displaystyle\frac{\Delta h(T_m)}{T_m}\left[T - \frac{T_g\,\Delta T}{T_m}\right]$ [2)]	$\displaystyle\frac{\Delta h(T_m)}{T_m}[2K - 1]$ [3)] with $\displaystyle K = \frac{-T_g T_m + T[T_g + 2T_m]}{T_m[T + T_m]}$	$\displaystyle\frac{\Delta h(T_m)\,\Delta T}{T_m}\,K$ [3)]
HSC	$\displaystyle\frac{\Delta h(T_m)}{T_m}\frac{2T}{T_m}$	$\displaystyle\frac{\Delta h(T_m)}{T_m}\frac{T^2}{T_m}$	$\displaystyle\frac{\Delta h(T_m)}{T_m}\frac{T - \Delta T}{T_m}$	$\displaystyle\frac{\Delta h(T_m)}{T_m}\frac{\Delta T}{T_m}\frac{T}{T_m}$
HT$_\infty$ = 0	$\displaystyle\frac{\Delta h(T_m)}{T_m}$	$\displaystyle\frac{\Delta h(T_m)}{T_m}T$ [4)]	$\displaystyle\frac{\Delta h(T_m)}{T_m}\left[1 - \frac{2\Delta T}{T + T_m}\right]$ [3)]	$\displaystyle\frac{\Delta h(T_m)T}{T_m}\frac{2\Delta T}{T + T_m}$ [3)]
SDM	$\displaystyle\frac{\Delta h(T_m)}{T_m}\frac{T}{T_m}$	$\displaystyle\frac{\Delta h(T_m)}{T_m}\frac{T^2 + T_m^2}{2T_m}$	$\displaystyle\frac{\Delta h(T_m)}{T_m}\frac{T}{T_m}$	$\displaystyle\frac{\Delta h(T_m)}{T_m}\frac{[T_m^2 - T^2]}{2T_m}$

[1] For explanation of code, see text.
[2] Using $T_\infty \simeq T_g T_m/(T_g + T_m)$.
[3] Using approximation $\ln T_m/T \simeq 2\Delta T/(T + T_m)$.
[4] Using $T_\infty \simeq 0$.

127

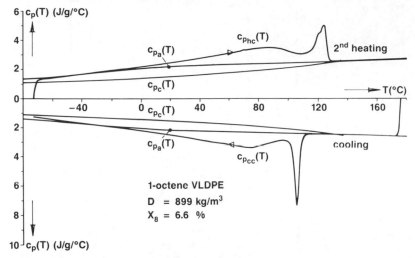

Fig. 5.10. DSC-2 continuous specific heat capacity curves for cooling $c_{pcc}(T)$ at 10 °C/min from 177 °C to −73 °C and subsequent heating $c_{phc}(T)$ at 10 °C/min for a VLDPE (7.321 mg): reference curves $c_{pc}(T)$ and $c_{pa}(T)$ as in Fig. 5.5, using polynomials

defined below T_m^o:

$$w^c(T) = \frac{h_a(T) - h(T)}{h_a(T) - h_c(T)} \tag{5.43}$$

Thus, with Eq. 5.39, $0 \leq w^c(T) \leq 1$. The notation for crystallinity expressed as a percentage is $W^c(T)$. By analogy, crystallinities can be defined via g and s.

Equation 5.43 is based on the assumptions [55, 58, 103, 104] made in the two-phase model [27], on the basis of which it can be stated that

$$h(T) \equiv w^c(T)h_c(T) + w^a(T)h_a(T) \tag{5.44}$$

with

$$w^c(T) + w^a(T) \equiv 1 \tag{5.45}$$

$h(T)$ follows from the experimental heat capacity values by integration; the reference functions $h_a(T)$ and $h_c(T)$ are discussed in section 5.2.1.

The crystallinity, defined in accordance with Eq. 5.44, is the mass crystallinity. In the two-phase model applied here, interface contributions are not explicitly taken into account for the samples discussed. In principle, the crystallinity thus calculated, based on enthalpy, should coincide with the crystallinities based on other techniques (NMR, X-ray, density, etc.) if interface contributions are negligible.

If the two-phase model *is* valid while there are non-negligible interface contributions, the enthalpy-based crystallinity will deviate from the results obtained using other techniques, because these other techniques either are not sensitive to interface contributions or show a different dependence. The modification of Eq. 5.43 to account—within the two-phase model—for the imperfection of the crystallites (folds, defects, finite crystallite thickness, etc.) contributing to the enthalpy [59, 68–70, 101, 105–111] may be seen in this light.

A similar line of reasoning holds for the case in which there are three (non-equilibrium) phases (for instance a third phase being a transition region between

128

the crystalline phase and the amorphous phase, such as a diffuse boundary layer [112–117]); see section 5.3.4 for an enthalpy-based mass crystallinity in a three-phase model without interface contributions being explicitly taken into account.

Generally, in real samples in which there are n (more than two) phases, which means that an n-phase model would have to be applied, the phases cannot be in perfect thermodynamic equilibrium. Any analytical technique will respond to this in its own characteristic way, and it is highly unlikely that the resulting calculated fractions for the phases will agree. No doubt this is also one of the main reasons for the differences in calculated crystallinity, based on measurements by different techniques, using the two-phase model.

In Fig. 5.11, the $h_{hc}(T)$ curve of the VLDPE from Fig. 5.10 is shown with the reference curves for LPE. $W^c_{hc}(T)$ was calculated according to Eq. 5.43. Figure 5.11 shows that the melting process is already in progress from $-60\,°C$ onwards, and is not finished until about $130\,°C$.

Figure 5.12 shows another example of enthalpy and crystallinity curves as obtained in cooling and heating runs, in this case for a polypropylene homopolymer ($M_w = 495$ kg/mol, $M_w/M_n = 5.4$ according to SEC, density after compression molding). The reference curves used are those of the ATHAS databank, 1990 version. This version assumes c_{p_a} and c_{p_c} to depend linearly on the temperature and gives the following values: $T_g = -3.2\,°C$; $T^o_m = 187.5\,°C$; $h_a(T^o_m) = 807$ J/g; $h_c(T^o_m) = 206.7$ J/g.

The c_p values for the melt are in good agreement with $c_{p_a}(T)$. Not surprisingly, the enthalpy curves for cooling and heating in Fig. 5.12 practically coincide with $h_a(T)$ in the temperature range concerned. The enthalpy curves were calibrated using the reference point $(T^o_m = 187.5\,°C, h_a(T^o_m) = 807$ J/g$)$. Also at low temperatures $h_{cc}(T)$ and $h_{hc}(T)$ are in good agreement, as expected for a good set of measurements in the absence of further crystallization during the isothermal waiting time and subsequent heating. The crystallinity curves, which, if there are differences between

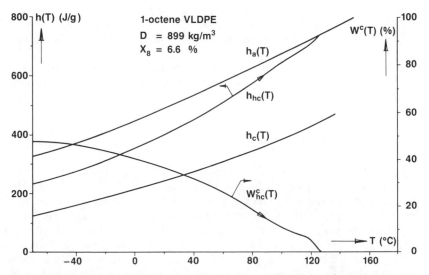

Fig. 5.11. Specific enthalpy heating curve $h_{hc}(T)$ for a VLDPE based on the measurements shown in Fig. 5.10, the reference curves $h_c(T)$, $h_a(T)$ as in Fig. 5.8(a) and the enthalpy-based mass percentage crystallinity heating curve $W^c_{hc}(T)$

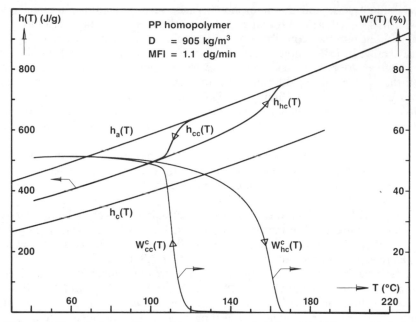

Fig. 5.12. Specific enthalpy cooling and heating curves $h_{cc}(T)$ and $h_{hc}(T)$ respectively for a PP based on DSC-7 continuous heat capacity measurements (actually based on the measurements shown in Fig. 5.15); reference curves $h_c(T)$, $h_a(T)$ from ATHAS databank: enthalpy-based mass percentage crystallinity curves for cooling $W^c_{cc}(T)$ and heating $W^c_{hc}(T)$ according to Eq. 5.43

the various experimental curves or between the experimental curves and the reference .curves, show these differences in greatly amplified form, exhibit only small discrepancies in the melting region: the deviation from the zero line is less than 1 %. These results agree well with the crystallinity curves obtained by alternative methods (see Fig. 5.16).

A similar formalism can be drawn up for a multiphase model in which the same assumptions are made with regard to additivity of some of the properties of the polymer regions and the absence of contributions from interfaces (see sections 5.2.7 and 5.3.4).

5.2.5 Transition enthalpy determination by extrapolation from melt heat capacities

In determining first-order transition enthalpies, the thermal analyst is often faced with the problem of where to draw the so-called base line [8, 24, 27, 47, 118–131]. He will usually choose two points before and after the peak, connect them by a straight line and determine the peak area included between the DSC curve and this line. The choice of these points is rather arbitrary, although most analysts will develop a certain 'feel'. Extrapolation from the melt is another commonly applied procedure. But what is the meaning of a point of intersection obtained in this manner? Depending on the specific shape of the curve, the temperature corresponding to the point of intersection, referred to below as T^*, will vary. Should the peak area thus obtained be labeled T^*? And what if there is no point of intersection? This is a

130

distinct possibility in the case of polymers with broad crystallization and melting ranges; see, for example, Fig. 5.10 and the LLDPE and VLDPE curves shown in chapter 9. Also, more insight should be obtained into the exact nature of the quantity evaluated when a certain 'base line' is drawn and the area above it is calculated. These questions clearly need to be answered, and an analyst-independent evaluation procedure needs to be developed.

Extrapolation from the melt to obtain the transition enthalpy [24, 27, 47, 88, 90] deserves closer study. Essentially (see section 5.1.6), an extrapolation from the melt of a heat capacity curve serves as an approximation of $c_{p_a}(T)$. As noted in section 5.1.4, $c_{p_l}(T)$ can usually be approximated well by a straight line with a positive temperature coefficient. This means that when the part of the c_p curve that describes the melt is extrapolated to lower temperatures according to a straight line, referred to as $c_{Pl,extr \to T}(T)$, a good approximation of $c_{p_a}(T)$ is obtained.

In the two-phase model, the area enclosed by the extrapolation curve and the DSC curve appears to be related to the numerator in the expression for $w^c(T)$ (see ref. 43), since if $h_{hc}(T)$ represents the enthalpy of the sample, calculated from $c_{p_{hc}}(T)$ (hc stands for heating curve), then

$$h_a(T) - h_{hc}(T) = h_a(T) - h_{hc}(T_r) - [h_{hc}(T) - h_{hc}(T_r)] \tag{5.46}$$

where T_r is a reference temperature in the melt—for example the equilibrium melt transition temperature— with $h_{hc}(T_r) = h_a(T_r)$. Rearrangement gives

$$h_a(T) - h_{hc}(T) = h_{hc}(T_r) - h_{hc}(T) - [h_{hc}(T_r) - h_a(T)] \tag{5.47}$$

For the first term of the right-hand member:

$$h_{hc}(T_r) - h_{hc}(T) = \int_T^{T_r} c_{p_{hc}}(T) \, dT \tag{5.48}$$

Assuming that $c_{p_a}(T)$ is approximated well by the extrapolation of $c_{p_{hc}}$ from the melt to T, referred to as $c_{Pl,extr \to T,hc}(T)$, we get for the second term of the right-hand member of (47):

$$h_{hc}(T_r) - h_a(T) \approx \int_T^{T_r} c_{Pl,extr \to T,hc}(T) \, dT \tag{5.49}$$

Thus, Eq. 5.47 can be rewritten as

$$h_a(T) - h_{hc}(T) \approx \int_T^{T_r} c_{p_{hc}}(T) \, dT - \int_T^{T_r} c_{Pl,extr \to T,hc}(T) \, dT \tag{5.50}$$

$$= [A_1 - A_2]_{hc,T} \tag{5.51}$$

(see Fig. 5.13). The reason only a dq/dT presentation is given in Fig. 5.13 is that it is not necessary to perform a c_p measurement in order to determine $[A_1 - A_2]_{hc,T}$ (see also section 5.2.6).

Although a straight-line extrapolation has been drawn in Fig. 5.13, in accordance with experience, this is not essential for the derivation given above: the analysis is also valid for a curve. The crucial point is whether or not the extrapolation, whatever form it takes, is a good approximation of $c_{p_a}(T)$ or $(dq/dT)_a$. Cooling curves (cc) are treated analogously.

131

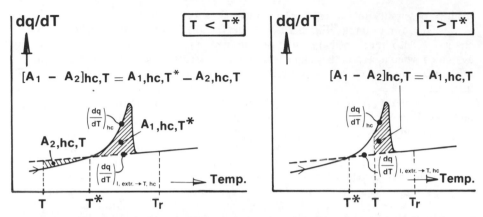

Fig. 5.13. Estimation of the transition enthalpy at T, $h_a(T) - h_{hc}(T)$, by calculation of $[A_1 - A_2]$ at T for a heating curve $(dq/dT)_{hc}$ (——) using melt extrapolation, referred to as $(dq/dT)_{l,extr \to T,hc}$ (– – –) in order to approximation $(dq/dT)_a$: similar definitions apply to a cooling curve

Figure 5.13 shows how the numerator in the expression for the crystallinity can be calculated via a simple area determination. Clearly, the calculation is valid for any temperature and for any curve shape. Extrapolation from the melt to the point of intersection with the heating curve (point T^*) and determination of the area A_{1,hc,T^*} yields $h_a(T^*) - h_{hc}(T^*)$. The analysis shows that this leads to a value of the numerator in Eq. 5.43, and hence also to, for example, a value for the crystallinity at T^*. Although peak-area determinations are often intuitively performed in this manner, no-one is really interested in information at T^*, of course; it is usually the value at a standard temperature used in comparisons, for example room temperature or application temperature, that is of interest. Figure 5.13 shows how to proceed when $T > T^*$ or $T < T^*$. It is also clear that when there is no point of intersection one should proceed as in the case $T > T^*$, so that even then a meaningful analysis is possible. Apparently, there are still excess phenomena at T in that case, since $c_{p_{hc}}(T) > c_{Pl,extr \to T,hc}(T) \approx c_{Pa}(T)$. After all, if there were only a base-line heat capacity, Eq. 5.42 would hold, with $c_{p_{hc}}(T) = c_{Pb,hc}(T) \le c_{Pa}(T)$.

A remarkable finding, at first glance, is that the transition enthalpy, defined as $h_a(T) - h(T)$ (numerator in Eq. 5.43), as a function of temperature exhibits a maximum at T^* if there is a point of intersection at T^*. This can be understood by working out what A_{1,hc,T^*} and $A_{2,hc,T}$ in Fig. 5.13 actually represent, because $h_a(T) - h(T) \approx [A_1 - A_2]_T$ (see Eq. 5.51). Figure 5.16 (see section 5.2.6) illustrates this for a polypropylene sample.

In the ultimate $w^c(T)$ this is of course no longer noticeable, because the temperature dependence of $\Delta h(T) = h_a(T) - h_c(T)$ (denominator in Eq. 5.43, see Fig. 5.16 for PP and Fig. 5.8(a) for PE) compensates the effect described above. The result is a $w^c(T)$ function which increases with decreasing temperature and may or may not become constant. On the basis of the preceding discussion, Eq. 5.43 can be written as

$$W^c(T) = \frac{[A_1 - A_2]_T}{\Delta h(T)} \cdot 100\,\% \qquad (5.52)$$

Figure 5.14 shows the resulting $W^c(T)$ curves for the LPE NBS SRM 1484, based on the measurements shown in Figs. 5.3 and 5.6 [24]. For this sample the temperatures

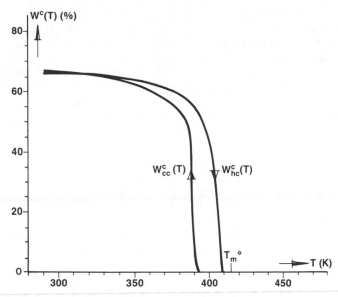

Fig. 5.14. Calculation of the crystallinity curves in cooling $W^c_{cc}(T)$ and subsequent heating $W^c_{hc}(T)$ above 290 K for an LPE according to Eqs. 5.52 and 5.53, based on the measurements shown in Fig. 5.3

corresponding to the points of intersection between the lines resulting from extrapolation of the melt data and the cooling and heating curves are $T^*_{cc} = 340$ K and $T^*_{hc} = 350$ K, respectively.

It should be noted that this calculation of $W^c(T)$ was performed using a near-perfect approximation of $\Delta h(T)$ for polyethylene (see Eq. 5.73), for the case 290 K $< T < T^o_m$:

$$\Delta h(T)_M = 293 - 0.3092 \times 10^{-5}(414.6 - T)^2(414.6 + 2T) \text{ J/g} \qquad (5.53)$$

This function is easy to use.

We have thus arrived at an analyst-independent transition-enthalpy determination with a sound foundation. It is clear, however, that the method differs from the universally used peak-area determination. We have also shown that the crystallinity determination is possible on the basis of this transition enthalpy determination. For transition enthalpy and crystallinity determinations, which are described in the next section, it is not necessary to perform actual heat capacity measurements. In section 5.2.7 the base-line determination is discussed, and it is shown that the base line is at any rate not the aforementioned line obtained by extrapolation from the melt.

5.2.6 Transition enthalpy and crystallinity from (non-heat capacity) DSC curves

Measurement of c_p curves is not strictly necessary for the analysis given in the preceding section [24, 47]. It suffices that temperature-dependent instrumental deviations are corrected for. In our experience, this can be effectively done by subtracting an empty-pan measurement from the actual measurement as a standard procedure. The same empty-pan measurement can normally be used for several measurements

133

and sometimes—depending on the stability of the DSC and possible fouling by the sample—for several days or longer.

The 'pseudo-c_p' character of such measurements makes it possible to obtain more insight into the quality of a measurement, because one can make comparisons with reference states and, in particular, check the temperature dependence. Simple checks are $dc_{p_s}(T)/dT$, $dc_{p_l}(T)/dT > 0$; $dc_{p_s}(T)/dT > dc_{p_l}(T)/dT$, etc.

In practice, however, it is advisable to avoid confusion with a real c_p measurement by standardizing the presentation, after checking, of such 'pseudo-c_p' measurements by, for example, plotting the curves 'horizontally'. By this we mean making the curve part that represents the signal from the melt more or less temperature-independent by adding to the measured signal a linearly temperature-dependent component. Clearly, this will not change the position of a specific point on the curve with respect to its temperature; the magnitude of areas will not be affected and therefore this procedure will not interfere with the determination of T^* and $[A_1 - A_2]_T$ (see preceding section).

In this way, samples can be quantitatively compared in terms of transition enthalpy without resorting to oversimplified peak evaluation on the one hand or being forced to perform c_p measurements on the other.

If data on $\Delta h(T) = h_a(T) - h_c(T)$ (the denominator in Eq. 5.43) are available, from the literature or from the previously mentioned ATHAS databank, then the

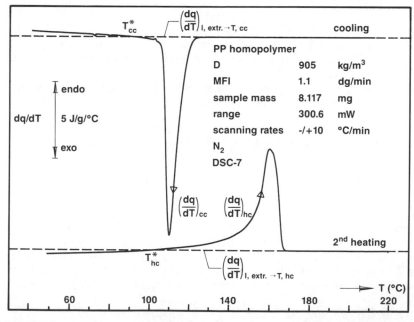

Fig. 5.15. DSC-7 'pseudo-c_p' measurements for a PP: the cooling curve $(dq/dT)_{cc}$ and the subsequent heating curve $(dq/dT)_{hc}$ were obtained by correcting the results of the measurements (combination of pan with sample and empty reference pan) by subtracting empty-pan measurements (two empty pans) and 'horizontalizing' the melt sections of the resulting curves; extrapolations from the melt along straight lines are referred to as $(dq/dT)_{l,extr \to T,cc}$ for cooling and $(dq/dT)_{l,extr \to T,hc}$ for heating; sample mass 8.117 mg; scanning rates $-/+10\,°C/min$ between 40 and 220 °C; isothermal stays 5 min (40 °C) and 10 min (220 °C)

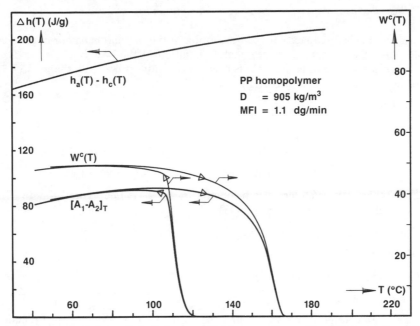

Fig. 5.16. Terms needed to calculate the crystallinities according to Eq. 5.52 for cooling $W^c_{cc}(T)$ and subsequent heating $W^c_{hc}(T)$ for the PP shown in Fig. 5.15: $[A_1 - A_2]_T$ in cooling and heating, determined by applying the procedure indicated in Fig. 5.13 to the 'pseudo-c_p' measurements shown in Fig. 5.15; $h_a(T) - h_c(T)$ from the ATHAS databank

crystallinity as a function of temperature can also be calculated from pseudo-c_p measurements, using Eq. 5.52.

Figures 5.15 and 5.16 illustrate the various aspects of the procedure for the polypropylene homopolymer mentioned earlier (see Fig. 5.12).

In Fig. 5.15 the melt sections of the cooling and heating curves that are characteristic of PP have been 'horizontalized'; in other words, they have been made temperature-independent, although this is not strictly necessary for the procedure. The extrapolations from the melt have been performed along straight lines, through the points on the curves that correspond to 140 and 180 °C (cooling curve, extrapolation referred to as $(dq/dT)_{l,extr \to T,cc}$, see section 5.2.5 for notation) and 180 and 200 °C (heating curve, extrapolation referred to as $(dq/dT)_{l,extr \to T,hc}$, see section 5.2.5 for notation). Although the extrapolations cover wide temperature ranges, the straight-line extrapolation is a plausible one, as Fig. 5.15 shows. This is confirmed by the ATHAS databank: data from this bank, too, show a linear change of $c_{p_a}(T)$ with T. The points of intersection with the cooling and heating curves are $T^*_{cc} = 92.5$ °C and $T^*_{hc} = 100.8$ °C respectively.

Figure 5.16 shows $[A_1 - A_2]_{cc,T}$ and $[A_1 - A_2]_{hc,T}$, as well as $\Delta h(T)$, based on the ATHAS data (1990 version). The resulting $W^c_{cc}(T)$ and $W^c_{hc}(T)$ are also shown in Fig. 5.16. The $W^c_{cc}(T)$ curve decreases slightly with decreasing temperature at the lowest temperatures. This can be due to several causes, which will have to be investigated by additional research. The results are in good agreement with those of the analysis given in section 5.2.4 (see Fig. 5.12); at low temperatures, the differences amount to a few per cent (absolute).

135

5.2.7 Base-line heat capacity, excess heat capacity and peak base line

We now use the heat-capacity-based description of the thermal behavior of semi-crystalline polymers to arrive at a meaningful definition of the base line.

To this end, we use the link between heat capacity and enthalpy [55, 132]. Differentiation of Eq. 5.44 according to

$$c_p(T) = (dh/dT)_p \qquad (5.54)$$

yields

$$c_p(T) = w^c(T)c_{p_c}(T) + h_c(T)\,dw^c(T)/dT + w^a(T)c_{p_a}(T) + h_a(T)\,dw^a(T)/dT \qquad (5.55)$$

$$= w^c(T)c_{p_c}(T) + [1 - w^c(T)]c_{p_a}(T) - [h_a(T) - h_c(T)]\,dw^c(T)/dT \qquad (5.56)$$

We shall regard the $w(T)$ terms as belonging to the 'base line' c_p and the $dw(T)/dT$ terms as belonging to the 'excess' c_p [47]:

$$c_{p_b}(T) \equiv w^c(T)c_{p_c}(T) + [1 - w^c(T)]c_{p_a}(T) \qquad (5.57)$$

(see Eq. 5.1), and

$$c_{p_e}(T) = -[h_a(T) - h_c(T)]\,dw^c(T)/dT \qquad (5.58)$$

$$c_p(T) = c_{p_b}(T) + c_{p_e}(T) \qquad (5.59)$$

If w^c does not change with temperature, then $c_{p_e}(T) = 0$ and only $c_{p_b}(T)$ remains in Eq. 5.59. This function lies between $c_{p_c}(T)$ and $c_{p_a}(T)$ (see Eq. 5.42). The discussion in sections 5.1.6–5.1.8 is based on this special situation. An example of a $c_p(T)$ measurement in which the excess c_p is not zero, and $c_p(T)$ does not lie between $c_{p_c}(T)$ and $c_{p_a}(T)$ in the whole temperature range between T_g and T_m^o, is shown in Fig. 5.10 for VLDPE.

For the same polymer we can now calculate $c_{p_b}(T)$ and $c_{p_e}(T)$. Figures 5.17(a) and (b) respectively show these for a heating curve [47]. The measurement was not performed on the same sample as the measurement shown in Fig. 5.10, and it is of lower quality, as seen from the systematic deviation in the melt part range with regard to $c_{p_a}(T)$. In the melt $c_{p_e}(T) = c_p(T) - c_{p_a}(T)$ because $w^c(T) = 0$. In this particular case $c_{p_e}(T)$ is not equal to zero, which it should be. This is because of the systematic deviation of $c_p(T)$ with respect to $c_{p_a}(T)$ just mentioned.

The outlined method clearly distinguishes between, on the one hand, the contribution of the temperature-dependent c_{p_c} and c_{p_a} in the experimental $c_p(T)$ through crystallinity and, on the other, the contribution of melting in the experimental $c_p(T)$. The same evaluation can be made, of course, for cooling and concomitant crystallization.

Clearly, the excess function is essential for the translation of a change of thermal properties of a material into a change of material structure or morphology [116, 124], and the determination of $c_{p_e}(T)$ is a real challenge to the thermal analyst.

A heat capacity measurement need not necessarily be made to calculate $c_{p_e}(T)$ from Eq. 5.58 [47]. The first part of the right-hand member, the enthalpy reference differential function $h_a(T) - h_c(T) = \Delta h(T)$, is known for many polymers. The change in crystallinity with temperature $dw^c(T)/dT$ can be numerically calculated from $w^c(T)$. The latter can be calculated, as shown in the preceding section, from a pseudo-c_p measurement, according to Eq. 5.52.

Figure 5.18 shows the result of such a calculation for the polypropylene homopolymer described above. The combination of dq/dT measurements of a 'pseudo-c_p'

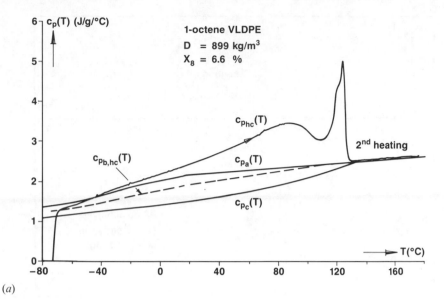

Fig. 5.17. (*a*) DSC-2 continuous specific heat capacity curves for heating $c_{Phc}(T)$ at 10 °C/min for a VLDPE (10.768 mg) after cooling at 10 °C/min from 177 °C to -73 °C (reference curves $c_{Pc}(T)$ and $c_{Pa}(T)$ as in Fig. 5.5, using polynomials; ($---$) base-line c_p for the heating curve $c_{Pb,hc}(T)$; (*b*) the resulting excess c_p for the heating curve $c_{Pe,hc}(T)$

character with extrapolations from the melt and with $\Delta h(T)$—the latter having been derived from the open literature (ATHAS databank) in this case—indeed appears to yield the c_p-excess functions, $c_{Pe,cc}(T)$ and $c_{Pe,hc}(T)$!

In principle, both excess functions are positive, since we have for the heating curve (Eq. 5.58):

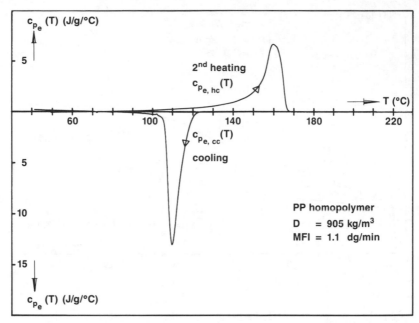

Fig. 5.18. Excess heat capacity curves for cooling $c_{\text{Pe,cc}}(T)$ and subsequent heating $c_{\text{Pe,hc}}(T)$ calculated according to Eq. 5.58 from the 'pseudo'-c_p measurements for the PP shown in Fig. 5.15: for $h_a(T) - h_c(T)$ the data from the ATHAS databank were used; $dw^c(T)/dT$ was calculated from $w^c(T)$ (see Fig. 5.16)

$$-[h_a(T) - h_c(T)] < 0; \qquad dw^c(T) < 0; \qquad dT > 0 \tag{5.60}$$

and for the cooling curve

$$-[h_a(T) - h_c(T)] < 0; \qquad dw^c(T) > 0; \qquad dT < 0 \tag{5.61}$$

$c_{\text{Pe,cc}}(T)$ turns out to be slightly negative at low temperatures (in other words, the cooling curve lies above the x-axis, see section 5.1.2 for the sign convention used here) because (see Fig. 5.16) $dw^c_{cc}(T)/dT > 0$ in that temperature range.

All this also means that the peak base line, which can now be defined as the base-line component in dq/dT obtained from a pseudo-c_p measurement, of a polymer can be calculated according to

$$(dq/dT)_b = (dq/dT)_{\text{measured}} - (dq/dT)_e$$
$$= (dq/dT)_{\text{measured}} - c_{\text{Pe}}(T) \tag{5.62}$$

In a pseudo-c_p measurement, any residual or deliberately introduced instrumental component in $(dq/dT)_b$ may be dependent on the temperature:

$$(dq/dT)_b = c_{\text{Pb}}(T) + f(T) \tag{5.63}$$

With regard to the accuracy of the determination of $c_{\text{Pe}}(T)$, everything depends on the magnitude of $f(T)$.

The approach we have followed within the two-phase model can be generalized to a multiphase model, provided that the assumptions with regard to additivity of a number of properties of regions and the absence of contributions from the interfaces between regions are valid. Analogously to Eqs. 5.44 and 5.45 we then get, for n phases:

$$h(T) \equiv \sum_{i=1}^{i=n} w^i(T)h_i(T) \tag{5.64}$$

with

$$\sum_{i=1}^{i=n} w^i(T) \equiv 1 \tag{5.65}$$

and

$$c_p(T) = c_{p_b}(T) + c_{p_e}(T) \tag{5.66}$$

with

$$c_{p_b}(T) \equiv \sum_{i=1}^{i=n} w^i(T)c_{p_i}(T) \tag{5.67}$$

$$c_{p_e}(T) \equiv \sum_{i=1}^{i=n} h_i(T)\,\mathrm{d}w^i(T)/\mathrm{d}T \tag{5.68}$$

Finally, we note that isothermal processes also can be described using the concepts discussed here. In the two-phase model used in this section, the change in measuring signal as a function of time at a temperature T is given by the excess function

$$\left(\frac{\mathrm{d}q}{\mathrm{d}t}\right)_{e,T} = \left(\frac{\mathrm{d}h}{\mathrm{d}t}\right)_{e,T} = -[h_a(T) - h_c(T)]\left(\frac{\mathrm{d}w^c}{\mathrm{d}t}\right)_T \tag{5.69}$$

The base-line signal is zero, $(\mathrm{d}q/\mathrm{d}t)_{b,T} = 0$, because $\mathrm{d}[h_c(T)]/\mathrm{d}t = \mathrm{d}[h_a(T)]/\mathrm{d}t = 0$, at least if a correction is made for instrument switch-off effects when the measuring temperature is reached and if there is no instrument drift. This means that crystallization and melting result in an exothermic (negative) and endothermic (positive) excess signal respectively. With the aid of the enthalpy differential function $\Delta h(T)$, $w^c(T, t)$ can simply be calculated from the measuring signal via integration if the integration constant (i.e. the initial or final crystallinity) has been determined from a dynamic experiment.

5.3 Applications and related subjects
5.3.1 The differential functions for LPE

As mentioned earlier (see section 5.2.3), the heat capacity differential function for polymers decreases with temperature in the region between T_g and T_m^o. In the case of polyethylene [6] (see Fig. 5.5, $\Delta c_p(T_m^o) \approx 0$) both the enthalpy differential function $\Delta h(T)$ and the entropy differential function $\Delta s(T)$ reach their maxima around T_m^o (see Fig. 5.9). This means that below T_m^o, $\Delta h(T)$ and $\Delta s(T)$ are smaller. It also means that, depending on the extent of supercooling in the crystallization range, $T_m^o - T_c$, and in the melting range, $T_m^o - T_m$, the temperature dependence of the differential functions must be taken into account in the evaluation of the measured data.

For a very small degree of supercooling, $T \lesssim T_m^o$, the CP0 approximation, $\Delta h(T) \approx \Delta h_{CP0} = \Delta h(T_m^o) = 293$ J/g (heat of fusion), can be used for $\Delta h(T)$, and likewise $\Delta s(T) \approx \Delta s_{CP0}$. As mentioned in section 5.2.3, $\Delta c_p(T_m^o) \approx 0$ also means that

139

the CP0 approximation for $\Delta g(T)$, the well-known expression $\Delta g(T) \approx \Delta g(T)_{CP0} = \Delta h(T_m^\circ)(T_m^\circ - T)/T_m^\circ$, can be used over a relatively large temperature range. All the data needed to assess the temperature dependence of the differential functions between 10 K and T_m° can be found in Table 4 of ref. 6.

For the temperature range between the temperature corresponding to the bend in $c_{p_a}(T)$ (for which different authors report different values; for the results reported here it is 290 K; see Fig. 5.5) and T_m° the temperature dependence of the differential functions can be taken into account [6, 24] without much difficulty. $\Delta c_p(T)$, $\Delta h(T)$, $\Delta s(T)$ and $\Delta g(T)$ can be mathematically represented very simply by use of the parabolic character of $\Delta c_p(T)$.

If we consider a parabola drawn through 0 K, 290 K and T_m° (see Fig. 5.5 and ref. 6) an experimental $\Delta c_p(T)$ function can be approximated with the aid of:

$$\Delta c_p(T)_M = f_{290}\, \Delta T \cdot T \tag{5.70}$$

where f_{290} is defined by taking the experimental Δc_p at 290 K:

$$f_{290} = \frac{\Delta c_p(290)}{(T_m^\circ - 290)\,290} \tag{5.71}$$

Reference 6 gives T_m° and f_{290} values with which, in the temperature range from 230 K to T_m°, various experimental differential functions for LPE can usually be approximated to within 1 %. In this manner new LPE experiments can also be characterized very simply and with a high degree of accuracy; the temperature 290 K can be used for every evaluation. As a rule of thumb one can take $f_{290} \approx 1.8 \times 10^{-5}$ (J/g per K^3) ($\sim 2.5 \times 10^{-4}$ (J/mol per K^3) or $\sim 6.0 \times 10^{-5}$ (cal/mol per K^3)), which leads to very simple shapes for the analytical expressions.

The various differential functions can now be approximated with the aid of simple analytical functions:

$$\Delta c_p(T)_M = 1.855 \times 10^{-5}(414.6 - T)T \quad \text{(J/g per K)} \tag{5.72}$$

$$\Delta h(T)_M = 293 - 0.3092 \times 10^{-5}(414.6 - T)^2(414.6 + 2T) \quad \text{(J/g)} \tag{5.73}$$

$$\Delta s(T)_M = 0.707 - 0.9275 \times 10^{-5}(414.6 - T)^2 \quad \text{(J/g per K)} \tag{5.74}$$

$$\Delta g(T)_M = 0.707(414.6 - T) - 0.3092 \times 10^{-5}(414.6 - T)^3 \quad \text{(J/g)} \tag{5.75}$$

The first term in each formula gives the approximation of the differential function for a very small degree of supercooling (the CP0 approximation, see Eqs. 5.24–5.27); the second term gives the correction required in the case of substantial supercooling. Equation 5.73 has been used before [24] (see Eq. 5.53) to approximate the denominator in the crystallinity expression (see Eqs. 5.43 and 5.52).

In the above equations 414.6 K and 1.855×10^{-5} (J/g per K^3) were used for T_m° and f_{290} respectively. Once again it should be stressed that in the case of LPE other values may be chosen for T_m° and f_{290} in the approximation formulae [6]; the approximation will still be near-perfect.

5.3.2 The differential functions for polystyrene

One of the reasons isotactic polystyrene is frequently used as a model polymer is that it is easily supercooled to a glass without crystallization. Thus, supercoolings of up to 150 K can be reached and it is therefore to be expected that the temperature dependence of the various differential functions becomes significant. Suzuki et al.

140

[133] and HOFFMAN et al. [134] have already paid much attention to this, and have particularly taken into account the temperature dependence of Δs and Δg.

Figure 5.19 shows $\Delta c_p(T)$ curves based on experimental data and curves based on the approximations given in Table 5.1. The polynomial coefficients of the KBO approximation are similar to those used by SUZUKI et al. The differences in $\Delta c_p(T)$ are not surprising, considering the differences in the samples and in the measuring procedures and techniques employed and, above all, considering the fact that c_p values that do not differ much (some of which have been obtained via extrapolation over a large temperature range) are subtracted from one another. The difference in $\Delta c_p(T_m^o)$ for WB, K and the other curves is not due to a difference in $c_{p_s}(T_m^o)$— obtained by extrapolation over 150 K—as expected, but to a difference of $\sim 6\%$ in $c_{p_l}(T_m^o)$. Considering, moreover, the uncertainty regarding $c_{p_s}(T_m^o)$, it is clear that $\Delta c_p(T_m^o)$ is in fact completely undetermined.

The approximations via HSC, $HT_\infty = 0$, SDM and CP0 fail with respect to $\Delta c_p(T_g)$, while there is 'incidental' agreement in the case of HT_∞. Furthermore, the temperature dependence of Δc_p in these approximations differs considerably from that of the curves based on experimental data, which is why these approximations should no longer be used.

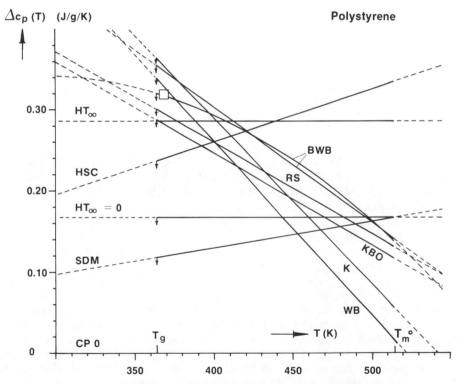

Fig. 5.19. Various approximations for the specific heat capacity differential function for polystyrene according to Table 5.1, based on experimental data from refs. 135 (KBO), 3 (WB), 136 and 3, first- and second-degree polynomial fits (BWB), refs. 82 (RS) and 137 (K): $\square$ = averages for $\Delta c_p(T_g)$ and T_g from Table 1 of ref. 6; arrows and interrupted lines as in Fig. 5.9

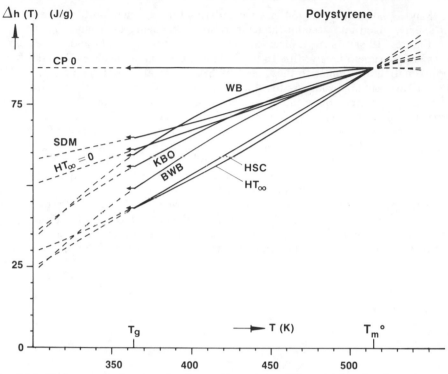

Fig. 5.20. Various approximations for the specific enthalpy differential function for polystyrene according to Table 5.1 and the experimental data in Fig. 5.19: $T_g = 364$ K, $T_m^o = 515.2$ K and $\Delta h(T_m^o) = 86.3$ J/g, according to ref. 133

In addition to the various approximations, Fig. 5.20 shows the curves with codes KBO, WB and BWB (first-degree polynomial for $\Delta c_p(T)$); the last two represent the possible variation in the experimental $\Delta c_p(T)$, as can be seen in Fig. 5.19, while KBO is often used in publications.

It is clear that at high degrees of supercooling the value at T_m^o cannot be used for $\Delta h(T)$ (and certainly cannot be used for $\Delta s(T)$). Moreover, at these high degrees of supercooling $\Delta g(T)$ differs considerably from the first-order CP0 approximation. As all experimental $\Delta c_p(T)$ functions are positive here, the CP0 approximation always constitutes an 'upper limit' (see section 5.2.3). In view of the positive temperature coefficient of $\Delta c_p(T)$ in the HSC approximation, the area under the curve is rather large compared to the areas under the other curves. In general, this approximation will hence give the lowest $\Delta h(T)$, $\Delta s(T)$ and $\Delta g(T)$ values. In comparison with the experimental curves, the HSC approximation is not a clear improvement on the CP0 approximation. Of course, it is preferable to use curves based on experimental evidence, such as BWB and WB curves (or better still, more recent data, see update [10] of the ATHAS 1980 databank [138]) for polystyrene, although the uncertainty introduced by the extrapolation of $c_{pg}(T)$ may lead to unexpected results. Incidentally, the HSC approximation indeed proves to be a convenient replacement for the HT_∞ approximation in the case of $\Delta h(T)$, $\Delta s(T)$ and $\Delta g(T)$, as noted in ref. 6. For $\Delta g(T)$ the $HT_\infty = 0$ approximation nearly coincides with the

KBO curve, which is why LAURITZEN et al. [139] and HOFFMAN et al. [134] use this approximation.

5.3.3 The δl catastrophe in crystallization theories

A striking example showing the importance of taking into account the temperature dependence of the driving force of crystallization $\Delta g(T)$ is the so-called δl catastrophe. In 1973 two articles [139, 140] were published on this topic. At that time a number of crystallization theories predicted a dramatic upswing in the fold length at high degrees of supercooling. However, experiments with isotactic polystyrene in dimethylphthalate [140] (see Fig. 5.21) showed the well-known inverse proportionality of the initial lamellar thickness l^* (see also the discussion in section 9.2.1) and ΔT at low degrees of supercooling but a levelling-off at high ΔTs (up to about 200 K). To account for the absence of a rapid increase in the growth rate and the initial lamellar thickness, LAURITZEN et al. derived the following expression for this thickness:

$$l^* = \frac{2\sigma_e}{\Delta g} + \delta l \tag{5.76}$$

with $2\sigma_e/\Delta g$ being the classical term if

$$\delta l = 0 \tag{5.77}$$

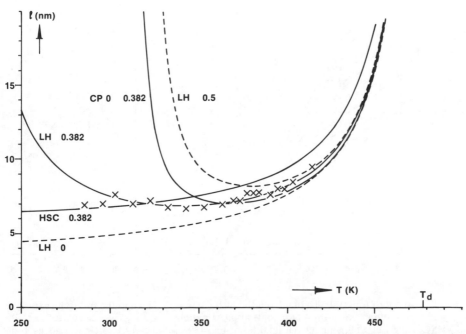

Fig. 5.21. The lamellar thickness l of isotactic polystyrene as a function of the crystallization temperature in a dilute solution: X from ref. 140; LH calculated by means of Eq. 5.76 according to ref. 139 for various Ψ values; fit of the experiment for $\Psi = 0.382$; pseudo-upswing for the same $\Psi = 0.382$ if the CP0 approximation is used for $\Delta g(T)$

143

and

$$\delta l = \frac{kT}{2b_0\sigma_s} \cdot \frac{2 + (1 - 2\Psi)a_0\,\Delta g/(2\sigma_s)}{[1 - a_0\,\Delta g\Psi/(2\sigma_s)][1 + a_0\,\Delta g(1 - \Psi)/(2\sigma_s)]} \tag{5.78}$$

where Δg stands for $\Delta g(T)$.

In Eq. 5.78 σ_e and σ_s are the end-surface and the side-surface free enthalpy for the crystal respectively, a_0 is the 'width' of a molecule, b_0 is the thickness of the growth layer, k is the Boltzmann constant and Ψ is a parameter between 0 and 1 indicating the degree to which the molecules are physically adsorbed on the substrate before being attached crystallographically (see also refs. 134, 141). The upswing in l^* that occurs when the value of δl becomes infinite would in this expression occur at the point where the denominator in Eq. 5.78 becomes zero, or at a temperature T_u, which is determined by

$$\Delta g(T_u) = \frac{2\sigma_s}{a_0}\frac{1}{\Psi} \tag{5.79}$$

Figure 5.22 shows the situation for crystallization of isotactic polystyrene from the melt and from a solution, on the basis of data given in ref. 139. Study of Eqs. 5.78

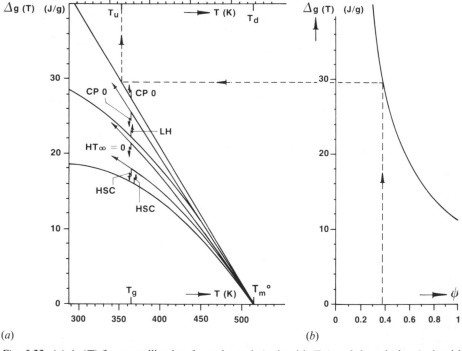

(a)   (b)

Fig. 5.22. (a) $\Delta g(T)$ for crystallization from the melt (axis with T_m°) and the solution (axis with T_m°) and the solution (axis with T_d) for some approximations: $\Delta g(T)_{\text{soln.}}$ was calculated for CP0 and HSC using Eq. 5.80, without the final term; ref. 139 used a $\Delta g(T)_{\text{soln.}}$ (indicated by code LH), which, up to a fairly high degree of supercooling, agrees with the $HT_\infty = 0$ and KBO approximations (parameter values as in ref. 139). (b) $\Delta g(T)$ as a function of Ψ according to Eq. 5.79, via multiplication by $v_c(T_m^\circ)$ expressed in J/g (ref. 139 uses $\Delta g(T)$, expressed in J/cm^3, arrived at by using the same value)

and 5.79 shows that $\Psi = 1$ always leads to an upswing whereas $\Psi = 0$ never does; there are a number of theories that can be seen as descriptions of these limiting cases [139]. From Fig. 5.22 it can easily be inferred at what value of Ψ the theory predicts an upswing. Figure 5.22(b), for instance, shows that for $\Psi = 1$ the corresponding $\Delta g(T)$ of about 11 J/g in Fig. 5.22(a) gives an upswing at $T \approx 440$ K for all the approximations of $\Delta g(T)$ shown here. For crystallization from the melt we see that, taking into account the various approximations for $\Delta g(T)$, a T_u value of $T_g < T_u < T_m^\circ$ would correspond to a Ψ value of between 0.45/0.65 and 1. As the growth rate experiments carried out [133] in the range between 385 K and 474 K give no indication whatever of a δl catastrophe, it must be concluded that in the picture presented by LAURITZEN et al., $\Psi \lesssim 0.55$. For temperatures below T_g, $\Delta g(T)$ is the continuation of the tangent in T_g to $\Delta g(T)$, which is represented by means of an arrow in Fig. 5.22. Vitrification, however, effectively prevents crystallization and a possible resulting upswing in this range.

For crystallization in a solution we may, according to ref. 94, approximate the driving force Δg_{soln} with the aid of

$$\Delta g(T)_{\text{soln}} \approx \Delta g(T) - \Delta g(T_d)\frac{T}{T_d} + \frac{\overline{\Delta h}\,\Delta T}{T_d} \tag{5.80}$$

where $\Delta g(T)$ is the driving force for crystallization from the melt used so far, T_d is the dissolution temperature, $\Delta T = T_d - T$ is the supercooling in solution and $\overline{\Delta h}$ is the mean value of the heat of solution in the temperature interval (T, T_d). On the basis of experimental $\Delta c_p(T)$ data, $\Delta g(T)$ is defined between T_g and T_m°; for polystyrene this implies a temperature range of ~ 150 K. As vitrification is suppressed in solutions, this does not suffice for a description of Jones' experiments covering a supercooling range of ~ 200 K. By analogy with LAURITZEN et al., we may envisage expansion of $\Delta g(T)$ over a range of 50 K, through extrapolation of $\Delta c_p(T)$ to lower temperatures. At an even greater degree of supercooling the extrapolated $\Delta s(T)$ functions will in a number of approximations tend towards zero at characteristic temperatures. Below such a characteristic temperature T_2 the following could apply [92, 142]:

$$\Delta g(T) = \Delta g(T_2)$$

$$= \Delta h(T_2) \tag{5.81}$$

However, the extrapolations are then conducted over such a wide range that the results obtained using the present experimental data will be purely speculative.

LAURITZEN et al. were able to describe the measurements by JONES et al. by assuming Ψ to be equal to 0.382 (see Fig. 5.21), as a result of which the upswing is shifted outside the experimental range for the $\Delta g(T)_{\text{soln}}$ used by them. In this case the use of $\Delta g(T)_{\text{soln,CP0}}$ would have led to a 'pseudo-upswing' at $T_u \approx 315$ K for the same $\Psi = 0.382$ value: follow the interrupted line in the direction indicated in Fig. 5.22.

Being based on so many assumptions, the above results are of course largely qualitative rather than quantitative, and moreover several new relevant theories have since been developed [143–148]. The above example is given mainly to illustrate the importance of recognizing the influence of the temperature dependence of the differential functions and the need to perform heat capacity measurements to establish these.

145

5.3.4 A rigid amorphous fraction: a three-phase model*

As already mentioned above, the two-phase model is a first-order approximation. For many polymers this approximation appears to lead to satisfactory results. In recent DSC research, deviations from the two-phase model have been established [5, 8, 85, 127, 149–153]. Deviations have been detected earlier in physical aging experiments [85, 153] and also by comparing results obtained with different analytical techniques, such as density measurements [154–158], dielectric relaxation [159, 160], dynamic mechanical analysis [161–163], electron microscopy [164, 165], NMR spectroscopy [166–173], Raman spectroscopy [112, 154, 156–158, 161–163, 174–180], small angle neutron and X-ray scattering [156] and theoretical modelling [181–186].

In particular in the glass transition temperature range, the connection described in section 5.1.7 between crystallinity and the jump in heat capacity does not always appear to apply: the increase is found to be smaller than would be expected on the basis of the crystallinity. It would seem that, on the one hand, a fraction of the material has not crystallized and hence does not contribute to the crystallinity, while of course it will not melt either, so that its contribution to the heat of fusion is also zero; on the other hand, it will not immediately and fully contribute to an increase in C_p during devitrification. The experimental data suggest that this fraction does have a (delayed) effect on the C_p when it is heated but that this process may cover a wide temperature range of a few dozen degrees, and may extend into the melting range. These phenomena, and also the results obtained with other analytical methods, suggest devitrification.

Since devitrification implies greater mobility of parts of molecules, which is reflected in an increase in the C_p, an explanation for the observed phenomena is likely to be found in the degree of mobility. It is thought that the observed deviations are caused by molecules whose mobility is somehow hindered, even though they are entirely or partially in the amorphous phase. Such a situation may well occur at the surfaces of crystals [85, 114, 153, 178, 187] and particles [188–190], and factors such as thermal history, annealing and recrystallization are probably influential. As a start, we distinguish between two states: a 'rigid amorphous' (or 'disturbed amorphous') state and a 'mobile amorphous' (or 'undisturbed amorphous') state.

A possible approach in the case of semi-crystalline polymers would be to change from a two-phase model to a three-phase model in which the rigid amorphous fraction [130] is taken into account. This third phase is from here on thought to be a transitional region located between the crystals and the mobile amorphous regions. Others (see ref. 178 and its references) have called this an interfacial or interphase region; here the term 'interface' is reserved for a transition between two phases which, perpendicular to the plane between the two phases, is of negligible dimensions.

The following discussion is again based on the assumption that a description is possible assuming additivity of properties, this time for three phases, while contributions by the interfaces to the thermal properties are not explicitly taken into account. In addition, it is assumed that rigid amorphous material behaves like vitrifying material [191, 192] during its formation.

---

*Co-author: J. van Ruiten, DSM Research, The Netherlands.

146

5.3.4.1 The change in c_p in the glass transition range

As a start, let us consider the situation in which the rigid amorphous and crystalline fractions do not change during the measurement, so that there are no excess contributions to the heat capacity. Such a situation may occur just above the glass transition temperature of mobile amorphous material (T_g refers hereafter to the glass transition temperature of this material). In that case ($T \gtrsim T_g$) (see Eqs. 5.67 and 5.68):

$$c_{p_e}(T) = 0 \tag{5.82}$$

$$c_{p_b}(T) = w^c(T)c_{p_c}(T) + w^r(T)c_{p_r}(T) + w^a(T)c_{p_a}(T) \tag{5.83}$$

where $w^r(T)$ is the rigid amorphous mass fraction ('rigidity') and $w^a(T)$ is the mobile amorphous mass fraction ('mobility') (see Eq. 5.65)

$$w^c(T) + w^r(T) + w^a(T) = 1 \tag{5.84}$$

If the rigid amorphous material behaves like vitrified material we can assume that, for $T \gtrsim T_g$:

$$c_{p_r}(T) \approx c_{p_g}(T) \tag{5.85}$$

Below the glass transition temperature the difference between the heat capacities of a vitrified phase and a crystalline phase is usually negligible, so that, extrapolating to temperatures above T_g, it is again assumed that:

$$c_{p_g}(T) \approx c_{p_c}(T) \tag{5.86}$$

For $T \gtrsim T_g$ it is further assumed that $c_{p_r}(T)$, $c_{p_g}(T)$ and $c_{p_c}(T)$ can be represented by $c_{p_s}(T)$.

In more general terms, we can define a mass fraction 'solidity' $w^s(T)$ with a heat capacity $c_{p_s}(T)$ as

$$w^s(T)c_{p_s}(T) \equiv w^c(T)c_{p_c}(T) + w^r(T)c_{p_r}(T) \tag{5.87}$$

where

$$w^s(T) \equiv w^c(T) + w^r(T) \tag{5.88}$$

$$w^s(T) + w^a(T) = 1 \tag{5.89}$$

Using this, in the absence of excess terms, the following applies just above the glass transition region:

$$c_{p_e}(T) = 0 \tag{5.90}$$

$$c_p(T) = c_{p_b}(T)$$
$$= w^s(T)c_{p_s}(T) + [1 - w^s(T)]c_{p_a}(T) \tag{5.91}$$

from which it follows, by analogy with Eq. 5.6, that at $T \gtrsim T_g$:

$$w^s(T) \approx 1 - \frac{c_{p_b}(T) - c_{p_s}(T)}{c_{p_a}(T) - c_{p_s}(T)} \tag{5.92}$$

Of course, if crystalline, rigid and mobile amorphous material are all present at the same time, $w^s(T) > w^c(T)$ and the jump in c_p at T_g is smaller than when there is only crystalline and mobile amorphous material.

Just below the glass transition temperature there is only crystalline and vitrified material or—on the assumption that vitrified material can also be regarded as rigid amorphous material—only crystalline and rigid material, which are together referred to as solid material. In that case $T \lesssim T_g$:

$$w^a(T) = 0; \qquad w^r(T) = 1 - w^c(T); \qquad w^s(T) = 1 \tag{5.93}$$

It is again assumed that $c_{p_r}(T)$, $c_{p_g}(T)$ and $c_{p_c}(T)$ can be represented by $c_{p_s}(T)$.

It is not true that no excess phenomena occur at the glass transition temperature: due to the transition from mobile amorphous into rigid amorphous material or vice versa, $dw^a(T_g) \neq 0$ and $dw^r(T_g) \neq 0$ at T_g. As will become apparent at the end of this chapter, the related contributions to c_{p_e} nullify one another, which means that there is no net excess term.

5.3.4.2 The semi-crystalline state, $T_g < T < T_c$ or T_m

Usually, allowance must be made for an 'excess' heat capacity $c_{p_e}(T)$ (see section 5.2.7), as well as for a 'base-line' heat capacity $c_{p_b}(T)$. This is particularly the case in the temperature range between T_g and T_m^o.

In a three-phase model, in which the material is assumed to consist of crystalline, 'rigid amorphous' and 'mobile amorphous' phases, the following applies, on the assumption of additivity of the enthalpies with respect to the regions concerned, while contributions by the interfaces between the regions are not explicitly taken into account (see Eq. 5.64):

$$h(T) \equiv w^c(T)h_c(T) + w^r(T)h_r(T) + w^a(T)h_a(T) \tag{5.94}$$

where (see Eq. 5.65)

$$w^c(T) + w^r(T) + w^a(T) \equiv 1 \tag{5.95}$$

The examples at the end of this section show that $h_r(T)$ and $c_{p_r}(T)$ are complex functions. In particular, $h_r(T)$ is a complex function since, unlike $h_c(T)$ and $h_a(T)$, it depends on the time and temperature history.

By analogy with Eqs. 5.65–5.68, it follows that

$$c_p(T) \equiv c_{p_b}(T) + c_{p_e}(T) \tag{5.96}$$

where

$$c_{p_b}(T) \equiv w^c(T)c_{p_c}(T) + w^r(T)c_{p_r}(T) + w^a(T)c_{p_a}(T) \tag{5.97}$$

$$c_{p_e}(T) \equiv h_c(T)\, dw^c(T)/dT + h_r(T)\, dw^r(T)/dT + h_a(T)\, dw^a(T)/dT \tag{5.98}$$

The excess heat capacity can also be expressed as

$$c_{p_e}(T) = -[h_a(T) - h_c(T)]\, dw^c(T)/dT - [h_a(T) - h_r(T)]\, dw^r(T)/dT \tag{5.99}$$

In Eq. 5.98 the excess function comprises the following components:

$$c_{p_e}^c(T) \equiv h_c(T)\, dw^c(T)/dT \tag{5.100}$$

$$c_{p_e}^r(T) \equiv h_r(T)\, dw^r(T)/dT \tag{5.101}$$

$$c_{p_e}^a(T) \equiv h_a(T)\, dw^a(T)/dT \tag{5.102}$$

These expressions show that the components are dependent on enthalpy functions. The options chosen for these functions, for example the reference point $h_a(T_m^o) = 0$ chosen for the examples given at the end of this section, affect the sign and the size

148

of the components. Because of this, these individual components are not very interesting; what really counts is the net result: $c_{pe}(T)$.

In the case of cooling it is only the changes that occur as a result of the transition from mobile amorphous to crystalline material that are relevant to the excess function, and not the changes associated with the transition from mobile amorphous to rigid amorphous material. This is because when, at T_j, a fraction, $\Delta w^r(T_j)$, of the mobile amorphous material changes into rigid amorphous material, the following applies (i.e. on the assumption that the process is a vitrification process):

$$w_j^r \equiv \Delta w^r(T_j) \text{ is constant for } T < T_j \text{ so} \tag{5.103a}$$

$$dw_j^r/dT = 0 \quad \text{ for } T \neq T_j \tag{5.103b}$$

For the corresponding $h_{r_j}(T)$ it follows

$$h_{r_j}(T_j) = h_a(T_j) \tag{5.103c}$$

and

$$h_{r_j}(T) = h_a(T) - \int_T^{T_j} [c_{p r_j}(T)]\, dT \tag{5.103d}$$

Eq. 5.103c implies that (in cooling, not in heating) at $T = T_j$ the contribution of $h_{r_j}(T)$ to the second term in Eq. 5.99 nullifies the contribution of $h_a(T)$ at T_j, which is why the second term is zero and only the first term contributes to the result:

$$c_{pe}(T) = -[h_a(T) - h_c(T)]\, dw^c(T)/dT \tag{5.104}$$

For $T < T_j$, $h_{r_j}(T)$ will not contribute to the second term because of (5.103b). Again, only the first term in (5.99) counts.

If $w^c(T)$ does not change at T_g, which is usually the case, then there is no net result: $c_{pe}(T_g) = 0$.

5.3.4.3 The mass fraction 'solidity'

The mass fraction 'solidity', which has already been introduced in the absence of excess phenomena just above the glass transition region, is based on the following general expression for $h_s(T)$ (see Eqs. 5.64 and 5.65)

$$h(T) = w^s(T)h_s(T) + w^a(T)h_a(T) \tag{5.105}$$

where $h_s(T)$ is defined by

$$w^s(T)h_s(T) \equiv w^c(T)h_c(T) + w^r(T)h_r(T) \tag{5.106}$$

where

$$w^s(T) \equiv w^c(T) + w^r(T) \tag{5.107}$$

$$w^s(T) + w^a(T) = 1 \tag{5.108}$$

The fraction 'solidity' thus represents material in a state of low mobility, i.e. in either a crystalline or a rigid amorphous state.

The mass fraction solidity is then, via Eq. 5.105 (see also Eq. 5.43):

$$w^s(T) = \frac{h_a(T) - h(T)}{h_a(T) - h_s(T)} \tag{5.109}$$

By analogy with Eqs. 5.55–5.59, it follows that

$$c_p(T) = c_{p_b}(T) + c_{p_e}(T) \tag{5.110}$$

where

$$c_{p_b}(T) = w^s(T)c_{p_s}(T) + [1 - w^s(T)]c_{p_a}(T) \tag{5.111}$$

$$c_{p_e}(T) = h_s(T)\, dw^s(T)/dT + h_a(T)\, dw^a(T)/dT \tag{5.112}$$

and also

$$c_{p_e}(T) = -[h_a(T) - h_s(T)]\, dw^s(T)/dT \tag{5.113}$$

If $c_{p_e}(T) = 0$ then these expressions lead to Eq. 5.91. If there is no 'rigid amorphous' fraction then the expressions of course reduce to those according to the two-phase model.

5.3.4.4 The mass fraction 'non-crystallinity'

In the same way, a mass fraction 'non-crystallinity', indicated by w^{nc}, and a $h_{nc}(T)$ can be introduced via

$$h(T) = w^c(T)h_c(T) + w^{nc}(T)h_{nc}(T) \tag{5.114}$$

where $h_{nc}(T)$ is defined by

$$w^{nc}(T)h_{nc}(T) \equiv w^a(T)h_a(T) + w^r(T)h_r(T) \tag{5.115}$$

where

$$w^{nc}(T) \equiv w^a(T) + w^r(T) \tag{5.116}$$

$$w^c(T) + w^{nc}(T) = 1 \tag{5.117}$$

The fraction 'non-crystallinity' requires no further explanation. $h_{nc}(T)$ and $c_{p_{nc}}(T)$ are not simple functions, as will become apparent from the examples at the end of this section.

The mass fraction crystallinity in the three-phase model given here follows from Eq. 5.114:

$$w^c(T) = \frac{h_{nc}(T) - h(T)}{h_{nc}(T) - h_c(T)} \tag{5.118}$$

The following applies:

$$c_p(T) = c_{p_b}(T) + c_{p_e}(T) \tag{5.119}$$

where

$$c_{p_b}(T) = w^c(T)c_{p_c}(T) + [1 - w^c(T)]c_{p_{nc}}(T) \tag{5.120}$$

$$c_{p_e}(T) = h_c(T)\, dw^c(T)/dT + h_{nc}(T)\, dw^{nc}(T)/dT \tag{5.121}$$

or

$$c_{p_e}(T) = -[h_{nc}(T) - h_c(T)]\, dw^c(T)/dT \tag{5.122}$$

If there is no rigid amorphous fraction, these expressions also reduce to the expressions according to the two-phase model.

150

5.3.4.5 Examples: rigid amorphous regarded as vitrified amorphous

We now illustrate the concepts discussed so far with the aid of two example calculations for hypothetical polymers that may occur in a state in which, besides amorphous and crystalline regions, rigid amorphous regions may be present. These rigid amorphous regions are here assumed to be directly coupled to crystalline regions, for example as a transitional layer at the crystal surface. Of course, a model can also be developed for an amorphous filled polymer, which was actually the system first reported in the literature [188–190]. What we are concerned with here is to show the connection between the various $c_p(T)$, $h(T)$ and $w(T)$ functions. In so doing we make a number of debatable assumptions; we do not claim to introduce an adequate material model. What we are doing here is showing that the following steps are the least that will have to be undertaken in order to arrive at an acceptable analysis of the thermal properties of a polymer in which rigid amorphous material is present.

The aim is to arrive at a total analysis. By this we mean the need to couple measuring results obtained in the glass transition range to results obtained in the temperature range in which crystallization and melting take place. As already explained in detail in this chapter, it is possible to arrive at consistent results only if due allowance is made for the temperature dependence of the required functions.

Let us consider the cooling of a (hypothetical) semi-crystalline polymer in which at T_c crystallization occurs while at the same time part of the mobile amorphous material changes into rigid amorphous material, the remaining part following at T_g. We assume the following characteristic values:

$$T_m^o = 500 \text{ K} \qquad h_a(T_m^o) = 0 \text{ J/g} \qquad c_{pa} = 2.1 \text{ J/g per K}$$

$$T_c = 450 \text{ K} \qquad \Delta h(T_m^o) = 200 \text{ J/g} \qquad c_{pc} = 1 \text{ J/g per K}$$

$$T_g = 300 \text{ K} \qquad\qquad\qquad\qquad\qquad c_{pg} = c_{pc}$$

To make things easier, we first take a c_{pc} and c_{pa} that are not temperature-dependent so that $h_c(T)$ and $h_a(T)$ are straight lines. ·

With respect to the rigid amorphous regions, we assume:

$$c_{pr} = c_{pg} = c_{pc} \tag{5.123}$$

$$w^r = fw^c \quad \text{and} \quad w^a = 1 - (1+f)w^c \qquad \text{if } 0 \leq w^c \leq 1/(1+f) \tag{5.124}$$

For f and the ultimate crystallinity, we take in this example

$$f = 0.3 \quad \text{and} \quad w^c = 0.7 \tag{5.125}$$

The assumption $c_{pr} = c_{pg}$ means that the rigid amorphous material is assumed to be in a kind of vitrified state. Between T_g and T_m^o it is assumed that c_{pg} can be approximated by c_{pc}.

A detailed material model could for example include a heat capacity that varies as a function of the distance to the crystal surface, from c_{pg} to c_{pa}. The assumption that the rigidity w^r is a constant factor of the crystallinity w^c for $w^c \leq 1/(1+f)$ is debatable. For $1/(1+f) \leq w^c \leq 1$ it is plausible that 'mobile' amorphous material (mobility) is absent; it is then assumed that in the case of extremely high crystallinity values the mobility of chain segments in non-crystalline regions is restricted to such an extent that there is no longer any question of mobility of the kind characterizing 'undisturbed' amorphous material [169]. One could picture for example lamellar crystals with interlamellar space widths equal to or less than the width of the transitional region required for a chain segment coming from a crystallite to lose its

151

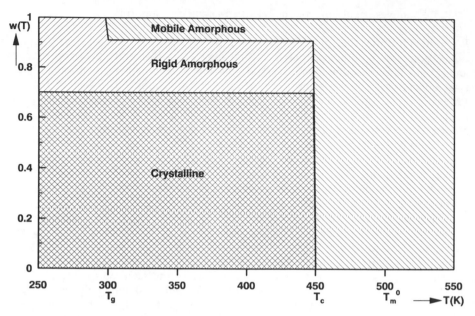

Fig. 5.23. The chosen mobile amorphous ('mobility'), rigid amorphous ('rigidity') and crystalline ('crystallinity') fractions as a function of temperature for a hypothetical polymer: see the text for the assumed polymer data

correlation with the crystal. If there is not enough space, then the correlation remains and so does the risk of restricted mobility, i.e. rigidity.

In the case of lamellar crystallites the above, debatable, assumptions (Eq. 5.125) imply that the width of the transitional region l_r between a crystallite and the mobile amorphous phase is related to the thickness of the lamella l according to $l_r = 0.3l$. Alternatively, it can easily be imagined that l_r is inversely proportional to the mobility of the chain segments at the crystal surface, which, in turn, depends on chain structure and actual temperature [169]. As mentioned above, our main concern in these examples is to illustrate a 'total analysis' approach rather than a material model.

Let us assume that crystallization from the melt takes place around T_c in a very narrow temperature range and that the crystalline mass fraction increases from 0 to 0.7. According to Eq. 5.124, a rigid amorphous fraction is formed at the same time. The fractions 'mobility' $w^a(T)$ and 'rigidity' $w^r(T)$ are thus determined and, of course, so are the derived fractions 'non-crystallinity' $w^{nc}(T)$ and 'solidity' $w^s(T)$. Figure 5.23 shows the w functions over the relevant temperature range. Between T_g and T_c, w^a, w^r, w^{nc}, w^c and w^s are 0.09, 0.21, 0.30, 0.70 and 0.91 respectively.

Figure 5.24 shows the curves of the $c_p(T)$ and $h(T)$ functions. In contrast to previous plots, the (positive) heat capacity cooling curve $c_{p_{cc}}(T)$ is plotted here in the positive y direction. Due to the presence of rigid amorphous material, the increase in specific heat capacity at T_g does not equal

$$c_{p_b}(T_g) - c_{p_c}(T_g) = [1 - w^c(T_g)][c_{p_a}(T_g) - c_{p_c}(T_g)]$$
$$= (1 - 0.7)(2.1 - 1)$$
$$= 0.33 \text{ J/g per K} \qquad (5.126)$$

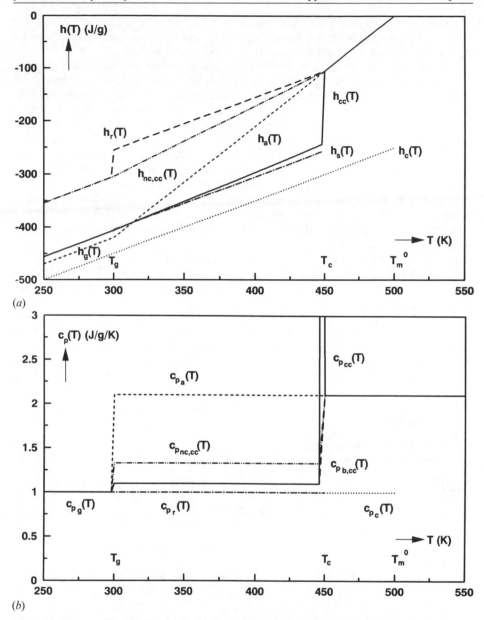

Fig. 5.24. The specific enthalpy (*a*) and heat capacity (*b*) functions for the hypothetical polymer corresponding to the various fractions present (see Fig. 5.23): symbols are explained in the text

but

$$c_{\text{pb}}(T_g) - c_{\text{ps}}(T_g) = [1 - w^s(T_g)][c_{\text{pa}}(T_g) - c_{\text{ps}}(T_g)]$$
$$= (1 - 0.91)(2.1 - 1)$$
$$= 0.099 \text{ J/g per K} \tag{5.127}$$

153

as mentioned earlier (see Eq. 5.92). (Just above T_g, $c_p(T) = c_{p_b}(T)$ because $c_{p_c}(T) = 0$. For $T \gtrsim T_g$ it is again assumed that c_{p_r}, c_{p_g} and c_{p_c} can be represented by c_{p_s}.) Between T_g and T_c, c_{p_b} and $c_{p_{nc}}$ are 1.099 and 1.33 J/g per K respectively.

Figure 5.24(a) also includes the $h_r(T)$ function. In cooling from the melt, at T_c 21 % of the total mass changes from mobile amorphous material into rigid amorphous material. The assumption that this material then behaves as vitrified material implies a second-order transition at T_c having the character of a glass transition. A bend is thus formed in the $h(T)$ curve for this material; the slope below the transition is then defined by c_{p_g}, which is in this example assumed to be equal to c_{p_c}. The $h_r(T)$ curve for this material then runs parallel to the $h_c(T)$ curve. At T_g the remaining mobile amorphous material, which is 9 % of the total mass, becomes rigid amorphous. The $h_r(T)$ function jumps to the $h_{nc,cc}(T)$ function and runs parallel to $h_g(T)$ (and in this case also to $h_c(T)$) below T_g. In this example, the $h_r(T)$ function below T_g is thus a result of two contributions: first from mobile amorphous material that vitrifies during crystallization at T_c and second from residual mobile amorphous material that vitrifies in the 'classical' manner at T_g.

In the temperature range $T_g \leftrightarrow T_c$ the $h_{nc}(T)$ function lies between $h_a(T)$ and $h_r(T)$ according to Eqs. 5.115 and 5.116, where w^a, w^r and w^{nc} are 0.09, 0.21 and 0.30 respectively. Below T_g, $h_{nc}(T)$ and $h_r(T)$ coincide (see Eqs. 5.115 and 5.116), and w^a, w^r and w^{nc} are 0, 0.30 and 0.30 respectively.

$h_{cc}(T)$, the specific enthalpy cooling curve, follows from Eq. 5.94; as it moves towards lower temperatures, it is the result of a first-order transition (crystallization) for a fraction (0.7) of the material at T_c, a second-order transition (vitrification) of another fraction (0.21) of the material at T_c, and finally another second-order transition (vitrification) of the remaining fraction of the material (0.09) at T_g, down

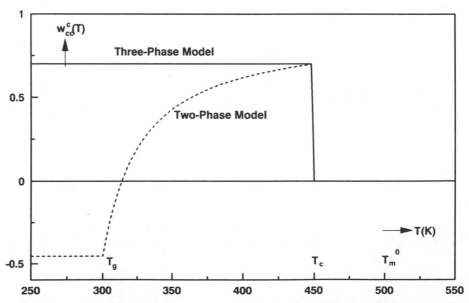

Fig. 5.25. The crystallinity as a function of the temperature obtained via cooling $w_{cc}^c(T)$ evaluated with the aid of the two-phase model and with the aid of a three-phase model: calculation according to the two-phase model (which is not valid in this case) in the extreme example considered here would even lead to negative crystallinities

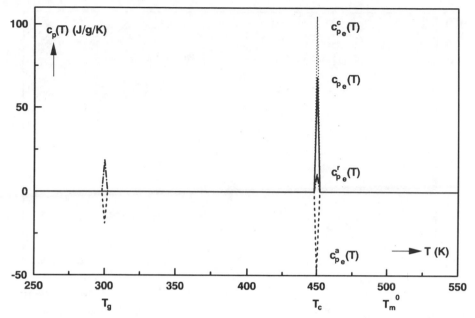

Fig. 5.26. The excess heat capacity $c_{p_e}(T)$ and its various components: only at T_c is the excess function not zero

to which this last fraction has remained unchanged. Figure 5.24 clearly shows that if there are rigid amorphous regions as well as mobile amorphous and crystalline regions, the experimental $h(T)$ curve may lie *above* the $h_a(T)$ in the model described here.

This is clear from Eqs. 5.114 and 5.117, in which $h(T)$ is formulated for the three-phase model as the weighted sum (i.e. weighted via the crystallinity) of $h_c(T)$ and $h_{nc}(T)$. In the two-phase model, on the other hand, the weighted sum (also via the crystallinity) of $h_c(T)$ and $h_a(T)$ is to be used. It is now also clear that Eq. 5.118 must be used to calculate the crystallinity. The use of the usual expression for $w^c(T)$ within the two-phase model (see Eq. 5.43) would lead to excessively low results. This is clearly shown in Fig. 5.25; in the example used here (which is non-realistic in terms of the c_p values), the use of the two-phase model would even lead to negative crystallinity values. Our example clearly shows the essentials of taking into account a rigid amorphous fraction.

Finally, the excess heat capacity can be calculated. Figure 5.26 gives $c_{p_e}(T)$ (see Eq. 5.98), as well as the various contributions to the excess function according to Eqs. 5.100–5.102. As mentioned earlier, the sign of the peaks is also determined by the choice of enthalpy calibration, and only changes in $w^c(T)$ are relevant to the excess function in the case of cooling. Because of this last fact, $c_{p_e}(T) = 0$ at T_g in Fig. 5.26.

By way of illustration, we now give another example of a hypothetical polymer becoming partially crystalline and partially rigid on cooling. This time, crystallization from the melt takes place between $T_{c,l}$ (l = lower) and $T_{c,u}$ (u = upper), while at the same time a portion of the mobile amorphous material changes into rigid amorphous material, the remainder following at T_g. The temperature dependence of the heat

155

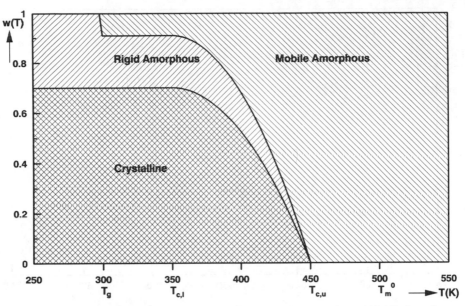

Fig. 5.27. The chosen mobile amorphous ('mobility'), rigid amorphous ('rigidity') and crystalline ('crystallinity') fractions as a function of temperature for a hypothetical polymer: see the text for the assumed polymer data

capacities is now more realistic:

$T_m^\circ = 500$ K $h_a(T_m^\circ) = 0$ J/g $c_{Pa} = 1 + 0.0025T$ J/g per K

$T_{c,u} = 450$ K $\Delta h(T_m^\circ) = 200$ J/g $c_{Pc} = 0.0040T$ J/g per K

$T_{c,l} = 350$ K $c_{Pg} = c_{Pc}$

$T_g = 300$ K

As a starting point we take a crystallinity cooling curve that is parabolic between $T_{c,l}$ and $T_{c,u}$, constant at and below $T_{c,l}$ (this value again being 0.7) and zero at and above $T_{c,u}$:

$$w^c(T) = 0.7 - 0.7 \times 10^{-4} \times (T - 350)^2 \tag{5.128}$$

With respect to the rigid amorphous regions we assume the same coupling between $w^r(T)$ and $w^c(T)$ as in Eqs. 5.124 and 5.125, where $f = 0.3$. With these data all $w(T)$ functions are known (see Fig. 5.27).

The specific heat capacity and specific enthalpy functions are shown in Fig. 5.28. The (positive) heat capacity cooling curve is here plotted in the positive y direction. As the c_p functions are linearly dependent on T, the $h(T)$ functions are now second-degree curves. In comparison with the first example, the rigid amorphous fraction now changes more gradually in cooling from $T_{c,u}$ to $T_{c,l}$. We assume that the rigid amorphous material behaves as vitrified material, which means that a rigid amorphous fraction is formed at every temperature $T_j, j = 1, 2, 3 \ldots$, see (5.103):

$w_j^r \equiv \Delta w^r(T_j)$ is constant for $T < T_j$ so $\qquad\qquad\qquad$ (5.129a)

$dw_j^r/dT = 0 \qquad$ for $T \neq T_j$ $\qquad\qquad\qquad\qquad\qquad\qquad$ (5.129b)

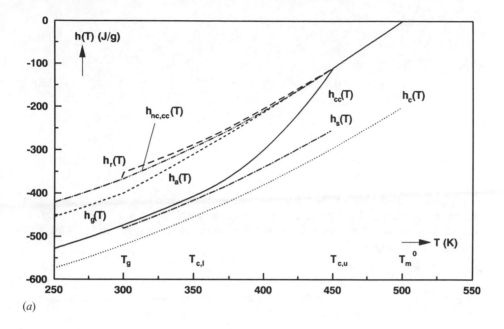

(a)

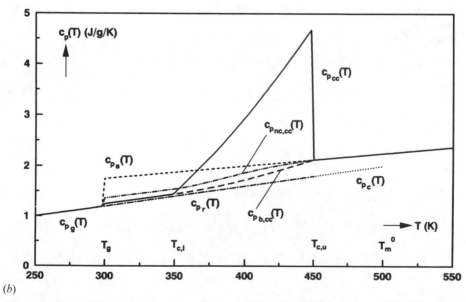

(b)

Fig. 5.28. The specific enthalpy (a) and heat capacity (b) functions for the hypothetical polymer, corresponding to the fractions present (see Fig. 5.27): symbols explained in the text

157

The related $h_{r_j}(T)$ is defined by

$$h_{r_j}(T_j) = h_a(T_j) \tag{5.130a}$$

and

$$h_{r_j}(T) = h_a(T_j) - \int_T^{T_j} [c_{p_{r_j}}(T)]\, dT \tag{5.130b}$$

or

$$h_{r_j}(T) = h_a(T) + \int_T^{T_j} [c_{p_a}(T) - c_{p_{r_j}}(T)]\, dT \tag{5.131}$$

Summation, see (5.134), of the various j contributions yields the totals for the rigid amorphous material. The functions comprising the total contribute each with different enthalpy, of which the explicit temperature dependence is given in (5.130b). For $w^r(T)$ the following applies:

$$w^r(T) = \sum_j w_j^r \tag{5.132}$$

Here the sum is over all vitrification events j which took place at temperatures higher than T. It is now also clear what $h_r(T)$ is in such a case:

$$w^r(T)h_r(T) \equiv \sum_j w_j^r h_{r_j}(T) \tag{5.133}$$

or

$$h_r(T) = \sum_j [w_j^r / w^r(T)]h_{r_j}(T) \tag{5.134a}$$

$$= [w_i^r / w^r(T)]h_{r_i}(T) + \sum_{j \neq i} [w_j^r / w^r(T)]h_{r_j}(T) \tag{5.134b}$$

Equation 5.94 can hence be rewritten as

$$h(T) = w^c(T)h_c(T) + \sum_j w_j^r h_{r_j}(T) + w^a(T)h_a(T) \tag{5.135a}$$

$$= w^c(T)h_c(T) + w_i^r h_{r_i}(T) + \sum_{j \neq i} w_j^r h_{r_j}(T) + w^a(T)h_a(T) \tag{5.135b}$$

and Eqs. 5.96–5.98 can be rewritten as

$$c_p(T) = c_{p_b}(T) + c_{p_e}(T) \tag{5.136}$$

where

$$c_{p_b}(T) = w^c(T)c_{p_c}(T) + \sum_j w_j^r c_{p_{r_j}}(T) + w^a(T)c_{p_a}(T) \tag{5.137}$$

$$c_{p_e}(T) = h_c(T)\, dw^c(T)/dT + \sum_j h_{r_j}(T)\, dw_j^r/dT + h_a(T)\, dw^a(T)/dT \tag{5.138a}$$

$$= h_c(T)\, dw^c(T)/dT + h_{r_i}(T)\, dw_i^r/dT + \sum_{j \neq i} h_{r_j}(T)\, dw_j^r/dT + h_a(T)\, dw^a(T)/dT \tag{5.138b}$$

the second way of formulation, (5.138b), clarifies that at $T = T_i$ only the specific fraction of material that becomes rigid at that specific temperature contributes to

158

$c_{p_e}(T)$ because, see (5.129b),

$$[dw^r_{j \neq i}/dT]_{T_i} = 0 \text{ so} \tag{5.139}$$

$$c_{p_e}(T_i) = h_c(T_i)[dw^c(T)/dT]_{T_i} + h_{r_i}(T_i)[dw^r_i/dT]_{T_i} + h_a(T_i)[dw^a(T)/dT]_{T_i} \tag{5.140}$$

and because

$h_{r_i}(T_i) = h_a(T_i)$ and, see (5.95),

$$[dw^a(T)/dT]_{T_i} = -[dw^r_i/dT]_{T_i} - [dw^c(T)/dT]_{T_i} \tag{5.141}$$

it follows

$$c_{p_e}(T_i) = [h_c(T_i) - h_a(T_i)][dw^c(T)/dT]_{T_i} \tag{5.142}$$

as was found before, see (5.104)

If the $c_{p_{r_j}}(T)$ are the same for every $j = 1, 2, 3, \ldots$, then

$$c_{p_{r_j}}(T) = c_{p_r}(T) \tag{5.143}$$

and Eq. 5.137 is reduced to Eq. 5.97, or to Eq. 5.111. If, as in the previous example

$$c_{p_r}(T) = c_{p_g}(T) = c_{p_c}(T) \tag{5.144}$$

then

$$c_{p_b}(T) = [1 - w^a(T)]c_{p_c}(T) + w^a(T)c_{p_a}(T) \tag{5.145}$$

It is clear that $h_r(T)$ is usually a complex function that depends on the way in which rigid amorphous material is formed or disappears.

Of course, what has been stated above also applies to the combination of melting and the disappearance of rigid amorphous material. However, in that case one should reckon with an excess enthalpy (enthalpy recovery) due to devitrification of rigid amorphous material when this devitrification takes place at a temperature above that temperature at which the rigid amorphous material was formed.

Figure 5.28(b) also shows again that the curve obtained via extrapolation from the melt bears no relation whatever to the base line (see sections 5.1.6 and 5.2.5).

Figure 5.29 shows the crystallinity function according to the three-phase model outlined here, and also shows what happens when the data are (incorrectly) evaluated according to the two-phase model.

Finally, the excess heat capacity and its components can be calculated. Figure 5.30 shows $c_{p_e}(T)$ (see Eq. 5.98) and also the various components (see Eqs. 5.100–5.102). As discussed above, $c_{p_e}(T)$ is not zero only in the range from $T_{c,l}$ to $T_{c,u}$ and, in cooling, in that range only the fraction of the mobile amorphous material that becomes crystalline is relevant.

The applicability of a three-phase formulation (leaving aside for the moment the validity of the assumptions used) will depend greatly on the possibility of arriving at a total analysis of the polymer in question. Important factors in the case of a semi-crystalline polymer are data in a temperature range from below the glass transition region to above the melting region, and the possibility of realizing different crystallinities, and states differing in the degree to which 'rigid amorphous' material is present (if at all).

159

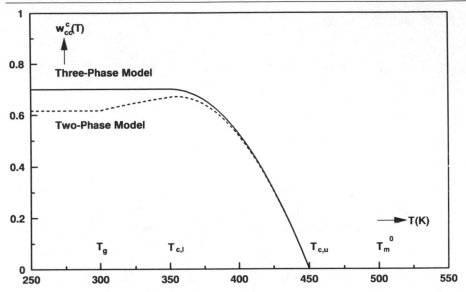

Fig. 5.29. The crystallinity as a function of the temperature obtained via cooling $w_{cc}^c(T)$ evaluated with the aid of the two-phase model and of a three-phase model: the use of the two-phase model (which is not valid here) would successively lead to an increase, a decrease and a stabilization of the crystallinity with decreasing temperature

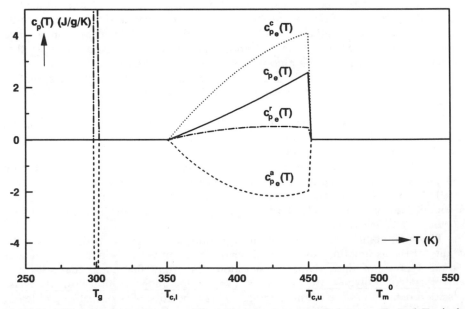

Fig. 5.30. The excess heat capacity $c_{p_e}(T)$ and its components: only between $T_{c,l}$ and $T_{c,u}$ is the excess function not zero

Acknowledgments

The author is grateful to M. F. J. Pijpers, who performed most of the experiments, to J. van Ruiten and L. C. E. Struik for critical reading and stimulating discussions, to G. Schuler for the drawings, and to H. J. Rhebergen, who provided the English translation of the original Dutch text. Thanks are also due to DSM for giving permission for publication of this work.

References

1 *Wunderlich, B., Cheng, S. Z. D.:* Gaz. Chim. Ital. 116, p. 345 (1986); *Wunderlich, B., Cheng, S. Z. D., Loufakis, K.,* in: Encyclopedia of Polymer Science and Engineering, p. 767, Wiley, New York, 1989

2 *Freire, E., van Osdol, W. W., Mayorga, O. L., Sanchez-Ruiz, J. M.:* Annu. Rev. Biophys. Biophys. Chem. 19, p. 159 (1990); *van Osdol, W. W., Mayorga, O. L., Freire, E.:* Biophys. J. 59, p. 48 (1991); *Mayorga, O. L., Navarro Rascon, A., Freire, E.,* in: Thermal Analysis and Calorimetry in Polymer Physics. *Mathot, V. B. F.* (Ed.), Special issue Thermochimica Acta, Elsevier Science Publishers, Amsterdam, 1994.

3 *Wunderlich, B., Baur, H.:* Adv. Polym. Sci. 7, p. 151 (1970)

4 The ATHAS Databank 1980. *Gaur, U., Lau, S.-F., Shu, H.-C., Wunderlich, B. B., Mehta, A., Wunderlich, B.:* J. Phys. Chem. Ref. Data 10, pp. 89, 119, 1001 (1981); 11, pp. 313, 1065 (1982); 12, pp. 29, 65, 91 (1983); Update, *Varma-Nair* et al. (1991)

5 *Wunderlich, B.,* in: Thermal Characterization of Polymeric Materials. *Turi, E. A.* (Ed.), p. 92, Academic Press, Orlando, FL, 1981

6 *Mathot, V. B. F.:* Polymer 25, p. 579 (1984). Errata: Polymer 27, p. 969 (1986)

7 *Pan, R., Cao, M.-Y., Wunderlich, B.,* in: Polymer Handbook. *Brandrup, J., Immergut, E. H.* (Eds.), 3rd Edn., VI/371, Wiley, New York, 1989

8 *Wunderlich, B.:* Thermal Analysis. Academic Press, San Diego, CA, 1990

9 *Darsey, J. A.:* Macromol. Chem. Rapid Commun. 12, p. 325 (1991)

10 *Varma-Nair, M., Wunderlich, B.:* J. Phys. Chem. Ref. Data 20(2), p. 349 (1991)

11 *Gaur, U., Wunderlich, B.:* J. Phys. Chem. Ref. Data 10(1), p. 119 (1981)

12 *Höhne, G. W. H., Glöggler, E.:* Thermochim. Acta 151, p. 295 (1989)

13 *Höhne, G. W. H., Cammenga, H. K., Eysel, W., Gmelin, E., Hemminger, W.:* Thermochim. Acta 160, p. 1 (1990)

14 *Höhne, G. W. H.:* J. Therm. Anal. 37, p. 1987 (1991)

15 *Callanan, J. E., McDermott, K. M.:* J. Chem. Thermodynamics 22, p. 225 (1990)

16 *Menzcel, J. D., Leslie, T. M.:* Thermochim. Acta 166, p. 309 (1990)

17 *Schick, C., Höhne, G. W. H.:* Thermochim. Acta 187, p. 351 (1991)

18 *Illers, K.-H.:* Eur. Polym. J. 10, p. 911 (1974)

19 *Jin, Y., Wunderlich, B.:* J. Therm. Anal. 36, p. 765 (1990)

20 *Furakawa, G. T.:* J. Am. Chem. Soc. 75, p. 522 (1953)

21 *Ditmars, D. A., Douglas, T. B.:* J. Res. Nat. Bur. Stand. 75A, p. 401 (1971)

22 *Mathot, V., Pijpers, M., Beulen, J., Graff, R., van der Velden, G.,* in: Proc. 2nd Eur. Symp. on Thermal Analysis 1981 (ESTA-2). *Dollimore, D.* (Ed.), p. 264, Heyden, London, 1981

23 *Gray, A. P.:* Polym. Prepr. (Am. Chem. Soc. Div. Polym. Chem.) 6(2), p. 956 (1965)

24 *Mathot, V. B. F., Pijpers, M. F. J.:* J. Therm. Anal. 28, p. 349 (1983)

25 *O'Neill, M. J.:* Anal. Chem. 38(10), p. 1331 (1966)

26 *Gray, A. P.:* Instrument News (Perkin Elmer Corp.) 20/2, p. 8 (1969)

27 *Gray, A. P.:* Thermochim. Acta 1, p. 563 (1970)

28 *Richardson, M. J.,* in: Developments in Polymer Characterization. *Dawkins, J.V.* (Ed.), Vol. 1, p. 205, Applied Science Publishers, London, 1978

29 *Mathot, V. B. F., Fabrie, Ch. C. M.:* J. Polym. Sci.: Part B: Polym. Phys. 28, p. 2487 (1990)

30 *Mathot, V. B. F., Fabrie, Ch. C. M., Tiemersma-Thoone, G. P. J. M., van der Velden, G. P. M.:* J. Polym. Sci.: Part B: Polym. Phys. 28, p. 2509 (1990)

31 *Bestul, A. B., Chang, S. S.:* J. Chem. Phys. 43, p. 4532 (1965)

161

32 *Lombardi, G.:* For Better Thermal Analysis. 2nd Edn., International Confederation for Thermal Analysis (ICTA), 1980. *Hill., J. O.* (Ed.): For Better Thermal Analysis and Calorimetry, 3rd Edn., ICTA, 1991

33 *Wunderlich, B.,* in: Physical Methods of Chemistry. *Weissberger, A. Rossiter B. W.* (Eds.), Vol. 1, Part 5, Chapter 8, p. 427, Wiley (Interscience), New York, 1971

34 *Mackenzie, R. C.:* Anal. Proc. (Lond.) 17, p. 217 (1980)

35 *Hemminger, W., Höhne, G.:* Calorimetry, Fundamentals and Practice. VCH Verlags-Gesellschaft Weinheim, Deerfield Beach (FL), Basel, 1984 (in German: 1979)

36 *Turi, E. A.* (Ed.): Thermal Characterization of Polymeric Materials. Academic Press, New York, 1981, Second, revised, Edn. in press, Academic Press, Boston

37 *Sesták J.:* Thermophysical Properties of Solids, Their Measurement and Theoretical Thermal Analysis. Vol. XIID of Thermal Analysis, Elsevier, Amsterdam; Academia, Prague, 1984

38 *Hemminger, W. F., Cammenga, H. K.:* Methoden der Thermischen Analyse. Springer-Verlag, Berlin, Heidelberg, New York, 1989

39 *Richardson, M. J.,* in: Comprehensive Polymer Science. *Booth, C., Price, C.* (Eds.), Vol. 1, p. 867, Pergamon, Oxford, 1989

40 *Bershtein, V. A., Egorov, V. M.:* Differential Scanning Calorimetry in Physicochemistry of Polymers (in Russian), Khimiya Leningradskoe (Chemistry Publ. House), Leningrad, 1990, English version in press; *Bershtein, V. A., Egorov, V. M., Egorova, L. M., Ryzhov, V. A.,* in: Thermal Analysis and Calorimetry in Polymer Physics. *Mathot, V. B. F.* (Ed.), Special issue Thermochimica Acta, Elsevier Science Publishers, Amsterdam, 1994

41 *Godovsky, Y. K.:* Thermophysical Properties of Polymers. Springer-Verlag, New York, Berlin, Heidelberg, 1992

42 Chemical Abstracts (CA) Selects: Thermal Analysis SVC 03B; Thermochemistry SVC 034, Chemical Abstracts Service, Columbus, OH

43 Journal of Thermal Analysis, *Simon, J.* (Ed.), Akademiai Kiado, Budapest; Wiley, Chichester

44 Thermal Analysis Reviews and Abstracts (TAR and A) (formerly Thermal Analysis Abstracts (TAA)), *Clark, G. M.* (Ed.), Official ICTA Publication, Interscience, Communications, London

45 Thermochimica Acta, *Wendlandt, W. W.* (Ed.), Elsevier Science Publishers, Amsterdam

46 Detailed information in: For Better Thermal Analysis and Calorimetry. *Hill, J. O.* (Ed.), 3rd Edn., International Confederation for Thermal Analysis (ICTA), 1991

47 *Mathot, V. B. F., Pijpers, M. F. J.:* Thermochim. Acta 151, p. 241 (1989)

48 *Mehta, A., Bopp, R. C., Gaur, U., Wunderlich, B.:* J. Therm. Anal. 13, p. 197 (1978)

49 *Gaur, U., Shu, H.-C., Mehta, A., Wunderlich, B.:* J. Phys. Chem. Ref. Data 10(1), p. 89, 1981

50 *Cheng, S. S., Horman, J. A., Bestul, A. A.:* J. Res. NBS 71A(4), p. 293 (1967)

51 *Wunderlich, B., Jones, L. D.:* J. Macromol. Sci.-Phys. B3(1), p. 67 (1969)

52 *Wunderlich, B., Gaur, U.,* in: Thermal Analysis. *Hemminger, W.* (Ed.), Vol. 2, p. 409, Birkhauser Verlag, Basel, 1980

53 *Meyer, H., Kilian, H.-G.:* Progr. Colloid Polym. Sci. 64, p. 154 (1978)

54 *Meyer, H., Kilian, H.-G.:* Progr. Colloid Polym. Sci. 64, p. 166 (1978)

55 *Wunderlich, B.:* Macromolecular Physics, Vol. 1: Crystal Structure, Morphology, Defects. Academic Press, New York, 1973

56 *Kast, W.,* in: Die Physik der Hochpolymeren. *Stuart, H. A.* (Ed.), p. 232, Springer-Verlag, Braunschweig, 1955

57 *Mandelkern, L.:* Crystallization of Polymers. McGraw-Hill, New York, 1964

58 *Miller, R. L.,* in: Encyclopedia of Polymer Science Technology. *Bikales, N. M.* (Ed.), Vol. 4, p. 449, Interscience, New York, 1966

59 *Dole, M.:* J. Polym. Sci., Part C 18, p. 57 (1967)

60 *Hobbs, S. Y., Mankin, G. I.:* J. Polym. Sci., Part A-2 9, p. 1907 (1971)

61 *Wunderlich, B.:* J. Phys. Chem. 69, p. 2078 (1965)

62 *Wunderlich, B.:* Macromolecular Physics, Vol. 3: Crystal Melting. Academic Press, New York, 1980

63 *Wegner, G., Fischer, W., Muñoz-Escalona, A.:* Makromol. Chem., Suppl. 1, p. 521 (1975)

64 *Wunderlich, B.:* Macromolecular Physics, Vol. 2: Crystal Nucleation, Growth, Annealing. Academic Press, New York, 1976
65 *Dröscher, M.:* Makromol. Chem. 178, p. 1195 (1977)
66 *Loufakis, K., Wunderlich, B.:* J. Phys. Chem. 92, p. 4205 (1988)
67 *Jin, Y., Wunderlich, B.:* J. Phys. Chem. 95, p. 9000 (1991)
68 *Fischer, E. W., Hinrichsen, G.:* Kolloid-Z. u. Z. Polymere 213(1–2), p. 28 (1966)
69 *Fischer, E. W., Hinrichsen, G.:* Kolloid-Z. u. Z. Polymere 213(1–2), p. 93 (1966)
70 *Fischer, E. W., Hinrichsen, G.:* Kolloid-Z. u. Z. Polymere 247, p. 858 (1971)
71 *Ruhemann, M., Simon, F.:* Z. Phys. Chem. A138, p. 1 (1928)
72 *Bekkedahl, N., Matheson, H.:* J. Res. NBS 15, 503 (1935)
73 *Bekkedahl, N., Scott, R. B.:* J. Res. NBS 29, 87 (1942)
74 *Sharonov, Y. A., Vol'kenshtein, M. V.:* Sov. Phys. Solid State 5(2), p. 429 (1963)
75 *Angell, C. A., Rao, K. J.:* J. Chem. Phys. 57(1), p. 470 (1972)
76 *Richardson, M. J., Savill, N. G.:* Brit. Polym. J. 11, p. 123 (1979)
77 *Wunderlich, B.:* J. Phys. Chem. 64, p. 1052 (1960)
78 *Peyser, P.,* in: Polymer Handbook. *Brandrup, J., Immergut, E. H.* (Eds.), 3rd Edn., VI/209, Wiley, New York, 1989
79 *Wolpert, S. M., Weitz, A., Wunderlich, B.:* J. Polym. Sci., Part A-2 9, p. 1887 (1971)
80 *Bourdariat, J., Berton, A., Chaussy, J., Isnard, R. Odin, J.:* Polymer 14, p. 167 (1973)
81 *Flynn, J. H.:* Thermochim. Acta 8, p. 69 (1974)
82 *Richardson, M. J., Savill, N. G.:* Polymer 16, p. 753 (1975)
83 *Brown, I. G., Wetton, R. E., Richardson, M. J., Savill, N. G.:* Polymer 19, p. 659 (1978)
84 *Ott, H.-J.:* Colloid Polym. Sci. 257, p. 486 (1979)
85 *Struik, L. C. E.,* in: Physical Aging in Amorphous Polymers and Other Materials. Elsevier, Amsterdam, 1978
86 *Bosma, M., ten Brinke, G., Ellis, T. S.:* Macromol. 21, p. 1465 (1988)
87 *Ellis, T. S.:* Polym. Prepr. (Am. Chem. Soc. Div. Polym. Chem.) 31(1), p. 285 (1990); *ten Brinke, G., Oudhuis, L., Ellis, T. S.,* in: Thermal Analysis and Calorimetry in Polymer Physics. *Mathot, V. B. F. (Ed.),* Special issue Thermochimica Acta, Elsevier Science Publishers, Amsterdam, 1994
88 *Raine, H. C., Richards, R. B., Ryder, H.:* Trans. Faraday Soc. 41, p. 56 (1945)
89 *Dole, M.:* Fortschr. Hochpolym.-Forsch. 2, p. 221 (1960)
90 *Richardson, M. J.:* J. Polym. Sci., Part C 38, p. 251 (1972)
91 *Kauzmann, W.:* Chem. Revs. 43, p. 219 (1948)
92 *Gibbs, J. H., DiMarzio, E. A.:* J. Chem. Phys. 28(3), p. 373 (1958)
93 *Suga, H., Seki, S.:* J. Non-Cryst. Solids 16, p. 171 (1974)
94 *Sanchez, I. C., DiMarzio, E. A.:* Macromol. 4, p. 677 (1971)
95 *Tammann, G.:* Verlag von J. A. Barth, Leipzig (1903)
96 *Thompson, C. V., Spaepen, F.:* Acta Metallurgica 27, p. 1855 (1979). The author recently became aware of this reference, and references therein, through J. D. Hoffman.
97 *Hoffman, J. D.:* J. Chem. Phys. 29, p. 1192 (1958)
98 *Hoffman, J. D., Weeks, J. J.:* J. Chem. Phys. 37, p. 1723 (1962)
99 *Dole, M., Hettinger, Jr., W. P., Larson, N. R., Wethington, Jr., J. A.:* J. Chem. Phys. 20(5), p. 781 (1952)
100 *Richardson, M. J.:* Br. Polym. J. 1, p. 132 (1969)
101 *Ver Strate, G., Wilchinsky, Z. W.:* J. Polym. Sci., Part A-2 9, p. 127 (1971)
102 *Blundell, D. J., Beckett, D. R., Willcocks, P. H.:* Polymer 22, p. 704 (1981)
103 *Baur, H.:* Prog. Colloid Polym. Sci. 66, p. 1 (1979)
104 *Baur, H.:* Pure & Appl. Chem. 52, p. 457 (1980)
105 *Hoffman, J. D.:* Soc. Plastics Engrs. Trans. 4, p. 315 (1964)
106 *Peterlin, A., Meinel, G.:* J. Appl. Phys. 36(10), p. 3028 (1965)
107 *Peterlin, A., Meinel, G.:* Appl. Polym. Symp. 2, p. 85 (1966)
108 *Hendus, H., Illers, K. H.:* Kunststoffe 57(3), p. 193 (1967)
109 *Mandelkern, L., Allou, Jr., A. L., Gopalan, M.:* J. Phys. Chem. 72, p. 309 (1968)
110 *Hara, K., Schonhorn, H.:* J. Appl. Phys. 42(12), p. 4549 (1971)
111 *Dosière, M.,* in: Handbook of Polymer Science and Technology, Vol. 2, *Cheremisinoff, N. P.* (Ed.), p. 367, Marcel Dekker, New York, 1989

112 *Strobl, G. R., Hagedorn, W.:* J. Polym. Sci.: Polym. Phys. Ed. 16, p. 1181 (1978)
113 *Strobl, G. R., Schneider, M.:* J. Polym. Sci.: Polym. Phys. Ed. 18, p. 1343 (1980)
114 *Vonk, C. G., Pijpers, A. P.:* J. Polym. Sci.: Polym. Phys. Ed. 23, p. 2517 (1985)
115 *Kilian, H.-G.:* Progr. Colloid & Polym. Sci. 72, p. 60 (1986)
116 *Rosenberger, B., Asbach, G. I., Kilian, H.-G., Wilke, W.:* Makromol. Chem. 189, p. 2627 (1988); *Kilian, H.-G.,* in: Thermal Analysis and Calorimetry in Polymer Physics. *Mathot, V. B. F.* (Ed.), Special issue Thermochimica Acta, Elsevier Science Publishers, Amsterdam, 1994
117 *Russell, T. P., Ito, H., Wignall, G. D.:* Macromol. 21, p. 1703 (1988)
118 *Brennan, W. P., Miller, B., Whitwell, J. C.* I&EC Fundamentals 8, p. 314 (1969)
119 *Brennan, W. P., Miller, B., Whitwell, J. C.,* in: Analytical Calorimetry. *Porter, R. S., Johnson, J. F.* (Eds.), Vol. 2, Plenum Press, New York–London, 1970
120 *Heuvel, H. M., Lind, K. C. J. B.:* Anal. Chem. 42(9), p. 1044 (1970)
121 *Guttman, C. M., Flynn, J. H.:* Anal. Chem. 45(2), p. 408 (1973)
122 *Richardson, M. J.:* Anal. Proc. (Lond.) 17(6), p. 228 (1980)
123 *Runt, J., Harrison, I. R.,* in: Methods of Experimental Physics, Vol. 16, Part B: Crystal Structure and Morphology. *Fava, R. A.* (Ed.), p. 287 (1980)
124 *Mandelkern, L., Glotin, M., Benson, R. A.:* Macromol. 14, p. 22 (1981)
125 *Varma-Nair, M.* (*Ed.*): The 3rd ATHAS Report. University of Tennessee, Dept. of Chem., Knoxville, 1985
126 *Bandara, U.:* J. Thermal Anal. 31, p. 1063 (1986)
127 *Cheng, S. Z. D., Wunderlich, B.:* Thermochim. Acta, 134, p. 161 (1988)
128 *Xiang, G., Yunliang, L.:* Thermochim. Acta 150, p. 53 (1989)
129 *Flammersheim, H.-J., Rudakoff, G., Eckhardt, N.:* Wiss. Z. Friedrich-Schiller-Univ. Jena: Naturwiss. Reihe 39(2/3), p. 231 (1990)
130 *Alsleben, M., Schick, C., Mischok, W.:* Thermochim. Acta 187, p. 261 (1991)
131 *Hemminger, W. F., Sarge, S. M.:* J. Thermal Anal. 37(7), p. 1455 (1991)
132 *Dole, M., Wunderlich, B.:* J. Polym. Sci. 24, p. 139 (1957)
133 *Suzuki, T., Kovacs, A. J.:* Polymer J. 1, p. 82 (1970)
134 *Hoffman, J. D., Davis, G. T., Lauritzen, Jr., J. I.,* in: Treatise on Solid State Chemistry. Vol. 3, p. 497, Plenum Press, New York–London, 1976
135 *Karasz, F. E., Bair, H. E., O'Reilly, J. M.:* J. Phys. Chem. 69, p. 2657 (1965)
136 *Bares, V., Wunderlich, B.:* J. Polym. Sci. 11, p. 861 (1973)
137 *Van Krevelen, D. W.:* Properties of Polymers. Elsevier, Amsterdam, 1972, 1976
138 *Gaur, U., Wunderlich, B.:* J. Phys. Chem. Ref. Data 11(2), p. 313 (1982)
139 *Lauritzen, Jr., J. I., Hoffman, J. D.:* J. Appl. Phys. 44, p. 4340 (1973)
140 *Jones, D. H., Latham, A. J., Keller, A., Girolamo, M.:* J. Polym. Sci.: Polym. Phys. Ed. 11, p. 1759 (1973)
141 *Hoffman, J. D., Miller, R. L.:* Macromol. 22, p. 3038 (1989)
142 *Adam, G., Gibbs, J. H.:* J. Chem. Phys. 43, p. 139 (1965)
143 *Point, J. J.:* Macromol. 12(4), p. 770 (1979)
144 *Point, J. J.:* Disc. Faraday Soc. 68, p. 167 (1979)
145 *Rault, J.* in: Preprints of Short Communications. Presented at IUPAC MAKRO MAINZ, Vol. 3, p. 1276 (1979)
146 *Hikosaka, M.:* Polymer 28, p. 1257 (1987)
147 *Hikosaka, M.:* Polymer 31, p. 458 (1990)
148 *Armitstead, K., Goldbeck-Wood, G.:* Adv. Polym. Sci. 100, p. 219 (1992)
149 *Cheng, S. Z. D., Cao, M.-Y., Wunderlich, B.:* Macromol. 19, p. 1868 (1986)
150 *Suzuki, H., Grebowicz, J., Wunderlich, B.:* Brit. Polym. J. 17, p. 1 (1986)
151 *Cheng, S. Z. D., Wunderlich, B.:* Macromol. 20, p. 1630 (1987)
152 *Cheng, S. Z. D., Pan, R., Wunderlich, B.:* Makromol. Chem. 189, p. 2443 (1988); see also a number of papers by *Cheng, S. Z. D., Wunderlich, B*
153 *Struik, L. C. E.:* Polymer 28, p. 1521 (1987); 28, p. 1534 (1987); 30, p. 799 (1989); 30, p. 815 (1989)
154 *Glotin, M., Mandelkern, L.:* Colloid Polym. Sci. 260, p. 182 (1982)
155 *Mandelkern, L.:* Polym. J. 17(1), p. 337 (1985)
156 *Clas, S.-D., Heyding, R. D., McFaddin, D. C., Russell, K. E., Scammell-Bullock, M. V., Kelusky, E. C., St-Cyr, D.:* J. Polym. Sci., Part B: Polym. Phys. 26, p. 1271 (1988)

157 Alamo, R. G., McLaughlin, K. W., Mandelkern, L.: Polym. Bull. 22, p. 299 (1989)
158 Mandelkern, L.: Adv. Chem. Ser. 227, p. 377 (1990)
159 Hahn, B., Wendorff, J., Yoon, D.Y.: Macromol. 18, p. 718 (1985); Hahn, B. R., Herrmann-Schönherr, O., Wendorff, J. H.: Polymer 28, p. 201 (1987)
160 Huo, P., Cebe, P.: Macromol. 25, p. 902 (1992). Huo, P., Cebe, P.: J. Polym. Sci. Part B: Polym. Phys. 30, p. 239 (1992). Huo, P., Cebe, P.: Colloid Polym. Sci., accepted for publication; Cebe, P., Huo, P. P., in: Thermal Analysis and Calorimetry in Polymer Physics. Mathot, V. B. F. (Ed.), Special issue Thermochimica Acta, Elsevier Science Publishers, Amsterdam, 1994
161 Popli, R., Glotin, M., Mandelkern, L., Benson, R. S.: J. Polym. Sci. Polym. Phys. Ed. 22, p. 407 (1984)
162 Clas, S.-D., McFaddin, D. C., Russell, K. E.: J. Polym. Sci., Part B: Polym. Phys. 25, p. 1057 (1987)
163 Mandelkern, L., Peacock, A. J., in: Studies in Physical and Theoretical Chemistry. Lacher, R. C. (Ed.), Vol. 54, p. 201, Elsevier Science Publishers, Amsterdam, 1988
164 Voigt-Martin, I. G., Mandelkern, L.: J. Polym. Sci. Polym. Phys. Ed. 22, p. 1901 (1984)
165 Kunz, M., Möller, M., Heinrich, U.-R., Cantow, H.-J.: Makromol. Chem., Makromol. Symp. 20/21, p. 147 (1988); 23, p. 57 (1989)
166 Bergmann, K., Nawotki, K.: Kolloid-Z. u. Z. Polymere 250, p. 1094 (1972)
167 Kitamaru, R., Horii, F., Hyon, S.-H.: J. Polym. Sci. Polym. Phys. Ed. 15, p. 821 (1977)
168 Bergmann, K.: J. Polym. Sci. Polym. Phys. Ed. 16, p. 1611 (1978)
169 Kitamaru, R., Horii, F.: Adv. Polym. Sci. 26, p. 139 (1978)
170 Axelson, D. E., Mandelkern, L., Popli, R., Mathieu, P.: J. Polym. Sci. Polym. Phys. Ed. 21, p. 2319 (1983)
171 Axelson, D. E., Russell, K. E.: Prog. Polym. Sci. 11, p. 221 (1985)
172 Kitamaru, R., Horii, F., Murayama, K.: Macromol. 19, p. 636 (1986)
173 McFaddin, D. C., Russell, K. E., Kelusky, E. C.: Polym. Commun. 29, p. 258 (1988)
174 Alamo, R., Domszy, R., Mandelkern, L.: J. Phys. Chem. 88, p. 6587 (1984)
175 Allen, R. C., Mandelkern, L.: Polym. Bull. 17, p. 473 (1987)
176 Richter, A., Carius, W., Hölzer, W., Schröter, O., Brandt, H., Hemmelmann, K.: Acta Polymerica 38(4), p. 220 (1987)
177 Alamo, R. G., Mandelkern, L.: Macromol. 22, p. 1273 (1989); Alamo, R. G., Mandelkern, L., in: Thermal Analysis and Calorimetry in Polymer Physics. Mathot, V. B. F. (Ed.), Special issue Thermochimica Acta, Elsevier Science Publishers, Amsterdam, 1994
178 Mandelkern, L., Alamo, R. G., Kennedy, M. A.: Macromol. 23, p. 4721 (1990)
179 Peacock, A. J., Mandelkern, L.: J. Polym. Sci., Part B: Polym. Phys. 28(11), p. 1917 (1990)
180 Wang, J., Pang, D., Huang, B.: Polym. Bull. 24, p. 241 (1990)
181 Mansfield, M. L.: Macromol. 16, p. 914 (1983)
182 Flory, P. J., Yoon, D. Y., Dill, K. A.: Macromol. 17, p. 862 (1984)
183 Yoon, D. Y., Flory, P. J.: Macromol. 17, p. 868 (1984)
184 Marqusee, J. A., Dill, K. A.: Macromol. 19, p. 2420 (1986)
185 Marqusee, J. A.: Macromol. 22, p. 472 (1989), and references therein
186 Kumar, S. K., Yoon, D. Y.: Macromol. 22, p. 3458 (1989), and references therein
187 Schick, C., Donth, E.: Physica Scr. 43, p. 423 (1991); Alsleben, M., Schick, C., in: Thermal Analysis and Calorimetry in Polymer Physics. Mathot, V. B. F. (Ed.), Special issue Thermochimica Acta, Elsevier Science Publishers, Amsterdam, 1994
188 Smit, P. P. A.: Rheol. Acta 5, p. 277 (1966)
189 Schwarzl, F. R.: Rheol. Acta 5, p. 270 (1966)
190 Kraus, G.: Adv. Polym. Sci. 8, p. 155 (1971)
191 Boyer, R. F.: Macromol. 6, p. 288 (1973)
192 Boyer, R. F.: J. Macromol. Sci. (Phys.) B8, p. 503 (1973)

Nomenclature

A	area belonging to a DSC curve
C_p	heat capacity at constant pressure
D	density

165

dQ	amount of heat transferred
G	free enthalpy or Gibbs free energy
H	enthalpy
l	lamellar thickness
l,extr $\rightarrow T$	extrapolation of melt part of DSC curve to T
m	mass
M	molar mass
S	entropy
T	temperature
t	time
T_c	crystallization temperature
T_d	dissolution temperature
T_g	glass transition temperature
T_m	melting temperature
T_m^o	equilibrium crystal–melt transition temperature
T_r	reference temperature
T^*	point of intersection of DSC curve with extrapolation from the melt part of the curve
w	mass fraction
W	mass percentage
$w^a(T)$	mobility
$w^c(T)$	crystallinity
$w^{nc}(T)$	non-crystallinity
$w^r(T)$	rigidity
$w^s(T)$	solidity
x	mole fraction
X	mole percentage
Δ	difference between reference states
$\Delta c_p(T)$	specific heat capacity differential function
$\Delta c_p(T_g)$	jump in specific heat capacity at T_g
$\Delta g(T)$	specific driving force of melting, specific free enthalpy differential function
$-\Delta g(T)$	specific driving force of crystallization
$\Delta h(T)$	specific heat of fusion, specific enthalpy differential function
$\Delta s(T)$	specific entropy differential function
ΔT	supercooling
σ_e	end-surface free enthalpy
σ_s	side-surface free enthalpy
	(Specific quantities are indicated by lower-case letters)

Abbreviations

DSC	differential scanning calorimetry/calorimeter
DTA	differential thermal analysis
LPE	linear polyethylene
LDPE	low-density polyethylene (high-pressure process)
LLDPE	linear low-density polyethylene
MFI	melt flow index
PP	polypropylene
SEC	size exclusion chromatography or gel permation chromatography
VLDPE	very low-density polyethylene

Subscripts and superscripts

a	amorphous
b	base line
c	crystal
cc	cooling curve
e	excess
g	glass

166

hc	heating curve
i	run number: $i = 1, 2, 3 \ldots$
j	run number: $j = 1, 2, 3 \ldots$
l	liquid, lower
m	melt
n	number-average
nc	non-crystalline
r	rigid
R	reference
s	solid
S	sample
u	upper, upswing
w	weight-average
z	z-average
8	1-octene as comonomer
*	initial

Chapter 6

The Glass Transition Region

Mike J. Richardson, National Physical Laboratory,
Teddington, Middlesex TW11 0LW, Great Britain

Contents

6.1 Introduction: glass formation

Although the viscosity of a liquid increases on cooling, the molecular or segmental mobility at the crystallization temperature T_c is usually sufficient to allow the formation of some degree of crystalline order. However, if crystallization is slow relative to the rate of cooling (glycerol and many isotactic polymers fall easily into this category) or is impossible for steric reasons (atactic polymers), further cooling leads to such high viscosities that molecular mobility is impaired and the material cannot relax to the equilibrium conformation characteristic of the supercooled liquid at that particular temperature. Instead it becomes an immobile glass with a frozen-in molecular conformation typical of some higher temperature liquid.

The specific heat capacity c_p in the glassy (subscript g) and liquid (subscript l) states may be thought of as having, respectively, a vibrational component c_{pg} plus, for a liquid, an additional relaxational or 'hole energy' term c_{pr} which is needed to generate the steady increase of free volume with temperature that characterizes this state. A c_p–temperature curve therefore shows a discontinuity in the glass transition T_g region as c_{pr} becomes active. Because the signal from a DSC is related to c_p, this technique is very convenient for the study of this important region which marks the maximum or minimum serivce temperature of many plastics and rubbers, respectively.

If the solid–liquid analogy is further developed [1] to consider the rate of response of a glass-forming material to a step change in temperature (a decrease ΔT is shown in Fig. 6.1(a)), the appropriate thermodynamic quantity, the enthalpy h, decreases instantaneously for both the vibrational (Δh_g) and relaxational (Δh_r) components provided that $T-\Delta T$ remains well above the T_g region (solid line, Fig. 6.1(b)). As mobility falls in the T_g region, relaxation becomes increasingly difficult

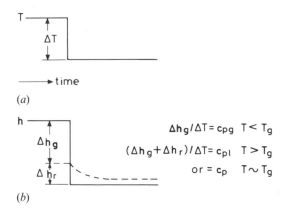

(a)

(b)

$\Delta h_g / \Delta T = c_{pg}$ $T < T_g$

$(\Delta h_g + \Delta h_r)/\Delta T = c_{pl}$ $T > T_g$

or $= c_p$ $T \sim T_g$

Fig. 6.1. The effect of a step change of temperature (ΔT, (a)) on the enthalpy of a liquid above (solid line, (b)) and within (broken line, (b)) the glass transition region

and a finite time is required for Δh_r to approach, but not necessarily attain (broken line, Fig. 6.1(b)) its equilibrium value; when the temperature is well below T_g, structural relaxation effectively ceases, $\Delta h_r = 0$ and the material is now a 'stable' glass. Passage through the T_g region therefore spans relaxation times from minute fractions of a second (in the liquid) to approaching infinity (in the glass). The intervening region, with relaxation times of the order of experimental time scales, can provide a wealth of structural information. The calorimetric procedures that are now available to exploit this are discussed in this chapter.

6.2 Experimental aspects

DSC curves for a variety of glasses are shown in Fig. 6.2. If the labels are removed it is impossible to distinguish (by shape or by the magnitude of the specific heat increment Δc_{pg} in the T_g region) between low molar mass (MM) and polymeric materials and even the glass formed from an oriented (nematic) liquid. Similar curves are also found for inorganic glasses and this universality of behavior in the T_g region has led to a very fruitful exchange of ideas between ceramicists, polymer scientists and theoreticians [2–4].

Although there is an obvious Δc_{pg} step in all the examples of Fig. 6.2, it is also clear that supplementary structure is possible. In fact the peaks to be seen in Fig. 6.2 can be manipulated at will, and the discussion below shows how they can be used to provide details of prior thermal history. At this point it is sufficient to indicate that the 'glass transition' is not a unique temperature, as is clear from the example of Fig. 6.3 which shows curves for polystyrene glasses that differ only in their rate of cooling from the liquid into the glass. The range of conditions is extreme, but extremes are very useful both in defining and in clarifying problems. Figure 6.3 illustrates this: if some (as yet undefined) point in the Δc_{pg} region is taken as T_g a paradox emerges — the well-annealed (slowly cooled) sample appears to have the highest T_g, whereas for many years it has been known from dilatometry that the reverse is true. Progress towards a deeper undestanding of the glass transition can be hoped for only if this rather basic problem can be overcome.

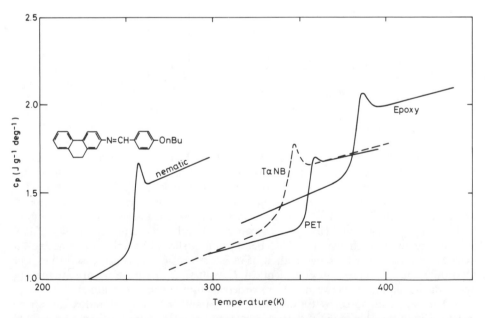

Fig. 6.2. The glass transition region of some low and high molar mass organic materials cooled at 10 K/min and reheated at 20 K/min: $T\alpha NB$ = tri-(α naphthyl) benzene; PET = polyethylene terephthalate

171

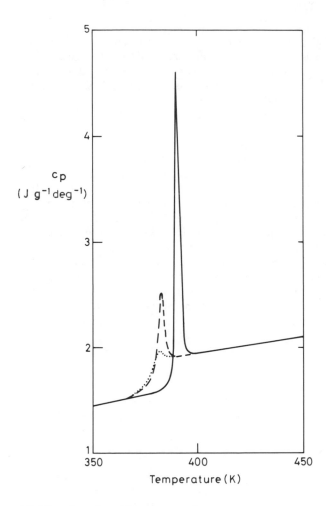

Fig. 6.3. The specific heat capacity of anionic polystyrene ($M_n = 36\,000$) glasses formed by cooling through T_g at 2.46 K/day (———), 0.3125 K/min (– – –) and 80 K/min (····): heating rate $= 20$ K/min

6.3 The glass transition temperature

Many points on the DSC curve have been used to 'define' T_g. A selection is given in Fig. 6.4, which shows both cooling (glass-formation or vitrification, Fig. 6.4(a)) and the subsequent reheating curve (devitrification, Fig. 6.4(b)). Although Figs. 6.4(a) and (b) are schematic, to make the points more obvious, they are based on real experience in the sense that only in exceptional circumstances do corresponding points on Figs. 6.4(a) and (b) agree. This must raise doubts as to the validity of *any* of the constructions shown: if T_g is to be of value it should be able to characterize a particular glass rather than the method of measurement (there is no reason to expect that different techniques will give identical T_g values; at this stage we are attempting to reconcile anomalies between data recorded using the same technique).

The problems posed by Figs. 6.4(a) and (b) and the earlier paradox can all be resolved by a suitable modification of a technique that was in use long before DSC was available. The technique was dilatometry, with T_g being defined by the point of intersection of specific volume, v, curves for the glassy and liquid phases [5]. The

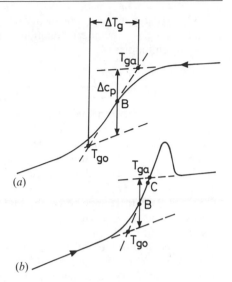

Fig. 6.4. The glass transition region in cooling (a) and subsequent reheating (b), showing some commonly used definitions of T_g: T_{go} is the extrapolated onset and T_{ga} the extrapolated end, the width ΔT_g is defined by $\Delta T_g = T_{ga} - T_{go}$, B is the point where half the specific heat increment has occurred $\equiv T_g(\frac{1}{2}\Delta c_p)$, C is the point of inflection

appropriate thermodynamic analogue of specific volume is specific enthalpy h, and changes in this quantity can be obtained by integration of the DSC heat capacity curves. It should be emphasized that the 'point of intersection' is for data from the 'stable' glass and liquid phases (Fig. 6.5) so that extrapolations from outside the glass transition region are required. In the T_g region itself the curve is much influenced by experimental conditions.

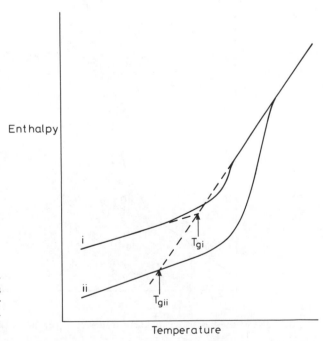

Fig. 6.5. Schematic enthalpy–temperature curves showing how T_g is defined for a (i) quenched and (ii) well-annealed glass

173

For most polymers, the heat capacities below (c_{pg}) and above (c_{pl}) the glass transition region are linear functions of temperature. Integration gives the corresponding enthalpies [6]

$$h_g(T) = aT + \tfrac{1}{2}bT^2 + P \tag{6.1}$$

$$h_1(T) = cT + \tfrac{1}{2}dT^2 + Q \tag{6.2}$$

The difference between the integration constants, $Q - P$, is obtained by subtracting Eq. 6.1 (with $T = T_1$) from Eq. 6.2 (with $T = T_2$) to give $h_1(T_2) - h_g(T_1)$, the energy increment through T_g which is represented by the shaded area of Fig. 6.6 (T_1 and T_2 are convenient temperatures in their respective 'stable' regions—well away from the transition itself). The glass temperature T_g then follows as the solution of the equation $h_g(T_g) = h_1(T_g)$.

Since c_{pg} is little affected by thermal history (at most, by a few tenths of one per cent [7, 8]) and c_{pl} is a true material property, the influence of thermal history on T_g is given by variations in P, and hence $Q - P$. These are in turn reflected by changes in the peak magnitude and the location of the Δc_{pg} step (Fig. 6.3): on an enthalpy curve they are represented by differences between (i) and (ii) in Fig. 6.5. The calculation of T_g as described here is equivalent to a construction (Fig. 6.6) on the $c_p - T$ curve which gives a step increment BC at T_g on an idealized path ABCD with T_g located so that areas $\kappa + \mu = \lambda$ [9, 10].

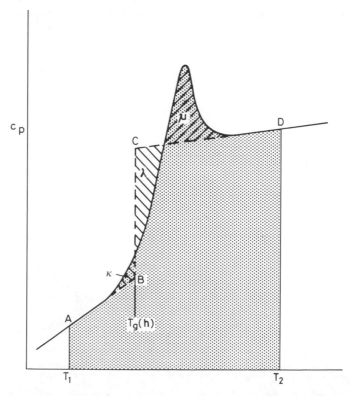

Fig. 6.6. Use of the heat capacity curve to derive $T_g(h)$: areas ($\kappa + \mu$) = area λ

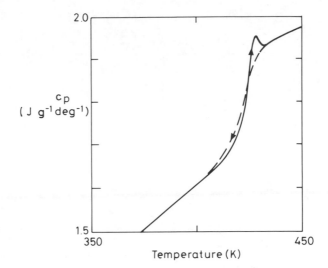

Fig. 6.7. Specific heat capacity curves in cooling and subsequent heating $(-/+20$ K/min for the polycarbonate of Table 6.1)

The consistency of the above method of determining T can be checked: measurements made in cooling at a rate q_- and on subsequent reheating (q_+) should agree [11] and T_g should shift to lower temperatures as annealing procedures are improved.

The formation and subsequent reheating of a polycarbonate glass are shown in Fig. 6.7. Values of T_g, calculated according to the enthalpic procedure discussed above $(T_g(h))$, are given in Table 6.1 together with data for two of the most popular definitions of Figs. 6.4(a) and (b). At first sight there seems to be some correlation between $T_g(h)$ and $T_g(\frac{1}{2}\Delta c_p)$, but this holds only when $|q_-|/q_+ > 1$, i.e. for glasses that are 'quenched' relative to subsequent DSC heating rates. This is clearly shown by the more general results for polyphenylene oxide in the second half of Table 6.1; here the data are for glasses formed by in situ cooling in the DSC (thermal treatments that are far less extreme than those shown in Fig. 6.3). There is a progressive deviation between $T_g(h)$ and other data, and only $T_g(h)$ removes the anomalous 'increase' in T_g (see, for example, Fig. 6.3 or Table 6.1) with improved annealing (lower cooling rates). There is, in fact, a simple relation between $T_g(h)$ and $\log q_-$ [6, 12]: Fig. 6.8 shows results from a series of experiments which includes the polyphenylene oxide data of Table 6.1. $T_g(h)$ is reproducible to ± 1 K and a curve

Table 6.1. Glass transition temperatures derived from DSC curves

Material	Thermal treatment K/min		Glass temperature K		
	Cooling	Heating	$T_g(h)$	T_{go}	$T_g(\frac{1}{2}\Delta c_p)$
Polycarbonate[1]	20	—	421.4	414.7	422.1
$(M_w/M_n = 38100/14400)$	20	20	421.7	419.1	422.7
Polyphenylene oxide	0.3125	20	479.6	494.4	493.8
$(M_w/M_n = 46400/17000)$	5	20	484.1	490.7	490.0
	80	20	487.6	486.3	489.6

[1]See Fig. 6.7.

175

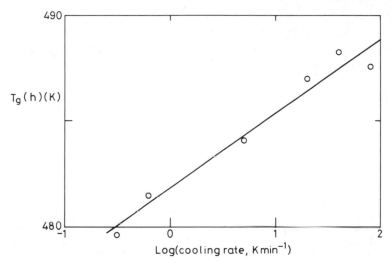

Fig. 6.8. Enthalpic glass temperature $T_g(h)$ for the polyphenylene oxide of Table 6.1

similar to Fig. 6.8 is useful in determining an unknown cooling rate; an injection molded material, for example, typically cools through T_g at hundreds or a few thousand degrees per minute (depending on size and shape), and these rates represent only short extrapolations on the logarithmic scale of Fig. 6.8.

Because of the discrepancies between $T_g(h)$ and other definitions of T_g, the value of much existing DSC work might be questioned. This is a reasonable criticism which is generally, and quite fortuitously, less serious than might be implied by some of the data of Table 6.1. The reason for this is that effects of thermal history have rarely been considered and most measurements have been made on materials that have been 'quenched'—either an as-received pellet or a molding—or after cooling in the DSC at a (generally unspecified) high cooling rate following some kind of 'conditioning' (generally, heating above T_g to minimize internal stress and/or to improve contact between sample and pan). Under these circumstances $|q_-|/q_+ > 1$ and $T_g(h) \approx T_g(\frac{1}{2}\Delta c_p)$. $T_g(\frac{1}{2}\Delta c_p)$, rather than T_{go}, is the preferred approximation to T_g here because it approaches an average, mid-point, value on the enthalpy curve, whereas T_{go} defines only the start of the transition. At this point it should be mentioned that *any* definition of a glass *temperature* ignores the fact that the transition covers a *range* of temperature: for some practical applications, such as indicating the upper service temperature of a polymer, it may be more appropriate to use T_{go} than $T_g(h)$ or $T_g(\frac{1}{2}\Delta c_p)$. Conventional DSC constructions badly overestimate T_g when a sample has been annealed—often unknowingly. If, for example, T_g of a material is only a few tens of degrees above ambient there may be extensive isothermal annealing. 'Ambient' may refer to the conventional room temperature or the service condition of a component that is 'run hot' on a routine basis; respective examples are polyvinyl acetate [13] or the many types of epoxy resins used in the electrical industry.

6.4 Calorimetry, dilatometry and other methods of determining T_g

It is useful to comment at this stage on the various methods used to measure T_g. The present calorimetric approach is essentially an equilibrium procedure that characterizes the original, metastable, glass. Any annealing effects in the T_g region (due to the finite heating rate in the DSC) are compensated via the $h_1(T_2) - h_g(T_1)$ term. An analogous approach is used in dilatometry which has the advantage of direct measurement of changes in the relevant thermodynamic quantity, the specific volume, through the movement of, for example, a column of mercury in a capillary tube. This technique requires relatively large samples, of the order of grams, so heating or cooling rates must be low—degrees per hour rather than per minute as with DSC.

Dilatometric data are easy to deal with visually because the expansion coefficient changes by a factor of two or three on passing from the glass to the liquid. The heat capacity, in contrast, increases by only 20%–40%, and it is very difficult to make a visual estimate of the point of intersection of enthalpy curves for the glass and liquid, especially as each has to be extrapolated—the computational procedure discussed above must be used. These comments might be taken to imply that the dilatometric technique is to be preferred, but the reverse is the case: dilatometric measurements are slow and must be preceded by lengthy filling and degassing operations; DSC is rapid, needs only milligrams of material, and a wide range of thermal histories can be examined. When results for the same glasses are compared $T_g(v) \approx T_g(h)$ for simple thermal histories (exceptions are discussed in section 6.5.1). There is no a priori reason that this should be so: the experiment is observing the response of a glass that forms because of incomplete relaxation from the melt (Fig. 6.5) and the rate of enthalpy relaxation need not equal that of volume [14]. It is fortunate that the two techniques appear to give similar results because it is common to group data from a variety of sources into general correlations (e.g. relations between T_g and molar mass).

No direct correlation should be expected between $T_g(h$ or $v)$ and T_g defined by a loss maximum in a mechanical or dielectric relaxation experiment. The latter T_g refers to a dynamic process for which T_g increases with frequency [15].

6.5 The influence of physical and compositional variables on T_g

Previous sections have discussed the appearance of the T_g region for samples with simple thermal histories, and these have been used to illustrate the procedures that lead to thermodynamically meaningful values of the glass temperature. The latter cannot, however, be unique since different physical or mechanical treatments may lead to the same value of $T_g(h)$: evidence to differentiate between the several routes to a common $T_g(h)$ may be provided by the shape of the c_p-T curve in the transition region. This section considers the influence of physical, mechanical and compositional variables.

6.5.1 Stress

Figure 6.9 shows results for a freeze-dried sample of high MM polystyrene. The curve is complex and a range of glasses can be obtained by varying the solvent power and the rate of precipitation. The product, especially when of high MM, is a light floc that is difficult to use as such; it is easily consolidated into a more tractable form by compression in a small hand press. However, unless minimal pressures are used, the initial material is transformed to a new high-energy state [16–18]—the compacted polymer of Fig. 6.9, for example, has a room temperature enthalpy that is 4 J/g higher than that of a reference state cooled through T_g at 40 K/min. Since $\Delta c_{pg} \approx 0.3$ J/g per degree K, this corresponds to a rise in T_g of ~13 K. Qualitatively similar behavior is

177

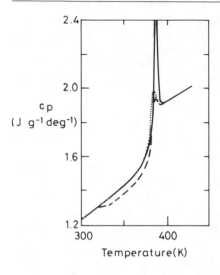

Fig. 6.9. Specific heat capacity of a freeze-dried polystyrene before (——) and after (– – –) densification at room temperature: the reference curve (···) is for the material after cooling at 20 K/min and rerunning (all, +20 K/min)

found for polymer that has been quenched: the Δc_{pg} step is preceded by a decrease in c_p as restricted molecular reorganization to a more stable conformation becomes possible with the approach to T_g.

The behavior of a glass that has been formed by the cooling of a molten polymer under pressure is quite different to that shown in Fig. 6.9. Sub-T_g peaks appear and these increase in size and shift to lower temperatures (Fig. 6.10) as the formation pressure is increased [19, 20]. Densification provides a clear example of conditions when $T_g(v) \neq T_g(h)$. It inevitably reduces the specific volume at room temperature and atmospheric pressure so that $T_g(v)$ decreases monotonously with increasing pressure. The enthalpy, on the other hand, initially decreases but then tends to increase as the sample stores energy. $T_g(h)$ therefore passes through a minimum before rising with further increase in pressure, at which stage $T_g(h) > T_g(v)$.

The conditions of Fig. 6.10 are not merely of academic interest; they simulate those encountered in normal injection molding conditions and it is useful to examine

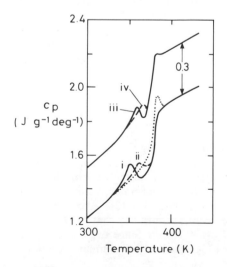

Fig. 6.10. Polystyrene glass formed under stress: (i), (ii) cooled at 1 K/min under (i) 310 and (ii) 270 MN/m², (iii), (iv) sections from the skin (iii) and core (iv) of an injection molded bar ((iii) and (iv) displaced +0.3 J/g per degree K) (reference and q_+ as in Fig. 6.9)

typical products prepared in this way. Curves (iii) and (iv) of Fig. 6.10 show thin sections representing, respectively, the skin and core of an injection molded bar of polystyrene. By analogy with curves (i) and (ii) the initial skin has formed under something approaching the full injection pressure with the core solidifying later (because of the finite rate of heat transfer) when the gate has closed and pressure decreased. It is possible to envisage this type of investigation giving an insight into the nature of flow in a mold and product dependence on distance from, and closure of, the gate.

    In all the above examples the polymer has undergone some form of stress. A further, and simpler, example of this is orientation. Little quantitative information is available to illustrate the effects of this because oriented material naturally retracts on heating through T_g and, unless restricted [21], movement of the sample can lead to gross artefacts. Evidence from other sources (dynamic mechanical analysis [22]) suggests that the specific units involved determine the direction in which T_g shifts with orientation—it would be valuable to have direct DSC evidence for this.

6.5.2 Molar mass

The simplest compositional change for an amorphous polymer is one due only to the effects of molar mass. This is conventionally modelled by a system composed of repeat and end units [23]—a formalism that allows the minor effects of MM on c_{pg} and c_{pl} to be brought into general equations for enthalpy changes in a homologous series [24]. The increase in T_g with MM to a limiting value, that is generally reached in the range $10^3 - 10^4$ g/mol, has long been known but the detailed shape of the T_g–MM curve is not clear because such curves are usually composites, using T_g derived via a variety of techniques and taking no account of the effects of thermal history. The DSC procedure is applicable to any polymer over the whole MM range and results should at least be able to differentiate between a single smooth T_g–MM curve and one made up of intersecting linear regions [25, 26]. Progress can then be made with more detailed theories that predict specific T_g–MM relations [27, 28]; these are currently restricted by a lack of both molecular data and accurate T_g values. An interesting application of DSC arises from this work—the determination of the MM of oligomers. In the low MM region T_g generally varies rapidly with MM, so the latter can be read from a suitable calibration curve.

    It was suggested some years ago that the product $\Delta\alpha T_g \approx$ constant, where $\Delta\alpha$ is the change in expansion coefficient on passing from the glass to the liquid. Like many generalizations this is now known to need many qualifications—mainly concerning chain stiffness. A comparable expression has been proposed [29] for the analogous quantity $\Delta c_p T_g$ and similar restrictions apply. However, a homologous series is a good test for such a relationship since molecular parameters are little changed while T_g may vary by a factor of 2 from high MM to monomer. For polystyrene the product is remarkably constant, decreasing by only two or three per cent as the degree of polymerization drops from ∞ to 4 [24].

6.5.3 Crosslinking

Crosslinking reduces molecular mobility, and this is regained only by appropriate thermal activation at higher temperatures through an increase in T_g. There is a parallel decrease in the magnitude of the Δc_{pg} step which vanishes for heavily

179

crosslinked materials such as dough molding compounds or phenolic resins. Examples are given in chapter 7 on the cure of thermosetting materials.

6.5.4 Fillers and liquid diluents

Previous comments have been concerned mainly with homopolymers. Crosslinking introduces a form of copolymerization but the range of compositional variables—co-units, blends, plasticizers, diluents, fillers, and other additives—covers several major technological areas and these can only be touched on here when DSC has important applications [30].

In any mixture of A (polymer) and B (low MM component), the effect of B on the location, breadth (ΔT_g, Fig. 6.4) and magnitude (Δc_{pg}) of T_{gA} must be considered [31]. Inert fillers might be expected to have a purely diluting effect on Δc_{pg} and leave T_{gA} and ΔT_g unchanged. In practice a variety of results has been reported, with changes of T_{gA} ascribed to stress due to differential contraction between A and (solid) B on cooling through the transition region [32]. In addition to the T_g region itself the effect on c_p is of interest; this is generally of the form $c_p = w_A c_{pA} + w_B c_{pB}$, where w_A and w_B are mass fractions of the respective components. Deviations from this simple relationship are found only for systems with specific interactions so that, in general, if c_{pA} and c_{pB} differ sufficiently—as with molten A and an inorganic filler—measurement of c_p may provide useful compositional detail, especially in conjunction with similar information derived from the magnitude of Δc_{pg}.

If A and B are compatible, as with polymer/plasticizer systems, T_g naturally covers the whole range from T_{gB} to T_{gA} and the shape of the T_g–composition curve can be used to discuss A–B interactions. This is particularly useful for PVC with its hierarchy of structural regimes, and careful work has been able to show how plasticizer is distributed between these regions [33].

A very active area of current research concerns polymer gels, which are polymer-diluent systems in which phase separation plays a key role. These systems are of such importance that chapter 8 is devoted entirely to them, and DSC aspects are considered there.

6.5.5 Copolymers and blends

The DSC behavior of individual random copolymers in the glass transition region is very similar to that of homopolymers. However, T_g–composition curves may show either monotonous changes between values for the two pure components or maxima or minima may appear (details depend on sequence distributions and interactions between co-units). These features are a severe test of any predictive theory [34, 35], so it is especially important to ensure that they use consistent and reliable $T_g(h)$ data. The transition to block copolymers raises the possibility of observing two T_g (for amorphous blocks) corresponding to the individual components. 'Possibility' should be emphasized because, although two T_g regions are commonly found, it is not clear what is the minimum 'probe' size that can be resolved in the DSC.

Similar comments apply to blends [36, 37] for which the appearance of one or two regions (again assuming two amorphous components) is taken to imply compatibility or incompatibility respectively. Although these are undoubtedly reasonable generalizations, it is always better to supplement them using the additional information provided by each T_g, ΔT_g and Δc_{pg}. Thus T_g of component A in the blend may not equal that of pure A (for equivalent thermal histories), implying some solubility of B in A. A useful additional procedure is to encourage the growth of peaks, as in Fig. 6.3, by annealing (by slow cooling or isothermally just below T_g), as this may confirm that a minor step in a DSC curve is associated with a glass transition rather than an unrelated phenomenon. A cautionary note is appropriate at this stage concerning the magnitude of peaks: those of Fig. 6.3 are almost inevitable in curves for homopolymers; they are much reduced in multicomponent systems and it may be necessary to resort to special annealing schedules to produce them. The effort is worthwhile if an otherwise obscure event can be unambiguously ascribed to a glass transition.

Annealing procedures are especially useful if the critical temperature of a blend is spanned by a DSC experiment. In this case quenching may well retain the single high-temperature phase with progressively slower cooling encouraging phase separation.

6.5.6 Crystallinity

DSC work on partially crystalline polymers has tended to concentrate on the melting region, with little attention given to the glass transition. This was understandable when data were represented by a chart recording and there was only a single opportunity for display. Now, with electronic data storage almost standard, multiple displays with scale expansion are routine and the complementary nature of the T_g region should be utilized.

Crystalline regions act as physical crosslinks; they limit molecular mobility but the overall effect on the *location* of T_g is small (a few degrees) relative to the effects of chemical crosslinking (see, for example, Fig. 7.11). The magnitude of the Δc_{pg} step is naturally reduced as the concentration of amorphous regions is lowered with increasing crystallinity [38] but there are many examples [39] where the decrease is much greater than would be anticipated on the basis of the degree of crystallinity ξ alone. In this particular case the ratio of the heat capacity increments in the T_g region for partially crystalline and amorphous material should equal $1 - \xi$. An independent estimate of ξ is available from the heat of fusion (see chapters 5 and 9), but the two methods rarely give comparable data. Applications when there *is* agreement have been discussed by YAGFAROV [40]. The more common result, when $\Delta c_{pg\xi}$ has almost vanished at normal values of ξ (i.e. obtained without any special thermal treatment), is ascribed to the breakdown of the two-phase model: a 'rigid amorphous' fraction is believed to persist to above the conventional T_g region, and is considered in more detail in section 5.3.4.

6.6 Thermodynamic behavior of glasses

When a liquid crystallizes (at T_c with heat of crystallization $\Delta h(T_c)$) there is a large stepwise decrease in entropy ($\Delta s(T_c) = \Delta h(T_c)/T_c$) followed, on further cooling, by a slow decrease that is related to the specific heat capacity of the crystalline phase c_{pc} (Eq. 6.3). If the liquid does not crystallize, but merely supercools, the entropy decreases more rapidly (because $c_{pl} > c_{pc}$) and at some temperature T_K entropy curves for the crystalline and liquid states intersect. T_K is given by

$$\int_{T_K}^{T_c} (c_{pc}/T)\,\mathrm{d}T + \Delta h_c/T_c = \int_{T_K}^{T_c} (c_{pl}/T)\,\mathrm{d}T \tag{6.3}$$

181

(for thermodynamic rigor, the components of Eq. 6.3 should refer to reversible conditions; the use of $\Delta h(T_c)$ and T_c, which imply some degree of supercooling, does not affect the general argument that there is a crossover of entropy curves). Below T_K the entropy of the liquid should be lower than that of the crystalline solid, but this physical absurd situation, the KAUZMANN paradox [41], is precluded by the intervention of the glass transition (below which $c_{pg} \approx c_{pc}$). For a broad range of glasses $T_K \approx T_g - 50$, and it has been suggested that T_K represents T_g at the limit of infinitely slow cooling. An experimental test of this is impossible, but the most successful theory of the glass transition of polymers, that of GIBBS and DI MARZIO [42] has a thermodynamic basis: the configurational entropy decreases to zero at a finite temperature to give a second-order transition of the Ehrenfest type. The thermodynamic implications of this are controversial and have been reviewed [43, 44]. It is sufficient to note here that, although the Gibbs–Di Marzio theory was at first received with caution, it is now widely accepted because of its success in predicting such diverse effects as the effect of molar mass, crosslinking and deformation on T_g and also the magnitude of the Δc_{pg} step in the transition region [45].

GOLDSTEIN [46] has suggested how calorimetric measurements may be used to prove some of the assumptions of the Gibbs–Di Marzio theory—that vibrational frequencies are insensitive to the configurational state and that secondary relaxation processes can be neglected. The approach considers the effect of temperature on the entropy difference between glasses of different thermal history. It requires very accurate measurements of heat capacity because, where the comparison can be made, c_{pg} is only one or two per cent greater than c_{pc} so that differences in c_{pg} due to thermal history effects alone would be expected to be small—and this is found to be the case. Very precise adiabatic calorimetric measurements indicate that the Goldstein approach is valid but this technique cannot accommodate a wide range of thermal histories. DSC is better in this respect and, used in a differential mode, can detect *differences* in c_p of the order of a few tenths of one per cent. Careful studies on polystyrene glasses formed by cooling through T_g at 0.3125 and 80 K/min (a ratio of 1:256) show that, for this polymer at least, virtually all this excess entropy at T_g is configurational in origin [47].

6.7 Physical aging and enthalpy relaxation

As molten polymer is cooled through T_g it falls out of equilibrium into a metastable state that has an effectively infinite life at low temperatures. If, however, the temperature is not too far below T_g (a few tens of degrees K) the metastable glass can slowly relax to a more stable state with accompanying decreases in specific volume or enthalpy and changes in mechanical properties. The last may be beneficial or deleterious; respective examples are the reduction in creep as the free volume is lowered and the well-known embrittlement of polycarbonate. These effects are manifestations of the process of physical aging—so-called because they are totally reversible on thermal cycling and can thus be distinguished from irreversible chemical changes.

The study of the 'enthalpy relaxation' that inevitably accompanies physical aging is an area of much current interest both for the kinetic information that it provides and for the characterization of a particular aged state. This is reflected in the structure that develops in the T_g region; correlation with impact properties, for example, then provides a much more convenient measurement (the DSC curve) than an individual impact test, which has notoriously wide limits of uncertainty.

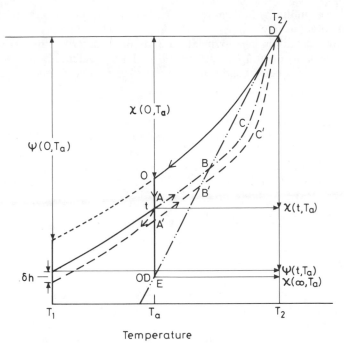

Temperature

Fig. 6.11. Enthalpy relaxation measurements showing the various quantities χ, ψ (see text for details): $-\cdot\cdot-\cdot\cdot$ is extrapolated liquid ($t = \infty$)

Figures 6.3 and 6.8 show the effect of cooling at constant rate on T_g. Cooling may be interrupted at any stage for an additional isothermal aging step of duration t. There are two alternatives at this stage (A, Fig. 6.11):

(a) an immediate rerun, ABCD, which gives $\chi(t, T_a) = h_l(T_2) - h_g(t, T_a)$, where (t, T_a) describes the isothermal annealing step and T_2 is any convenient arbitrary reference temperature in the liquid state [48], or

(b) further cooling to T_1, which is sufficiently low for the glass at this temperature to be considered stable on the experimental timescale, followed by a rerun [49] which leads to $\psi(t, T_a) = h_l(T_2) - h_g(t, T_a; T_1)$.

Ideally the later part of $\psi(t, T_a)$ would traverse the path ABCD of $\chi(t, T_a)$ but, depending on the location of T_a, there may be additional relaxation δh on cooling to T_1. The subsequent heating curve will therefore be in error by this amount (Fig. 6.11; ABC and A′B′C′ are shown parallel, this will be a good approximation at low temperatures but there may be deviations on approaching the liquid state). Enthalpy relaxation measurements are designed to produce

$$\Delta h(t, T_a) = h_g(0, T_a) - h_g(t, T_a) = \chi(t, T_a) - \chi(0, T_a) \tag{6.4}$$

$$= \text{(ideally)} \ \psi(t, T_a) - \psi(0, T_a) \tag{6.5}$$

If $\psi(t, T_a)$ is in error by δh this will be compensated to some extent by a corresponding (but not necessarily identical) error in $\psi(0, T_a)$. The alternative approach using $\chi(t, T_a)$ cannot be given unqualified approval because, depending on T_a and the rate of enthalpy relaxation, the isothermal base line may be perturbed leading to uncertainties that are at a maximum for the important quantity $\chi(0, T_a)$. For both χ and

183

ψ there is some ambiguity in defining $t = 0$, since the sample always requires a finite time to equilibrate at T_a.

Enthalpy relaxation measurements are used to describe the kinetics of the approach to equilibrium at $h_g(\infty, T_a) \equiv h_l(T_a)$ (this equivalence is almost universally assumed, but see [52] below) and it is often useful to normalize $\Delta h(t, T_a)$ in the form

$$z(t, T_a) = \Delta h(t, T_a)/\Delta h(\infty, T_a) \qquad (0 \leq z \leq 1 \text{ for } 0 \leq t \leq \infty) \tag{6.6}$$

so that $1 - z$ represents the *departure* from equilibrium. The quantity $\chi(\infty, T_a)$ can be measured directly for high T_a (when equilibrium can be attained on a reasonable timescale) or calculated using the almost universally observed linear heat capacity–temperature relation (Eq. 6.2) for the molten polymer to make the relatively short extrapolation from the liquid region (DE, Fig. 6.11). The approach to equilibrium is generally taken to be of the form

$$\Delta h(t, T_a) = \Delta h(\infty, T_a)[1 - \phi(t)] \tag{6.7}$$

with $\quad \phi(t) = \exp[-(t/\tau)] \tag{6.8}$

or $\qquad = \exp[-(t/\tau_0)^\beta] \qquad 0 < \beta \leq 1 \tag{6.9}$

Equation 6.8 was proposed by MARSHALL and PETRIE [50], the single relaxation time τ is a function of t; Eq. 6.9, the WILLIAMS–WATTS function [51], implies a distribution of relaxation time with the distribution narrowing as $\beta \to 1$. Equation 6.9 has been used recently by COWIE and FERGUSSON [52], who treat τ_0, β, and also $\Delta h(\infty, T_a)$ (for several T_a), as unknowns that are obtained by curve fitting of data to Eqs. 6.7 and 6.9. The resultant $\Delta h(\infty, T_a)$ show no correlation with 'extrapolated' (via DE, Fig. 6.11) values; this discrepancy must be resolved if Eqs. 6.7–6.9 are to have predictive validity.

6.8 Kinetic behavior in the glass transition region

Detailed theoretical descriptions of the thermal response of a glass-forming system to both cooling and heating cycles have now been developed to such a stage that they can predict and/or explain the complex fine structure that is observed on the DSC curve of a material following a wide range of thermal or mechanical treatment. Developments were pioneered by schools associated with the names of KOVACS (the most recent version has become known as the KAHR model after the authors' initials [53]) and MOYNIHAN [54]. These groups have independently arrived at similar equations which differ mainly in the mathematical formalisms used rather than in the underlying physical assumptions. Both approaches can in principle be applied to any relaxing quantity, although that of KAHR has been mainly expressed in terms of specific volume, with Moynihan's using examples from enthalpy relaxation. (HUTCHINSON [55] has, however, recently discussed the derivation of some KAHR parameters from DSC measurements.)

The Moynihan procedure is briefly summarized here because of its direct correlation with DSC results. An important quantity that appears in this treatment is the fictive temperature T_f which is used to characterize the structure of a given glass. T_f is the temperature on the equilibrium curve that corresponds to the structure at a given temperature T on the (for this particular case) enthalpy–temperature curve. It is obtained (Fig. 6.12) by extrapolation of the glassy curve from T to the equilibrium line so that T_f is defined by [10]

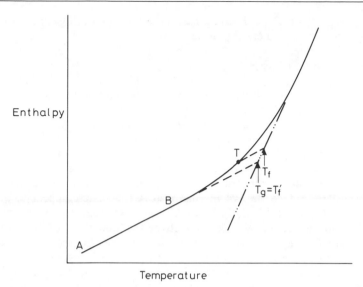

Fig. 6.12. The use of the fictive temperature T_f to define the structural state of a glass at T: T'_f is the limiting value reached at low temperatures and $T'_f = T_g$

$$h(T) + \int_{T}^{T_f} c_{pg}\, dT = h_1(T_f) \tag{6.10}$$

The temperature coefficient is given by

$$\frac{dT_f}{dT} = \frac{|c_p - c_{pg}|_T}{|c_{pl} - c_{pg}|_{T_f}} \tag{6.11}$$

which is equal to one in the liquid state, where $T_f = T$, and zero at low temperatures (all points in the region AB (Fig. 6.12) are of constant structure and the limiting low temperature fictive temperature $T'_f = T_g$) but may be considerably greater than unity for well-annealed glasses (e.g. Fig. 6.3).

Relaxation kinetics are both non-exponential and non-linear [54]. They are non-exponential in the sense that for small departures from equilibrium, $\partial \ln \Phi / \partial t$ is not constant, where $\Phi(t)$ is the recovery or decay function, which is usually taken to have the form of Eq. 6.9. The non-linear qualification refers to the dependence of $\Phi(t)$ on both temperature and structure T_f, as reflected in the NARAYANASWAMY equation [56]:

$$\tau_0 = A\, \exp\left[\left(\frac{x\, \Delta h}{RT}\right) + \frac{(1-x)\, \Delta h}{RT_f}\right] \qquad 0 \le x \le 1 \tag{6.12}$$

where A, x and Δh are constants and R is the gas constant; x is a parameter that partitions the effects of temperature and structure, and is a direct measure of non-linearity—there is linear behavior when $x = 1$. Δh is the activation enthalpy for structural relaxation, and is given [10] by

$$\partial \ln q_- / \partial(1/T'_f) = -\Delta h/R \tag{6.13}$$

where q_- is the cooling rate for glass formation. Δh can therefore be obtained directly

185

from the effect of q_- on T_f' ($= T_g(h)$). Structural relaxation during continuous cooling or heating is modelled numerically using [54]

$$T_f(T) = T_0 + \sum_{j=1}^{m} \Delta T_j \left[1 - \exp\left(- \sum_{k=j}^{m} \Delta T_k / q_k \tau_{0k} \right)^\beta \right] \tag{6.14}$$

T_0 is an equilibrium temperature and the ΔT are differential temperature steps. A, x and β are obtained by fitting the observed and calculated T_f vs. T or $\partial T_f / \partial T$ vs. T curves. It is possible to reproduce quite a wide range of conditions using a particular set of A, x, β and Δh values, but these depend to some extent on the conditions to be modelled [57]. This may be a result of the choice of Eq. 6.9 for the recovery function [58]—it is suggested that the function used by KAHR is more representative in the sense that, for their model, the structural parameter x, for example, behaves as a true materials constant, independent of thermal history [55]. There appear to be correlations between Δh and β [59] and between Δh and x [60], but it is by no means clear whether these are genuine effects or artefacts of the model chosen. It is relevant to mention here that there is not particularly good agreement between different authors, even for the 'directly measured' quantity Δh, let alone the fitting parameters A, x and β [24, 59–63]. Perhaps some experimental refinement is needed to match theoretical sophistication. What are essentially points of detail should not, however, be allowed mask an impressive ability to model the complex DSC curves that reflect a wide range of histories. As confidence in the procedures grows, the models will be used to predict long-term aging behavior, with obvious implications for related properties [64].

References

1 *DeBolt, M. A., Easteal, A. J., Macedo, P. B., Moynihan, C. T.*: J. Am. Ceram. Soc. 59, p. 16 (1976)
2 *Goldstein, M., Simha, R.* (Eds.): Ann. NY Acad. Sci. 279 (1976)
3 *O'Reilly, J. M., Goldstein, M.* (Eds.): Ann. NY Acad. Sci. 371 (1981)
4 *Angell, C. A., Goldstein, M.* (Eds.): Ann. NY Acad. Sci. 484 (1986)
5 *Kovacs, A. J.*: Fortschr. Hochpolym.-Forsch. 3, p. 394 (1963)
6 *Richardson, M. J., Savill, N. G.*: Polymer 16, p. 753 (1975)
7 *Chang, S. S., Bestul, A. B.*: J. Chem. Phys. 56, p. 503 (1972)
8 *Easteal, A. J., Wilder, J. A., Mohr, R. K., Moynihan, C. T.*: J. Am. Ceram. Soc. 60, p. 134 (1977)
9 *Flynn, J. H.*: Thermochim. Acta 8, p. 69 (1974)
10 *Moyhnihan, C. T., Esteal, A. J., DeBolt, M. A., Tucker, J.*: J. Am. Ceram. Soc. 59, p. 12 (1976)
11 *Peyser, P., Bascom, W. D.*: J. Macromol. Sci. Phys. 13, p. 597 (1977)
12 *Richardson, M. J., Savill, N. G.*: Br. Polym. J. 11, p. 123 (1979)
13 *Richardson, M. J.*, in: Comprehensive Polymer Science. *Booth, C., Price, C.* (Eds.), Vol. 1, p. 867, Pergamon, Oxford, 1989
14 *Moynihan, C. T., Gupta, P. K.*: J. Non-Cryst. Solids 29, p. 143 (1978)
15 *Read, B. E.*, in: Structure and Properties of Ionomers. *Pineri, M., Eisenberg, A.* (Eds.), p. 255, Reidel, Dordrecht, 1987
16 *Chang, S. S.*: J. Chem. Thermodyn. 9, p. 189 (1977)
17 *Bershtein, V. A., Yegorov, V. M., Razgulyayeva, L. G., Stepanov, V. A.*: Polym. Sci. USSR (Engl. Transl.) 20, p. 2560 (1978)
18 *Chang, B. T., Li, J. C. M.*: Polym. Eng. Sci. 28, p. 1198 (1988)
19 *Wunderlich, B., Weitz, A.*: J. Polym. Sci. Polym. Phys. Ed. 12, p. 2473 (1974)
20 *Brown, I. G., Wetton, R. E., Richardson, M. J., Savill, N. G.*: Polymer 19, p. 659 (1978)
21 *Latrous, K., Cavrot, J. P., Rietsch, F.*: Eur. Polym. J. 17, p. 1205 (1981)
22 *Wetton, R. E., Foster, G. M., Blok, M. de*, in: Proc. German Rubber Conference, Nurnberg, 4–7 June, 1988

23 *Ueberreiter, K., Otto-Laupenmuhlen, E.:* Z. Naturforsch. 8A, p. 664 (1953)
24 *Aras, L., Richardson, M. J.:* Polymer 30, p. 2246 (1989)
25 *Turner, D. T.:* Polymer 19, p. 789 (1978)
26 *Cowie, J. M. G.:* Eur. Polym. J. 11, p. 297 (1975)
27 *Couchman, P. R.:* Polym. Eng. Sci. 21, p. 377 (1981)
28 *Greenberg, A. R., Kusy, R. P.:* Polymer 25, p. 927 (1984)
29 *Boyer, R. F.:* J. Macromol. Sci. Phys. 7, p. 487 (1973)
30 *Turi, E. A. (Ed.):* Thermal Characterization of Polymeric Materials. Academic Press, New York, 1981
31 *Bair, H. E.,* in: Thermal Characterization of Polymeric Materials. *Turi, E. A.* (Ed.), p. 845, Academic Press, New York, 1981
32 *Alfthan, E., Ruvo, A. de, Rigdahl, M.:* Int. J. Polym. Mater. 7, p. 163 (1979)
33 *Bair, H. E., Warren, P. C.:* J. Macromol. Sci. Phys. 20, p. 381 (1981)
34 *Suzuki, H., Miyamoto, T.:* Bull. Inst. Chem. Res. Kyoto 66, p. 297 (1988)
35 *Suzuki, H., Mathot, V. B. F.:* Macromolecules 22, p. 1380 (1989)
36 *MacKnight, W. J., Karasz, F. E., Fried, J. R.,* in: Polymer Blends. *Paul, D. R., Newman, S.* (Eds.), Vol. 1, p. 224, Academic Press, New York, 1978
37 *Shalaby, S. W., Bair, H. E.,* in: Thermal Characterization of Polymeric Materials. *Turi, E. A.* (Ed.), p. 365, Academic Press, New York, 1981
38 *Wunderlich, B.:* Macromolecular Physics, Vol. 3, p. 348, Academic Press, New York, 1976
39 *Menczel, J., Wunderlich, B.:* Polym. Prep. Am. Chem. Soc. Div. Polym. Sci. 27(1), p. 255 (1986)
40 *Yagfarov, M. Sh.:* Polym. Sci. USSR (Engl. Transl.) 21, p. 2631 (1980)
41 *Kauzmann, W.:* Chem. Rev. 43, p. 219 (1948)
42 *Gibbs, J. H., Di Marzio, E. A.:* J. Chem. Phys. 28, p. 373 (1958)
43 *Goldstein, M.:* Macromolecules 10, p. 1407 (1977)
44 *Di Marzio, E. A.:* Macromolecules 10, 1407 (1977)
45 *McKenna, G. B.,* in: Comprehensive Polymer Science. *Booth, C., Price C.* (Eds.), Vol. 2, p. 311, Pergamon, Oxford, 1989
46 *Goldstein, M.:* J. Chem. Phys. 64, p. 4767 (1976)
47 *Goldstein, M., Richardson, M. J.,* in: Proc. 12th North American Thermal Analysis Society Conf., Williamsburg, VA. *Buck, J. C.* (Ed.), p. 244 1983
48 *Lagasse, R. R.:* J. Polym. Sci. Polym. Phys. Ed. 20, p. 279 (1982)
49 *Cowie, J. M. G., Fergusson, R. R.:* Polym. Comm. 27, p. 258 (1986)
50 *Marshall, A. S., Petrie, S. E. B.:* J. Appl. Phys. 46, p. 4223 (1975)
51 *Williams, G., Watts, D. C.:* Trans. Faraday Soc. 66, p. 80 (1970)
52 *Cowie, J. M. G., Fergusson, R. R.:* Macromolecules 22, p. 2307 (1989)
53 *Kovacs, A. J., Aklonis, J. J., Hutchinson, J. M., Ramos, A. R.:* J. Polym. Sci., Polym. Phys. Ed. 17, p. 1097 (1979)
54 *Goldstein, M., Simha, R.* (Eds.): Ann. NY Acad. Sci. 279, p. 15 (1976)
55 *Hutchinson, J. M., Ruddy, M.:* J. Polym. Sci. Polym. Phys. Ed. 26, p. 2341 (1988)
56 *Narayanaswamy, O. S.:* J. Am. Ceram. Soc. 54, p. 491 (1971)
57 *Stevens, G. C., Richardson, M. J.:* Polym. Comm. 26, p. 77 (1985)
58 *Hodge, I. M.:* Macromolecules 19, p. 936 (1986)
59 *Privalko, V. P., Demchenko, S. S., Lipatov, Y. S.:* Macromolecules 19, p. 901 (1986)
60 *Hodge, I. M.:* Macromolecules 16, p. 898 (1983)
61 *Chen, H. S., Wong, T. T.:* J. Appl. Phys. 52, p. 5898 (1981)
62 *Hodge, I. M., Huvard, G. S.:* Macromolecules 16, p. 371 (1983)
63 *Hutchinson, J. M., Ruddy, M., Wilson, M. R.:* Polymer 29, p. 152 (1988)
64 *Moynihan, C. T., Bruce, A. J., Gavin, D. L., Loehr, S. R., Opalka, S. M., Drexhage, M. G.:* Polym. Eng. Sci. 24, p. 1117 (1984)

Nomenclature

$a, b; c, d$	coefficients in enthalpy–temperature Eqs. 6.1, 6.2
A	pre-exponential term in Eq. 6.12
c_p	specific heat capacity at constant pressure; additional subscripts refer to glassy (g), crystalline (c), or liquid (l) states or to the relaxational (r) contribution to c_p; subscripts A, B indicate components of a binary mixture
$h_g(t, T_a)$	specific enthalpy after annealing for time t at temperature T_a (section 6.7)
$h_g(t, T_a; T_1)$	as above, but followed by cooling to T_1 (section 6.7)
$h_x(T)$	specific enthalpy at temperature T in the glassy ($x = g$) or liquid ($x = l$) state
MM	molar mass
P, Q	integration constants in Eqs. 6.1, 6.2
q	heating or cooling rate (when necessary the sign is shown as a subscript)
R	gas constant
t	annealing time (section 6.7)
T	temperature; subscripts may denote annealing (a), crystallization (c), fictive (f, Fig. 6.12) or glass transition (g) temperatures
$T_g(y)$	T_g determined via enthalpy ($y = h$) or specific volume ($y = v$) curves
$T_{go}, T_g(\frac{1}{2}\Delta cp)$	definitions of T_g commonly used in thermal analysis (Fig. 6.4, Table 6.1)
T_K	Kauzmann temperature (section 6.6)
T_1, T_2	temperatures in stable glass and liquid regions respectively (Fig. 6.6)
v	specific volume
w	mass fraction (subscripts may refer to components A, B of a binary mixture) (section 6.5.4)
x	'structural' parameter that expresses the relative effects of temperature and structure (through T_f) in Eq. 6.12
$z(t, T_a)$	term expressing the relative approach of a glass to equilibrium (Eq. 6.6)
β	parameter characterizing the distribution of relaxation times (Eq. 6.9)
δh	additional enthalpy relaxation on cooling from T_a to T_1 (Fig. 6.11)
Δc_{pg}	specific heat capacity increment at T_g for an amorphous polymer
$\Delta c_{pg\xi}$	as Δc_{pg}, but for a polymer of degree of crystallinity ξ (section 6.5.6)
Δh	activation enthalpy for structural relaxation (Eqs. 6.12, 6.13)
$\Delta h(t, T_a)$	decrease in specific enthalpy after annealing for time t at temperature T_a (Eqs. 6.4, 6.5)
$\Delta h(T_c)$	specific enthalpy of crystallization at T_c
$\Delta s(T_c)$	specific entropy of crystallization at T_c
ΔT_g	breadth of T_g region (Fig. 6.4)
$\Delta \alpha$	increase in expansion coefficient at T_g for an amorphous polymer
κ, λ, μ	areas defined in Fig. 6.6
ξ	degree of crystallinity
τ	relaxation time
$\Phi(t)$	recovery function (Eqs. 6.8, 6.9)
$\chi(t_1 T_1), \psi(t_1 T_a)$	changes in specific enthalpy defined in Fig. 6.11

Chapter 7

Curing of Thermosets

Mike J. Richardson, National Physical Laboratory,
Teddington, Middlesex TW11 0LW, Great Britain

Contents

7.1 Introduction

The cure of a resin is an exothermic event that can be followed using DSC in either the scanning or the isothermal mode. It was initially thought that the derivation of partial (Δh_t, the heat evolved up to time t) and total (Δh_∞) heats of reaction from a single scanning experiment held great promise for a subsequent kinetic analysis of the reaction: in reality few systems show the idealized behavior required and pitfalls await the uncritical user of much of the commercial 'kinetics' software now available. This chapter discusses physical aspects of the curing process with emphasis on the precautions that must be taken to obtain data that have thermodynamic validity. Chemical aspects are not considered in any detail, excellent summaries having already been given [1, 2]. Conference proceedings (e.g. refs. 3, 4) give a useful overall picture of the applicability of thermal analysis to kinetics in general: there are many common factors in the treatment of rates of polymerization, degradation (of both organic and inorganic materials) and crystallization. From the purely *physical* point of view the polymerization of a liquid system is one of the simplest cases to deal with—there are none of the effects of particle size, crystallography or partial pressure that can so confuse studies of the kinetics of decomposition or crystallization. Against this must be set the chemical complexity of most curing systems, and the added physical problem of increasing viscosity and the resultant potential change to diffusion control of the reaction.

The curing process describes the transformation of reactants into a crosslinked product (these are described using subscripts R and P respectively) and DSC can play an important role in the whole sequence—from compounding the initial materials, through the reaction and into the characterization of the final product. All these aspects are illustrated, although most emphasis is naturally placed on the curing stage.

7.2 Reactants

Homogeneity of the initial reaction mixture is best achieved for an all-liquid system and this may require the dissolution of a solid component: subsequent storage of the resultant mixture may in turn lead to phase separation. Conditions for dissolution can be found using DSC. Figure 7.1 shows the start of the endotherm characteristic of the dissolution of Ciba Geigy HT901 (phthalic anhydride) that has crystallized from solution in CT200 (a diglycidyl ether of bisphenol A, DGEBA). Some idea of the initial extent of solution can be gained from both the specific heat capacity c_p and the glass transition temperature T_g. The former can be regarded as an additive function of the two components A and B, where B represents the anhydride. The assumption of additivity is easily checked by direct measurements on the pure components, etc.; it holds very well for this system. Under these circumstances

$$c_p = w_A c_{plA} + w_B c_{pyB} \tag{7.1}$$

where w_A, w_B are mass fractions and the subscript l refers to liquid. The system is a simple mixture of liquid DGEBA and crystalline anhydride when $y = c$ (crystal) (dotted line, Fig. 7.1), and is a homogeneous liquid when $y = 1$ (solid line, $T > 300$ K, Fig. 7.1). It is clear from Fig. 7.1 that phase separation, as judged from the heat capacity at X, is incomplete and this is confirmed by the glass temperature, which is lower than that of the pure DGEBA component. The large plasticizing effect of phthalic anhydride is clear from the further 20 K depression of T_g on full solution

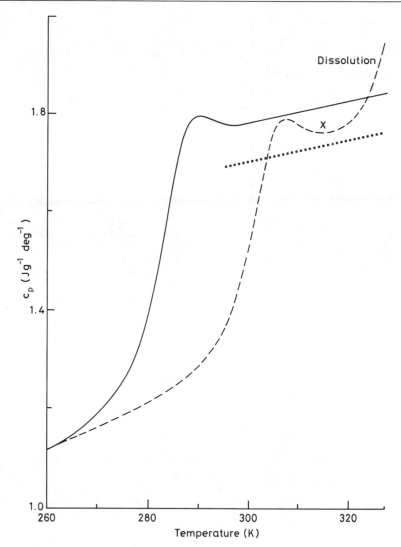

Fig. 7.1. DSC heating curves of phase separated (– – –) and homogeneous (——) epoxy/anhy-
dride systems: the dotted line is the additivity calculation for total phase separation (liquid
epoxy + crystalline anhydride); the small peaks that follow the steep rises in c_p are discussed
in chapter 6; cooling/heating rates 20 K/min

that is shown in Fig. 7.1. It is seen in section 7.3.2 that it is important to know the
glass temperature of the original homogeneous mixture T_{g0} when deriving an extent-
of-reaction parameter.

The properties of fully-cured DGEBA/anhydride resins can be changed by
varying the molar mass (MM) and hence the hydroxyl content of the initial DGEBA
prepolymer. Degrees of polymerization up to perhaps 10 are involved; this is just the
region where T_g is often a very sensitive function of the number average MM M_n.
Figure 7.2 shows that this is the case for DGEBA, and the (limited) data can be
represented by

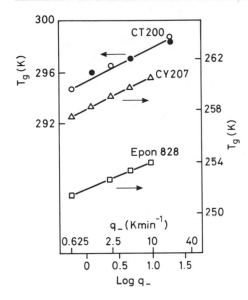

Fig. 7.2. The glass temperature T_g of some DGEBA prepolymers as a function of cooling rate q_- and number average molar mass (g/mol): 377 (Epon 828), 444 (CY207), 833 (CT200); different symbols indicate different batches

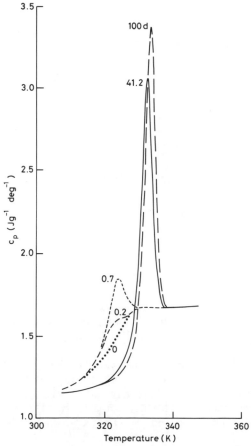

Fig. 7.3. DSC curves after isothermal aging (for the times, in days, shown) of a B-staged resin at room temperature: heating rate 10 K/min

$$T_g = 333.4 - 3.12 \times 10^4/M_n \quad \text{(K)} \tag{7.2}$$

for the glass formed by cooling through T_g at 5 K/min (note that it is important always to compare glasses formed using a common thermal history). A calibration curve such as this is very helpful for the characterization of an unknown DGEBA prepolymer. The apparent $T_{g\infty} = 333.4$ K ($T_{g\infty}$ is T_g for a linear, very long chain DGEBA) has little physical significance because, at these low M_n values, the parameters of the Flory–Fox equation (Eq. 7.2) show considerable deviations from their high M_n values [5].

For practical convenience—especially in the handling and transportation of resin-impregnated cloth—it is common to partially cure a system to give a 'B staged' resin that is still below the gel point (at which an infinite three-dimensional network is formed and above which flow and processability are lost) but has a T_g above ambient. Subsequent physical aging [6] is inevitable on storage at room temperature; this can be turned to advantage to characterize the age of the B staged material, which has a finite shelf life. Figure 7.3 shows how fine structure develops in the DSC curve of a prepreg that is used in the manufacture of printed circuit boards: for clarity curves are shown only for aging periods of fractions of a day and for many days. A complete description of isothermal annealing is complex, but for the room temperature aging of a B-staged resin with $T_g \approx 50\ °C$ there is an approximately logarithmic dependence on time (Fig. 7.4), although the detailed shape is naturally influenced by

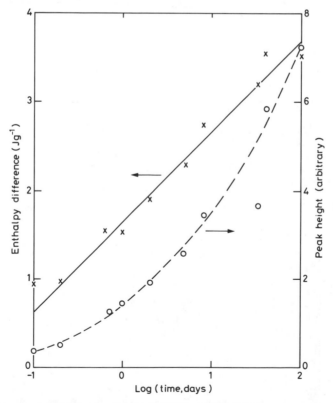

Fig. 7.4. Some measurements of aging for the resin of Fig. 7.3

the parameter chosen. Figure 7.4 shows the peak height of the aged material relative to that of the unaged, and also the enthalpy difference between the two at room temperature. The former is a purely empirical quantity which is influenced to some extent by sample mass and geometry; the latter reflects genuine material differences that better characterize the aging process. The 'peak height' is adequate for quality control purposes because here standardized procedures can be developed to give near-identical experimental conditions for all samples of a given material: the 'area' procedure is relevant to more basic work covering a range of different materials.

7.3 Cure

The cure of a resin is a thermally activated process that may be followed either isothermally or by scanning through the cure region (Figs. 7.5 and 7.6 respectively). There are practical and theoretical arguments for and against both procedures. The isothermal experiment clearly shows differences between two resins, but a closer examination reveals possible errors in both the initial and final stages of reaction:

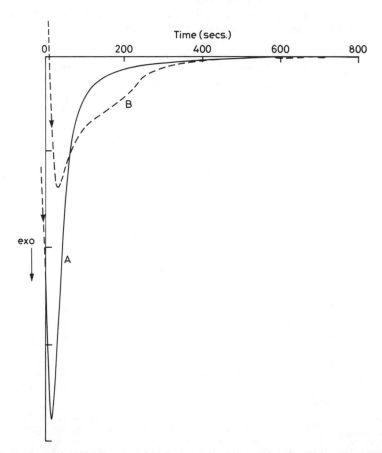

Fig. 7.5. Isothermal (120 °C) cure of high-reactivity (A) and low-reactivity (B) DMCs

194

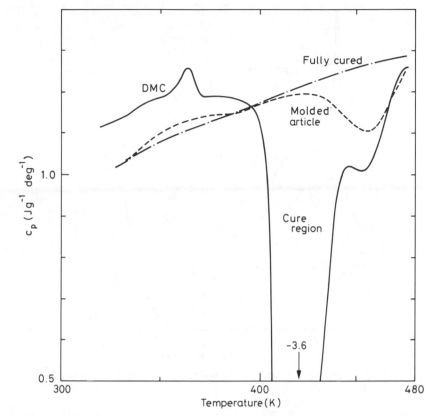

Fig. 7.6. Cure of a DMC (——) and of a molded article (– – –): heating rate 20 K/min

(i) In the examples of Fig. 7.5 cold material has been inserted into a preheated calorimeter, which naturally then requires some time to regain thermal control — the exact period is a function of instrumental design and the time constants of the individual apparatus. Because of this, it is difficult to define $t = 0$; in addition, for reactions described by nth order kinetics (for which the rate is a maximum at $t = 0$) some heat of reaction will go unrecorded in the initial equilibration period. The problem can be partially overcome by repeating the experiment (cold sample into hot calorimeter) with fully cured material [7, 8] to give a 'no reaction' base line; no instrumental manipulation, however, can overcome a sluggish response time.

(ii) The fully cured state is approached asymptotically (Fig. 7.5) and it becomes difficult to separate residual reaction from instrumental instability. It is always useful to let any reaction that is apparently complete in a time t to continue for a time $2t$ or more (or to run a fresh sample for this time), and note what effect this has on the apparent overall heat of reaction (see also Fig. 7.10 and section 7.3.1).

Assuming that the problems of (i) (see section 7.3.1) and (ii) can be overcome, there is a temperature range for any reaction over which useful data can be taken. This corresponds to experimental timescales from a few minutes up to perhaps two hours,

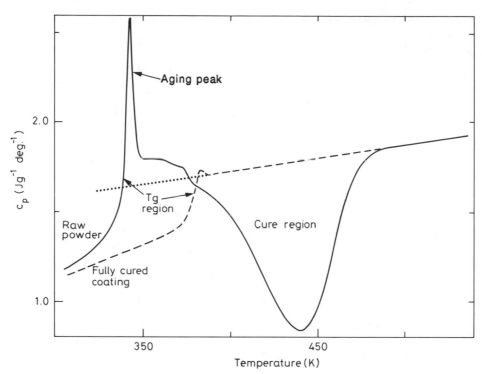

Fig. 7.7. DSC curves showing the cure of an epoxy powder coating (——) and the fully cured product (– – –): heating rate 10 K/min

with the latter figure depending very much on the performance of the individual instrument.

Because of the complexity of cure chemistry there may well be parallel reactions at a particular isothermal temperature. It is most unlikely that these have the same activation energies, so the relative rates of reactions will change with temperature. A set of isothermal experiments (that makes full use of the temperature 'window' discussed above) therefore gives information about the cure process that is not available from a single scan. Because of this, fundamental kinetic studies should always be based on isothermal, rather than scanning, experiments.

For the latter there is no difficulty in recording data throughout the cure region (Fig. 7.7) unless final cure is at such a high temperature that it encounters the start of thermal degradation. Under these circumstances the maximum temperature must be restricted, with final cure taking place isothermally (Fig. 7.10)—again this can lead to the problems discussed in (ii) above. There is a hint of this behavior for the dough molding compound (DMC, Fig. 7.6) for which the run was terminated at 200 °C, not because of thermal degradation, but to simulate post-curing conditions for the article previously molded at 160 °C. For both the initial DMC and the molded product there is an obvious secondary reaction at about 180 °C, which is only completed after post-cure at 200 °C when the heat capacity attains that of the 'fully cured' product. This is a very highly crosslinked material and, as a result (section 6.5.3), no glass transition region is observable on the 'rerun' curve (compare Fig. 7.7). On a more quantitative level there is a difference in heat capacity between the

196

reactants and products of Fig. 7.6 of ~0.1 J/g per degree K, and it is seen in section 7.3.1 that this leads to difficulties in calculating the heat of reaction.

The curve for the as-molded article shows a diffuse low-temperature endothermic event that probably corresponds to the loss of absorbed water—there is a mass loss of a few tenths of one per cent that is complete by about 400 K. Although not serious for this particular example, the endothermic loss of volatile materials can sometimes be sufficiently large to nearly compensate for the exothermic heat of reaction. Figure 7.8 shows the cure of a phenolic material in both normal (loose fitting lid) and sealed pans—the latter are made of stainless steel to contain the high internal pressure generated. 'Sealed pan' data are reproducible and give consistent heats of reaction,

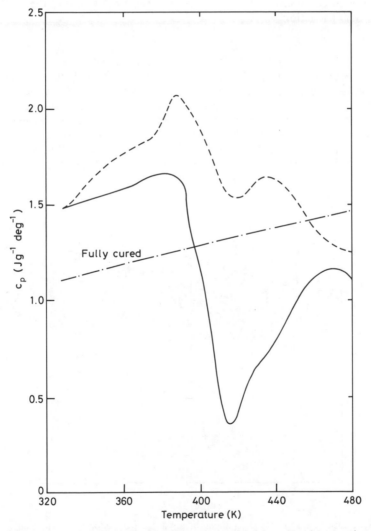

Fig. 7.8. DSC curves for the cure of a phenolic resin in sealed (——) and conventional (– – –) pans: there is a mass loss of 5.0 % from the conventional pan and the endothermic heat of vaporization tends to cancel the exothermic heat of cure; heating rate 20 K/min

but there is no uniformity in the 'normal pan' results—presumably reflecting differences in the rate of vapor loss due to variable efficiencies in the conventional crimping procedure. This emphasizes an important point—reactions that lead to volatile products *must* be carried out in sealed pans. It is impossible to obtain quantitative data for systems in which the mass is changing. Even with sealed pans there are problems: the vapor space is of uncertain volume and the internal pressure conditions are ill-defined; however, the resultant errors are small—they are essentially second-order with respect to the overall heat of reaction.

7.3.1 The heat of reaction Δh_∞

The extent of cure α of any system at a time t is given by the ratio of the partial (Δh_t) to total heats of reaction.

$$\alpha = \Delta h_t / \Delta h_\infty$$

Δh_∞ is therefore an important quantity, and will be discussed here in more detail than is usual for the generalized base-line procedures that are normally used to derive Δh_∞. An idealized scanning experiment will be considered first—one in which 'full cure' can be attained before the onset of thermal degradation and for which the high-temperature portion of the curve merges smoothly with that for the rerun (Figs. 7.7, 7.9). For this situation the DSC should give an overall heat of reaction, but how this relates to $\Delta h_\infty(T)$ (the heat of reaction at a temperature T) has still to be determined. In Fig. 7.9 the difference between the heat capacities of the products and the reactants

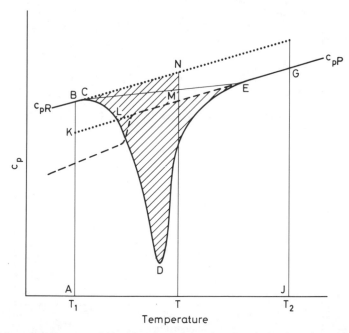

Fig. 7.9. Schematic diagram of the changes in apparent heat capacity of a resin as it cures (——) and the crosslinked product (– – –): dotted lines show extrapolations; the difference in heat capacity between reactants c_{pR} and products c_{pP} is greatly exaggerated; the hatched area corresponds to the heat of reaction at T

has been grossly distorted and, as drawn, Δc_p $(=c_{pP} - c_{pR})$ is a negative quantity—the usual sign if $\Delta c_p \neq 0$. To have any thermodynamic validity, Δh_∞ must refer to a particular temperature. For example:

$$\Delta h_\infty(T_1) = h_P(T_1) - h_R(T_1) \tag{7.3}$$

$$= [h_P(T_2) - h_R(T_1)] - [h_P(T_2) - h_P(T_1)] \tag{7.4}$$

$$= F(\text{ABCDGJ} - \text{AKGJ}) \tag{7.5}$$

$$= F(\text{BCLK} - \text{LED}) \tag{7.6}$$

where the alphabetical sequences define the relevant areas in Fig. 7.9 and F is the area-to-enthalpy conversion factor (section 4.4.2) for raw DSC data (i.e. before conversion to the $c_p - T$ curves of Fig. 7.7). It is implicitly assumed in the development of Eqs. 7.3–7.10 that reactants and products form homogeneous systems; additional terms can be introduced if there is a need to take account of phase changes. In Fig. 7.9, for example, the region corresponding to the second square-bracketed term of Eq. 7.4 shows a glass transition, and this is overcome in Eq. 7.5 by the extrapolation (to K) of data for the rubbery state. Extension to any other temperature follows from the basic definition

$$\Delta h_\infty(T) = \Delta h_\infty(T_1) + [h_P(T) - h_P(T_1)] - [h_R(T) - h_R(T_1)] \tag{7.7}$$

$$= \Delta h_\infty(T_1) + \Delta c_p \Delta T \tag{7.8}$$

$$= F(\text{BCLK} - \text{LED} - \text{BNMK}) \tag{7.9}$$

$$= F(\text{CNMED}) \tag{7.10}$$

where $\Delta T = T - T_1$, the final term in Eq. 7.9 is negative because $\Delta c_p < 0$. The area of Eq. 7.10 is hatched in Fig. 7.9. If $\Delta c_p = 0$, Δh_∞ is independent of temperature, lines BN and KG coincide and a simple linear base line defines Δh_∞. In all other cases the construction of Eq. 7.10 is needed. The conventional procedure, which joins the start and finish (CE, Fig. 7.9) of the reaction region, gives a thermodynamically meaningless quantity that approximates to $\Delta h_\infty(T_D)$ for a symmetrical cure curve (T_D is the temperature at which the rate of reaction is a maximum). In many cases points C and E are difficult to define, and the consequent implied appeal to the discretion of individual operators (in the choice of their location) cannot be a satisfactory basis for anything other than 'in-house' quality control.

Increasingly complex cure situations can be superimposed on the idealized conditions of Fig. 7.9. If, for example, the scan must be terminated because of potential thermal degradation, residual cure must be allowed to take place isothermally at T_f (Fig. 7.10). The relevant (negative) enthalpy must then be included in the overall enthalpy change [9]—the first term in square brackets in Eq. 7.4. A conventional rerun follows. Figure 7.10 indicates how the problems of asymptotic behavior ((ii) in section 7.3) can be overcome. Cure *appears* to be complete after $\frac{1}{2}$ h at 170 °C, but data have been recorded for a further $1\frac{1}{2}$ h. In the subsequent data treatment the reaction is assumed to be complete after $\frac{1}{2}$, 1, $1\frac{1}{2}$ and 2 h cure at 170 °C, and the corresponding final data points define the isothermal base line. Figure 7.10 clearly shows that $\frac{1}{2}$ h at 170 °C is inadequate. However curves for 1, $1\frac{1}{2}$ and 2 h are virtually indistinguishable, so that for practical purposes cure is complete after 1 h at 170 °C.

The combination of scanning and isothermal procedures shown in Fig. 7.10 suggests one way of overcoming the 'unrecorded enthalpy' problem encountered in conventional isothermal experiments ((i) in section 7.3). The introduction of a cold

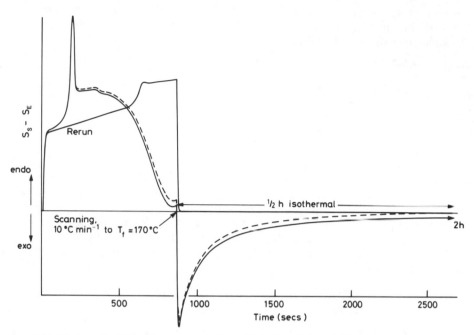

Fig. 7.10. Composite 'full-minus-empty' curve ($S_S - S_E$, section 4.1) for an epoxy resin: the sample was given an additional 2 h cure at 170 °C; the broken line shows the effect if reaction is assumed to be complete after $\frac{1}{2}$ h at 170 °C (the scanning curve is also affected because an incorrect isothermal base line is used); curves for 1 h and $1\frac{1}{2}$ h can be superimposed on that shown for 2 h (heating rate 10 K/min)

sample into a preheated DSC is simulated by using a high heating rate q_+ prior to the isothermal cure period. Data are recorded at all stages so that the extent of any non-isothermal prereaction is now quantifiable and, if q_+ is sufficiently large, is only a small fraction of Δh_∞. This type of experiment is best carried out using a power-compensation DSC because q_+ is well defined over the whole programmed temperature range. In a heat-flux DSC, at high q_+, the nominal q_+ is reduced as the desired isothermal temperature T_f is approached, but the actual temperature may still temporarily overshoot T_f. A useful experimental test of any instrument in this respect can be made by noting how closely the melting temperature (429.8 K) of indium can be approached without actually melting a sample of this material, see 4.4.1. The instrument is programmed at q_+ to $T_f = 429.8 - \delta T$, where $\delta T = 0.1, 1, 2, 5$ K, etc. and allowed to stabilize at this temperature. If 429.8 K is exceeded, and the indium melts, this may not be apparent in the generally large instrumental response but subsequent crystallization will be obvious. Crystallization may require a cooling run to overcome supercooling effects or, if there is a gross temperature overshoot, this may be overcome in the equilibration period when T_f is finally attained. A Perkin Elmer DSC2 may be heated at 160 K/min to 429.7 K without melting a small sample of indium.

The various heats of reaction can now be compared:

(i) $\Delta h_\infty(T)$ via scanning experiments
(ii*a*) $\Delta h_\infty(T)$ via conventional isothermal experiments
(ii*b*) $\Delta h_\infty(T)$ via corrected isothermal experiments
(iii) $\Delta h_\infty(T)$ via a combination of (i) and (ii) (e.g. as in Fig. 7.10).

Ideally all would agree, but (ii*a*) is expected to have the lowest value for reasons already given. Agreement can of course be obtained only if the curing mechanisms are unaffected by temperature and, especially, if there are no new reactions at higher temperatures. For complex curing systems these conditions are generally the exception rather than the rule and secondary reactions, for example, may be obvious (Figs. 7.5, 7.6). Systematic differences, especially between $\Delta h_\infty(T)$ via (i) and via (ii*b*), suggest changing mechanisms—similar conclusions should be arrived at when more detailed kinetic data are studied (section 7.5).

Many empirical quality control procedures for resins are based on measurements that are essentially those of (ii*a*) and these, while satisfactory for 'in-house' control, are inadequate for inter-laboratory comparisons—for these it is essential to know how (ii*a*) relates to (ii*b*).

Although the equivalence (or not) of the various $\Delta h_\infty(T)$ gives useful information about the path of a cure reaction, its main use is in defining the extent of cure α_t at any time t through the ratio of partial to total heats:

$$\alpha_t = \Delta h_t(T)/\Delta h_\infty(T) \tag{7.11}$$

$\Delta h_t(T)$ is calculated using the same procedures as those described above for $\Delta h_\infty(T)$; in particular in a scanning experiment appropriate allowances must again be made for any difference in heat capacity between reactants and products. α_t is, in turn, used to discuss reaction rates, etc. (section 7.5).

7.3.2 Other thermal properties

'In situ' DSC measurements leading to Δh_∞ can be made for reactions that are 'complete' on a timescale of up to perhaps two hours. For longer times it is difficult to distinguish between a slowly varying (and small) signal and instrumental drift, and in any case it is inefficient to tie up equipment for lengthy periods. Slow-to-cure systems should therefore be cured externally and samples taken at suitable intervals for subsequent examination using the DSC. The most convenient parameter to measure is T_g although, as already noted, the relevant Δc_{pg} step decreases with increasing crosslink density [10] and DSC cannot be used to detect T_g of cured DMC or phenolics, for example.

Figure 7.11 shows c_p–T curves for an epoxy resin system that has been cured at 398 K for the times shown [11]. The associated T_g data are plotted in Fig. 7.12 which also shows Δc_{pg}—the heat capacity is in principle another parameter (Y) that can be used to define $\alpha_t(Y)$:

$$\alpha_t(Y) = (Y_t - Y_0)/(Y_\infty - Y_0) \tag{7.12}$$

(in Eq. 7.11 $Y = h$), where subscripts refer to time. When $Y = T_g$, the initial value T_{g0} refers to the homogeneous reaction mixture (Fig. 7.1)—hence the emphasis placed on this quantity in section 7.2.

7.4 Products

The 'rerun' curves shown in Figs. 7.6 and 7.7 provide essential data for the reduction of the overall enthalpy change to the heat of reaction at a specified temperature. They are particularly valuable in cases such as that shown in Fig. 7.10, where an isothermal period is needed to complete the reaction. The 'product' curve shows how far the system was from full cure at the start of the final isothermal period.

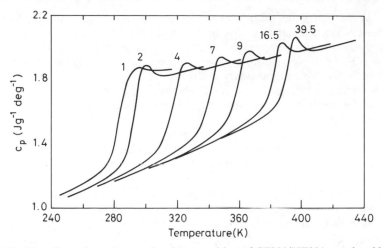

Fig. 7.11. The effect of reaction on the glass transition of CT200/HT901 cured at 398 K for the times (h) shown

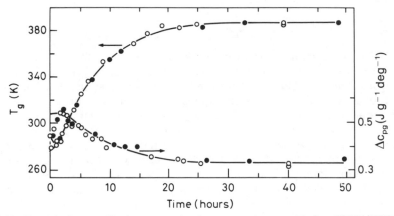

Fig. 7.12. T_g and the specific heat capacity increment Δc_{pg} at T_g for CT200/HT901 as a function of cure time at 398 K: some data for phase-separated materials are shown at low cure times ($\bigcirc$, $\bullet$ different preparations)

As with any glass (a full discussion is given in chapter 6) a cured epoxy, for example, can be annealed by cooling through T_g at various q_- or, isothermally, somewhat below T_g (Fig. 7.13), and the effect on the latter is very useful in characterizing the physically aged state of a material—a parameter that can then be correlated with changes in, say, impact properties.

The heat capacities of both precursor and cured material have useful analytical properties in that at 450 K, for example, a typical liquid or rubbery hydrocarbon polymer has $c_p \approx 2 \pm 0.2$ J/g per degree K (e.g. Figs. 7.7, 7.11). Lower values may be caused by the diluting effect of inorganic fillers (as in Fig. 7.6) or halogenation; the latter is especially relevant for many fire retardant (brominated) grades of epoxies.

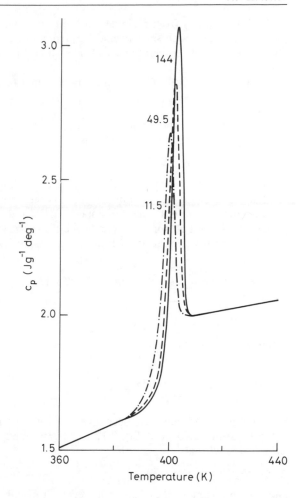

Fig. 7.13. DSC curves for fully cured CT200/HT901 annealed at 373 K for the times (h) shown: heating rate 20 K/min

7.5 Kinetics

Previous sections have considered the precautions necessary to obtain enthalpy–time–temperature data for polymerizing systems. The next stage is to establish general correlations—the production of 'kinetic data'. There are two very different ultimate objectives—predictive, which may be feasible, and the detailed discussion of a reaction mechanism, which must be accompanied by many qualifications. A very useful summary of the precautions needed to obtain kinetic data from DSC experiments has been given by FLYNN [12], in which the approach is pragmatic and consistency tests are emphasized (does the reaction change with α and/or temperature?).

The almost universal starting point for any discussion of DSC kinetics is the empirical assumption represented by

$$\mathrm{d}\alpha/\mathrm{d}t = f(\alpha)k(T) \tag{7.13}$$

i.e. the reaction rate can be split into variables of degree of conversion and temperature. The early hope, referred to in section 7.1, was that it would be possible to derive

203

both $f(\alpha)$ and $k(T)$ from a single scan, but experience has shown that this is rarely possible and it is much safer to find $f(\alpha)$ from an isothermal experiment. Comparable data at a series of temperatures will then show the general validity, or otherwise, of $f(\alpha)$.

The rate constant $k(T)$ is generally assumed to follow the Arrhenius equation

$$k(T) = A \exp(-E/RT) \tag{7.14}$$

where A is the pre-exponential (frequency) factor, E is an activation energy and R is the gas constant. Combination of Eqs. 7.13 and 7.14 gives, in logarithmic form:

$$\ln(d\alpha/dt)_i = Af(\alpha_i) - E/RT_i \tag{7.15}$$

where subscript i refers to a chosen degree of conversion. Thus a graph of the left-hand side of Eq. 7.15 against T_i^{-1} (T_i is the temperature at which α_i is attained) should give a straight line of slope $-E/R$ and intercept $Af(\alpha_i)$ [13, 14]. This procedure is repeated for i_1, i_2, etc. to see if E varies with α and/or temperature [15].

An equivalent equation in heating rates q_+ has been derived:

$$F(\alpha) = \int_0^\alpha d\alpha/f(\alpha) = (A/q_+) \int_{T_0}^T \exp(-E/RT)\,dT \tag{7.16}$$

$$= (AE/q_+ R)p(x) \tag{7.17}$$

where $x = E/RT$ and

$$p(x) = -\int_{x_0}^x [\exp(-x)/x^2]\,dx \tag{7.18}$$

Several approximations have been given for the exponential integral $p(x)$ [16, 17], and if the logarithmic form of Eq. 7.17 is differentiated, again at constant α, it may be shown [18, 19] that

$$E = 4.35\, d(\log q_+)/d(1/T) \tag{7.19}$$

and the approximation can be improved by a rapidly converging iterative procedure. Equation 7.19 is the basis of an ASTM method for determining Arrhenius constants [20]. This uses 'peak' (or 'trough') temperatures in Eq. 7.19, following the observation [21] that the maximum rate of reaction occurs at a constant value of α. All the procedures discussed above have the advantage that they do not need a specific analytical form for $f(\alpha)$; the resultant value of E can be used to predict the rate of reaction at a temperature T_k if it is known at another temperature T_j [12].

$$\ln(d\alpha/dT_k) = \ln(d\alpha/dT_j) + (E/R)(T_k - T_j)/(T_j T_k) \tag{7.20}$$

If DSC methods are to be used as a guide to $f(\alpha)$, special care must be taken, particularly in scanning experiments. As mentioned above, polymerization mechanisms are generally complex and simultaneous reactions are common. If the system is assumed to follow nth order kinetics, Eq. 7.15 be written

$$\ln(d\alpha/dt) = \ln A - E/RT + n\ln(1 - \alpha)$$

so that for an isothermal reaction a graph of ln(rate) against ln(residual fraction) should be a straight line of slope n. A scanning run can be fitted using a multiple regression analysis to solve for A, E and n, but the resultant parameters may simply characterize some 'average reaction' and have little fundamental significance (three parameters, whatever their meaning, are sufficient to describe rather complex curves). Any parameters derived from scanning experiments should therefore be checked by comparison with isothermal data [22] — either with the parameters themselves or by a prediction of, say, the time to half reaction, the latter being subsequently tested experimentally. The last procedure is recommended in ref. 20, although this implicitly assumes first-order kinetics in the calculation of the pre-exponential factor A as $(q_+ E/RT^2) \exp(E/RT)$ [23].

For more complex systems the chemistry must serve as a guide towards the rate-determining step. Many examples based on $\alpha(h)$ and/or the time or temperature derivative are given in refs. 1 and 2. Other parameters may be useful; for example it has been shown [11] that the rate constant obtained from $\alpha(T_g)$ is very similar to that for etherification (obtained from IR measurements) for DGEBA/anhydride systems.

Over the years many equations (that are now often named after their originators) have been developed on the basis of specific aspects of Eq. 7.13. FLYNN and WALL [24] and OZAWA [25] have given comprehensive reviews that show the assumptions and approximations inherent in many of these equations. There is no 'best' approach other than an open mind that treats each system on its own merits, emphasizing, where possible, tests for consistency of data.

References

1 *Prime, R. B.*, in: Thermal Characterization of Polymeric Materials. *Turi, E. A.* (Ed.), p. 435, Academic Press, New York, 1981
2 *Barton, J. M.:* Adv. Polym. Sci. 72, p. 111 (1985)
3 J. Thermal Anal. 5, pp. 179–354 (1973)
4 Thermochim. Acta 110 (1987)
5 *Aras, L., Richardson, M. J.:* Polymer 30, p. 2246 (1989)
6 See section 6.7
7 *Widmann, G.*, in Thermal Analysis. *Buzas. I.* (Ed.), Vol. 3, p. 359, Heyden, London, 1975
8 *Barton, J. M.:* Thermochim. Acta 71, p. 337 (1983)
9 *Chang, S.-S.:* J. Thermal Anal. 34, p. 135 (1988)
10 *Judovits, L. H., Bopp, R. C., Gaur, U., Wunderlich, B.:* J. Polym. Sci. Polym. Phys. Ed. 24, p. 2725 (1986)
11 *Stevens, G. C., Richardson, M. J.:* Polymer 24, p. 851 (1983)
12 *Flynn, J. H.:* J. Thermal Anal. 34, p. 367 (1988)
13 *Friedman, H. L.:* J. Polym. Sci. C6, p. 183 (1964)
14 *Ozawa, T.:* J. Thermal Anal. 31, p. 547 (1986)
15 *Richardson, M. J.:* Pure Appl. Chem. 64, p. 1789 (1992)
16 *Doyle, C. D.:* J. Appl. Polym. Sci. 5, p. 285 (1961)
17 *Doyle, C. D.:* J. Appl. Polym. Sci. 6, p. 639 (1962)
18 *Ozawa, T.:* Bull. Chem. Soc. Jpn. 38, p. 1881 (1965)
19 *Flynn, J. H, Wall, L.A.:* J. Polym. Sci. Polym. Letters Ed. 4, p. 323 (1966)
20 Standard Test Method for Arrhenius Kinetic Constants for Thermally Unstable Materials. E698, Vol. 14.02, American Society for Testing and Materials, Philadelphia, PA
21 *Horowitz, N. H., Metzger, G.:* Anal. Chem. 35, p. 1464 (1963)
22 *Duswalt, A. A.:* Thermochim. Acta 8, p. 57 (1974)
23 *Rogers, R. N., Smith, L. C.:* Anal. Chem. 39, p. 1024 (1967)
24 *Flynn, J. H., Wall, L. A.:* J. Res. Nat. Bur. Stds 70A, p. 487 (1966)
25 *Ozawa, T.:* J. Thermal Anal. 2, p. 301 (1970)

Nomenclature

A	pre-exponential term in Arrhenius equation (Eq. 7.14)
c_p	specific heat capacity at constant pressure: additional subscripts refer to the phase (c = crystal, g = glass, l = liquid), the component (A, B in binary mixture) or to reactants (R) or products (P).
DGEBA	diglycidyl ether of bisphenol A
DMC	dough molding compound
E	activation energy (Eq. 7.14)
$h_P(T), h_R(T)$	specific enthalpy of products (P) or reactants (R) at temperature T
$k(T)$	rate constant (Eq. 7.14)
M_n	number average molar mass
MM	molar mass
n	order of reaction
$p(x)$	exponential integral (Eq. 7.18)
q	heating or cooling rate (when necessary the sign is shown as a subscript)
R	gas constant
t	time
T	temperature
T_D	temperature at which rate of curing is at a maximum
T_f	final isothermal temperature following a (generally rapid) programmed heating
T_g, T_{g0}	glass temperature, initial ($t = 0$) T_g of a reacting system
T_j, T_k	temperatures used to demonstrate calculation of reaction rate (Eq. 7.20)
T_1, T_2	arbitrary temperatures in region before and after cure has occurred (Fig. 7.9)
w_A, w_B	mass fractions of components A, B, in binary mixture
Y	parameter used to define α_t (Y_0, Y_t, Y_∞ are values at start, after time t and at end of reaction)
α_t	extent of reaction after time t
$\alpha_t(Y)$	extent of reaction determined by parameter Y (T_g or enthalpy)
Δc_p	$c_{pP} - c_{pR}$
Δc_{pg}	specific heat capacity increment at T_g
Δh_t	partial specific enthalpy of reaction up to time t (relevant temperature may be shown)
Δh_∞	total specific enthalpy of reaction (relevant temperature may be shown)

Chapter 8

Thermal Transitions and Gelation in Polymer Solutions

Hugo Berghmans, Laboratory for Polymer Research, Katholieke Universiteit
Leuven, Celestijnenlaan 200F, B-3001 Heverlee, Belgium

Contents

8.1 Introduction

Moderately concentrated solutions of many synthetic and biological polymers solidify on cooling. This phenomenon, known as thermoreversible gelation, results from the formation of some kind of interconnectivity throughout the solution. It is fully reversible, as heating restores the original solution [1–4].

The term 'gel' originates from polymer chemistry, where these structures are obtained by chemical cross-linking to form a permanent, molecular network that can be swollen with solvent. The formation of these chemical gels, however, is not reversible unless the network is decomposed by heating, radiation, etc.

Another basic difference between chemical and thermoreversible gels is their mechanical behavior. Chemical gels show very characteristic elastic properties, generally not encountered with thermoreversible gels. These solid-like characteristics are observed only under well-defined conditions of stress and strain, and differ from system to system, making the exact definition of a thermoreversible gel rather difficult.

Consequently, a thermoreversible gel is defined here as a moderately concentrated polymer solution that solidifies on cooling. Such gels show elastic characteristics under well-defined conditions of stress or strain. They can be classified according to their formation mechanism, which determines their properties to a large extent.

The different formation mechanisms result from a thermal transition that can occur on cooling and heating of a polymer solution. The most frequently encountered mechanism is a liquid–solid demixing or crystallization (L–S demixing). Because one is working in solution, liquid–liquid demixing (L–L demixing) can also take place. In itself, this L–L demixing cannot be responsible for the formation of a rigid gel. Interference with an L–S demixing or a vitrification process, however, can create the necessary conditions for the solidification that is needed. On cooling of a solution of a stereoregular vinyl polymer, a change from a random coil to a helical conformation can also take place. These regular conformations can be stabilized by intramolecular interactions and/or by interaction with the solvent. In a second step, these regular chains will associate to form a three-dimensional network. This mechanism is called conformational gelation.

Calorimetry plays an important role in the study of the mechanisms of thermoreversible gelation as it is a very elegant way to detect the occurrence of thermal transitions. This is illustrated below for gelation induced by L–S demixing, L-L demixing in combination with an L–S demixing or a glass transition. Special attention is paid to the experimental aspects of the observation of these transitions by dynamic calorimetry.

8.2 The use of dynamic techniques in the study of crystallization and melting
8.2.1 Crystallization and melting of a low molecular mass substance

This is an equilibrium phenomenon. At constant pressure it represents an invariant situation: crystallization and melting take place at the same temperature, as illustrated by the temperature–time ($T-t$) relations shown in Fig. 8.1. The experiment that leads to these graphs consists of transferring the sample from a thermostatic bath at temperature T_2 to another one at temperature T_1 or vice versa ($T_2 > T_1$). The volume of these baths is suffiiciently large that the release or absorption of heat by the sample does not influence their temperature. The change in temperature inside the

208

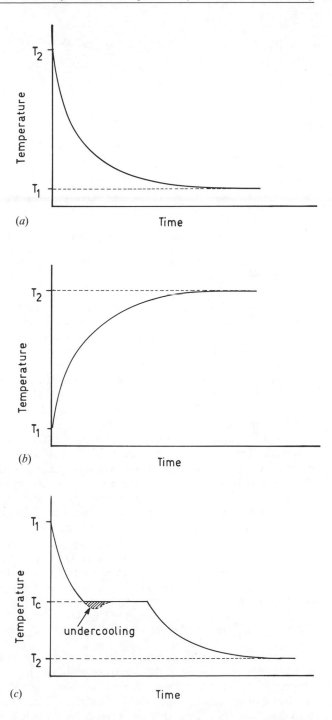

Fig. 8.1. T−t relations for cooling and heating of a substance: (*a*), (*b*) without thermal transition; (*c*) crystallization

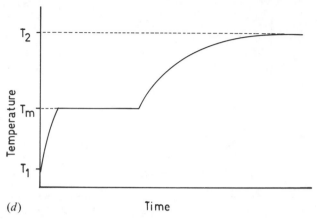

(*d*)   Time

Fig. 8.1 (continued) $T-t$ relations for cooling and heating of a substance: (*d*) melting

sample T is followed as a function of time. A temperature homogeneity must be realized throughout the sample.

If no thermal transition takes place during cooling or heating, $T-t$ relations like those represented in Fig. 8.1(a) and (b) are to be expected. When crystallization sets in on cooling, T will remain constant at T_c until the sample has been transformed completely into the crystalline phase (Fig. 8.1(c)).

The opposite phenomenon is observed when the sample is transferred from T_1 to T_2. The increase of T with time is interrupted by the melting of the crystalline phase at constant temperature. This represents the melting point T_m of the pure substance (Fig. 8.1(d)). Undercooling is very often observed (broken line, Fig. 8.1(c)), but once crystallization has started, the temperature inside the sample must return to the equilibrium temperature.

In a dynamic calorimetric experiment, this equilibrium situation cannot be observed. The sample holders are cooled or heated at a constant rate and no indication of the temperature T inside the sample is given. A second inconvenience is the rate of diffusion of the heat released or absorbed by the sample and the resulting temperature inhomogeneity inside the sample. Consequently, a more or less broad crystallization and melting peak is obtained. The shape of these signals depends on experimental conditions such as heating and cooling rate and the amount of sample. The temperature at the onset of the melting peak is therefore taken as the melting point T_m. The temperature at the onset of the exotherm observed on cooling corresponds to the crystallization temperature, T_c. The difference between T_m and T_c is the degree of undercooling ΔT. A simulation of melting and crystallization, as can be observed in this type of experiment, is shown in Fig. 8.2.

8.2.2 *Crystallization and melting of a polymer*

The problem becomes more complicated when a polymer is involved. Since this topic is dealt with abundantly in the literature, only the basic ideas are developed in this chapter [5]. The molecular condition for the crystallization of a polymer is the regularity of the chain. In case of vinyl polymers it means that only those with a very high degree of stereoregularity can crystallize. Only a few exceptions to this general statement are known. This problem is not encountered with condensation polymers such as polyesters of polyamides, as most of them have a regular chain structure. The

210

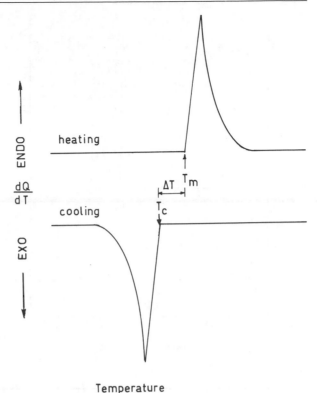

Fig. 8.2. Simulation of the curve obtained in a dynamic calorimetric experiment with a substance that melts and crystallizes

crystallization of copolymers is more complex and depends strongly on the chemical nature of the composing units and their distribution along the chain.

The chain-like nature of polymer molecules is responsible for different crystal morphologies. The most frequently encountered is the lamellar morphology. The thickness of the lamellae lies within the nanometer range, but their lateral dimensions are of the order of micrometers. The molecules are incorporated perpendicular to the largest surface by a mechanism of regular or irregular chain folding (Fig. 8.3). The formation of these crystals necessitates a certain degree of undercooling, as the rate of crystallization at the equilibrium melting temperature, T_m^o is zero. This degree of

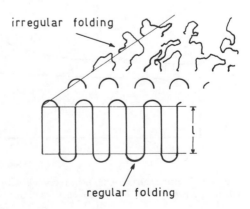

Fig. 8.3. Schematic representation of lamellar polymer crystals with regular and irregular chain folding

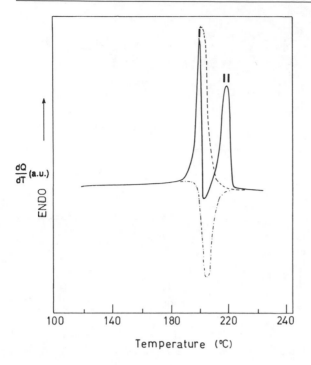

Fig. 8.4. Complex melting of isotactic polystyrene

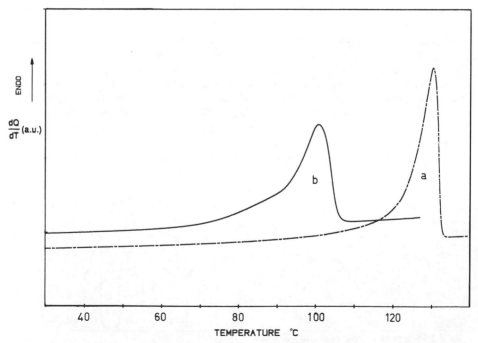

Fig. 8.5. Melting curve of HDPE, recorded with a Perkin-Elmer DSC-2: (a) without solvent; (b) in the presence of *o*-xylene (mass fraction polymer = 0.34)

212

undercooling also determines the thickness l of these crystallites and their melting point T_m. Annealing at temperatures above the crystallization temperature results in an increase in this thickness and therefore in an increase in the melting point. This transformation can take place during a heating experiment in a dynamic calorimeter, as illustrated in Fig. 8.4 for isotactic polystyrene, crystallized at 180 °C and heated at 20 °C/min. The low-temperature melting endotherm (I) represents the melting of the crystallites formed at the crystallization temperature T_c. The high-temperature endotherm (II) corresponds to the melting of recrystallized crystallites. The small exotherm between the two endotherms results from the overlap of the melting of the crystallites formed at T_c (broken-line endotherm) and the exotherm, resulting from their recrystallization (broken-line exotherm). Endotherm I therefore represents the melting of the originally formed crystallites with the lowest melting point. It is clear from these observations that the melting point of a crystalline polymer is not a constant. It depends strongly on the crystallization conditions and can be altered by any further treatment after the crystallization. The equilibrium melting temperature can be obtained only indirectly.

Melting peaks of polymers are always much broader than those of low molecular mass substances. This is due to the distribution in crystal thickness, perfection and size. Therefore the melting point is never taken as the temperature at the onset of the melting peak, but as the temperature at the maximum of the melting endotherm. This is shown in Fig. 8.5(a) for high density polyethylene (HDPE).

8.2.3 Crystallization and melting of a binary system of two low molecular mass substances

The melting point of a low molecular mass substance is depressed when a second substance, compatible with the first one in the melt, is added. Eutectic melting is observed when the melting points of the two substances do not differ greatly. A classic example is the system bismuth–cadmium; its phase diagram is schematically represented in Fig. 8.6. Adding Bi to Cd results in a melting point depression of Cd, and vice versa. The melting lines meet in the lowest melting point or eutectic point T_{eu}. In this point, the melt, a homogeneous solution of Bi and Cd, is in equilibrium with solid Bi and Cd. It is an invariant point at constant pressure, like the melting points of the pure substances. In between these invariant points, the melting point depends on the concentration.

This can also be illustrated by a $T–t$ relation experimentally obtained as described above. The pure substances and the system with an eutectic composition will show the same relation as the one represented in Fig. 8.1. Between these invariant points, the temperature inside the sample will not remain constant when crystallization occurs on cooling or when melting is induced by heating. Crystallization and melting will slow down the rate of cooling and heating respectively, but no plateau value can be observed (Fig. 8.7(a)). This necessitates a different interpretation of the melting curve obtained with a dynamic calorimetric experiment. The melting points that must be reported in the phase diagram for the construction of the liquidus are the temperatures not at the onset of the melting endotherm, but at the end of the endotherm. This of course is valid only when no other factors can be responsible for the broadening of the signal. Most of the errors will result from the rate of diffusion of heat through the sample. This can be eliminated by extrapolating these final melting points to zero scanning rate. A simulation of the melting curve of a system with a composition between the invariant points is shown in Fig. 8.7(b). The eutectic

213

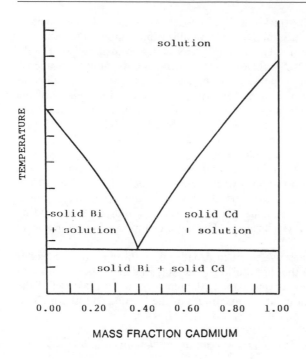

solution

TEMPERATURE

solid Bi
+ solution

solid Cd
+ solution

solid Bi + solid Cd

0.00   0.20   0.40   0.60   0.80   1.00

MASS FRACTION CADMIUM

Fig. 8.6. T–concentration phase diagram for the system Bi–Cd

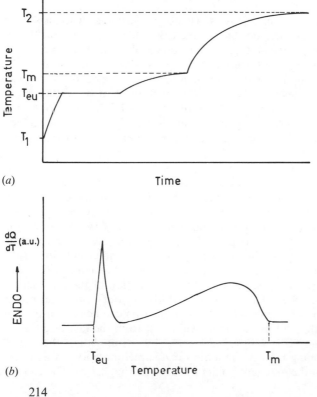

Temperature

T_2

T_m

T_{eu}

T_1

(*a*)   Time

$\frac{dQ}{dT}$ (a.u.)

ENDO

T_{eu} T_m

(*b*)   Temperature

Fig. 8.7. Melting of a two-component system at a concentration between that of a pure component and the eutectic composition: (*a*) T–t relation for melting; (*b*) simulation of the melting curve in a dynamic experiment

214

melting point T_{eu} corresponds to the temperature at the onset of the sharp peak. The temperature to be reported on the liquidus is T_m, the temperature at the end of the broad melting endotherm. A typical example, obtained with a DSC-2 Perkin-Elmer, is shown in Fig. 8.8 for the system Bi–Cd, with a Cd mass fraction of 0.19. The invariant eutectic melting occurs at 417 K as expected. The melting of Bi in equilibrium with the Bi–Cd melt extends over a much wider temperature domain. The final melting point of 498 K is still 9 K too high, due mainly to the rather high scanning rate of 5 K/min.

An increase of the difference between the melting points of the two substances will shift the eutectic point towards the axis of the lower melting substance. It will coincide with this axis if the difference in melting point becomes sufficiently large. This is realized when a solid is dissolved in a low melting solvent. The melting of the solvent will appear as an endotherm. The solid substance will melt or redissolve in the solvent over a temperature range that extends from the melting point of the solvent to the liquidus. This cannot be observed, however, with the calorimeters currently used for these studies. In the temperature range between the melting point of the solvent and the experimentally observable onset of the deviation of the liquidus from the pure solvent axis, only a very small fraction of the solid will melt. The smearing-out of this melting over a considerable temperature range will further limit the experimental observation of this initial melting. As a consequence, the melting endotherm can be integrated without any appreciable error between this temperature of 'first deviation' and the final temperature. This kind of melting is shown in Fig. 8.9 for the system octacosane (C28)–heptane (C7) [6]. The data were obtained with a Setaram BT 200 dynamic calorimeter. The melting of heptane is situated at the expected low temperature. The first indication of a melting of octacosane is observed

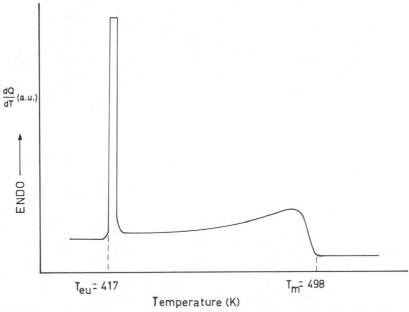

$Fig.$ $8.8.$ Melting curve of a Bi–Cd system with a Cd mass fraction of 0.19 (DSC-2 Perkin-Elmer, 5 K/min)

215

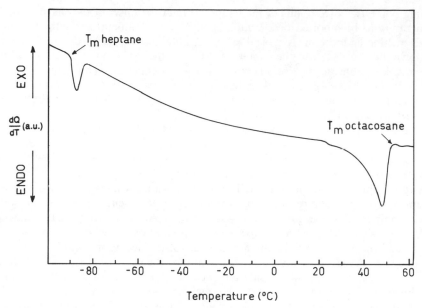

Fig. 8.9. Melting curve of the system octacosane (C$_{28}$)–heptane (C$_7$) (Setaram BT 200, 0.3 °C/min)

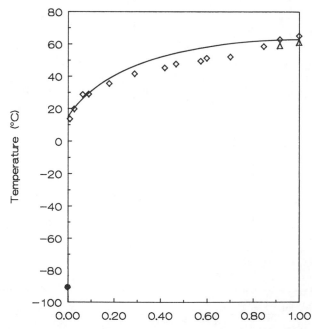

Fig. 8.10. *T*–concentration phase diagram for the system C$_{28}$–C$_7$: △, ◇, T_m of C$_{28}$; ● T_m of C$_7$

at $\sim -5\,°C$. It corresponds to the temperature at which the first deviation of the liquidus from the solvent axis can be detected by this method. This is illustrated in Fig. 8.10, which represents the complete phase diagram. The pure octacosane and the highly concentrated solutions show two melting points, due to the presence of two crystal forms. The continuous line in Fig. 8.10 represents the melting point depression for the ideal situation and was calculated using:

$$\frac{1}{T_m^o} - \frac{1}{T_m} = \frac{R}{\Delta H} \ln X_A$$

where the melting point of the pure substance and of the substance in the presence of the second substance are represented by T_m^o and T_m respectively. The ideal gas constant is R, the change in enthalpy on melting is ΔH, and X_A is to the mole fraction of the crystallizing substance.

Although the correspondence is not perfect, it is clear that the temperatures at the end of the melting endotherm represent rather well the melting points of the mixtures. They were not corrected for scanning rates. It has been shown that extrapolation to the zero scanning rate is necessary to obtain more exact values.

8.2.4 Crystallization and melting of a binary system composed of a polymer and a low molecular mass substance

Two different systems are discussed below. Attention is first paid to the melting of a polymer in the presence of a low molecular mass substance with about the same melting point: this is referred to as eutectic melting. The behavior of polymer solutions is then discussed.

8.2.4.1 Eutectic melting

The melting point of a polymer is reduced by the presence of a small molecule, and eutectic melting has frequently been observed [5]. Recently, the melting of the system antracene-poly(2,6-dimethylphenyleneoxide) (PPO) has been reported [6]. A typical melting curve is shown in Fig. 8.11. The samples were obtained by evaporation of a benzene solution. The eutectic melting endotherm is broadened by the presence of the polymer. The temperature at the maximum of the peak is therefore taken as the eutectic temperature, and is shown in Fig. 8.12. The broad melting of antracene starts at the eutectic temperature and ends at the melting point indicated. These different melting points constitute the liquidus in Fig. 8.12. In this phase diagram, the decrease of the melting point (temperature at the maximum of the melting endotherm) of PPO with increasing concentration of antracene is also shown. The non-equilibrium nature of these melting points was not taken into account. The eutectic point is situated at a PPO mass fraction of 0.70 and at 172 °C.

The importance of the error that can be introduced by the use of non-equilibrium values is illustrated by the phase behavior of the system poly(ethylene oxide)(PEO)-glutaric acid [7]: its phase diagram is shown in Fig. 8.13. The two substances have analogous melting points. The liquidus corresponding to the melting of glutaric acid can be regarded as an equilibrium line. A more complex behavior, however, is observed at the polymer side. Depending on the crystallization conditions, different curves are obtained. This results in different apparent eutectic points, with their specific concentration and melting temperature. The main reason for this apparent complexity

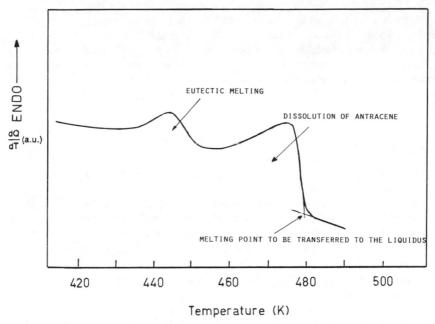

EUTECTIC MELTING

DISSOLUTION OF ANTRACENE

MELTING POINT TO BE TRANSFERRED TO THE LIQUIDUS

Temperature (K)

Fig. 8.11. Melting curve of the system PPO–antracene, mass fraction PPO = 0.55 (DSC-2 Perkin-Elmer, 5 K/min)

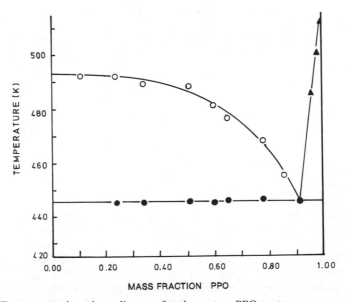

MASS FRACTION PPO

Fig. 8.12. *T*–concentration phase diagram for the system PPO–antracene

218

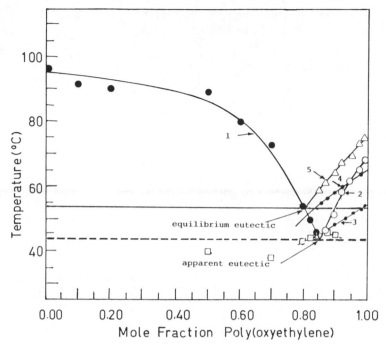

Fig. 8.13. T–concentration phase diagram for the system PEO–glutaric acid: ● T_m of glutaric acid (1); △ T_m^o of PEO (5); ○ T_m of PEO from dynamic crystallization (-10 °C/min) (2); • T_m from $T_c = 53$ °C (4); • T_m from $T_c = 24$ °C (3); □ T_{eu}

is the non-equilibrium value of melting points, obtained after isothermal or dynamic crystallization. A good approximation of the equilibrium melting temperature can be obtained by extrapolation of T_m, obtained after crystallization at different temperatures, to an equilibrium crystallization temperature $T_c = T_m$. The corresponding melting point represents the equilibrium value. When these equilibrium melting points are plotted as a function of the polymer concentration, an equilibrium liquidus and eutectic point are obtained. This eutectic point is situated at a molar fraction PEO of 0.80, and a melting temperature of 53 °C.

8.2.4.2 Polymer solutions

The phase behavior of polymer solutions is analogous to that observed with solutions of low molecular mass substances. The melting point of the polymer decreases with increasing solvent content, and the liquidus will tend to the pure solvent axis far above the melting point of the solvent.

However, an important difference arises from the specific crystallization mechanism of a polymer, which is a nucleation controlled phenomenon that proceeds at a measurable rate at a certain degree of undercooling. Fast crystallizing polymers such as polyethylene already crystallize from solution during a cooling operation in a calorimeter. The degree of undercooling at which this crystallization sets in is determined by the cooling rate and the rate of nucleation and crystallization. The melting of the sample is much less affected by this scanning rate. A shift to higher temperatures results mainly from superheating effects.

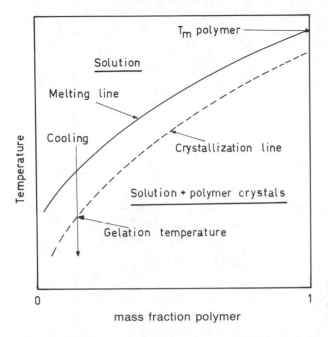

Fig. 8.14. Schematic representation of the depression, by a solvent, of T_m and T_c of a polymer

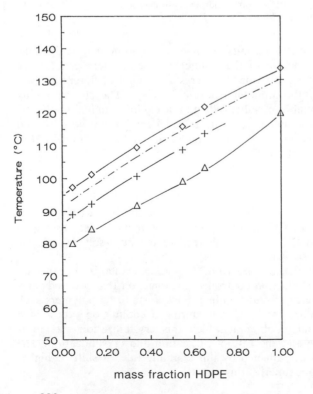

Fig. 8.15. Depression of T_m and T_c of HDPE by *o*-xylene (DSC-2 Perkin-Elmer, 5 °C/min): T_m^{end} ($\diamond$); T_m^{max} (+); T_c ($\triangle$); T_{corr} (−·−)

Consequently, if gelation of these solutions is studied and the whole temperature–concentration diagram is investigated, two lines must be considered. The melting line will represent the decrease of T_m with increasing solvent content; the second line, at lower temperatures, represents the onset of crystallization and will be called the crystallization line. This is schematically shown in Fig. 8.14. The temperature– concentration diagram for the system HDPE–o-xylene is shown in Fig. 8.15 [8]. The formation of crystalline thermoreversible gels results from the cooling to room temperature of a solution with a polymer mass fraction of 0.05–0.20. Gelation sets in when the crystallization line is crossed and the gelation temperature decreases with increasing scanning rate.

A further point of discussion is the exact location of the melting temperature. In Fig. 8.15, both the temperature at the maximum of the melting endotherm T_m^{max} and the final melting point T_m^{end} are represented. Neither represents the correct value, even if superheating effects are taken into account. The most exact value is the final melting point, corrected for the shape of the melting peak of the polymer in the absence of the solvent. This can be realized by subtracting the difference ΔT^{corr} between the T_m^{end} and T_m^{max} of the melting endotherm of the polymer in the absence of the solvent from T_m^{end}. This is shown in Fig. 8.16. The melting of the polymer in the absence of the solvent is represented in Fig. 8.16(a). This endotherm is shown as a broken curve in Fig. 8.16(b), and represents the melting of the polymer–solvent system and the correction of T_m^{end} of this system for the distribution of the melting of the polymer by itself. This 'corrected' melting point T_m^{corr} is also shown in Fig. 8.15. In principle, these melting points cannot be used in thermodynamic calculations, as they do not represent equilibrium values. They are obtained after crystallization at a certain degree of undercooling during a dynamic experiment or during an isothermal crystallization. The influence of this non-equilibrium situation has to be eliminated by, for example, an extrapolation to equilibrium conditions.

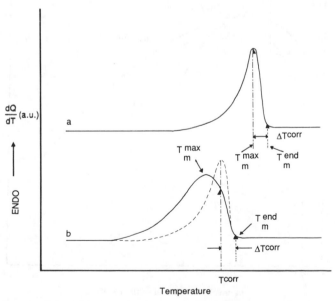

Fig. 8.16. Schematic illustration of the correction of T_m^{end} of a polymer–solvent system (b) for the distribution in melting of the polymer (a) as a consequence of its specific morphology

8.3 L–L demixing
8.3.1 L–L demixing in a binary system of low molecular mass substances
8.3.1.1 General considerations

A change in temperature at constant pressure can introduce L–L demixing in a homogeneous two-component system. This results in the formation of two liquid phases of different concentration in equilibrium with each other. The temperature dependence of the concentration of these coexisting phases can be shown by a phase diagram like the one shown in Fig. 8.17. This represents only one possible case with an upper critical solution temperature (UCST). It is most frequently encountered in systems that lead to the formation of thermoreversible gels. The opposite situation, a lower critical solution temperature (LCST), is also frequently observed.

The formation of these coexisting phases can occur by spinodal or binodal demixing. The latter can be observed when a solution with a concentration different from the critical one is cooled slowly. It is a nucleation controlled process, which

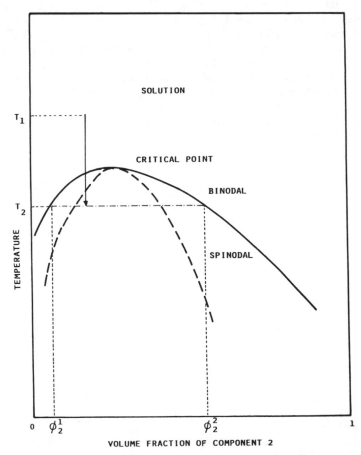

Fig. 8.17. Schematic representation of the *T*–concentration phase diagram for L–L demixing of a two-component system with a UCST

takes place at a certain rate. The former is a much faster process that sets in when a solution with a critical concentration is cooled into the demixing region. Solutions with a different concentration from the critical one can also follow this path when they are quenched through the binodal into the spinodal region. Because of the kinetic aspect of the binodal demixing, it can be suppressed by this very fast cooling. The final state that has to be reached, independent of the demixing mechanism, is one of two macroscopic layers of different concentration, in equilibrium with each other. This evolution toward an equilibrium situation can easily be followed by optical observations. The only condition is a sufficient difference between the refractive indices of the two substances. In the first moments of the demixing, the transparent solution becomes opaque as a consequence of the formation of microdomains of different concentration. These domains then collapse into larger ones, and this finally leads to the formation of the two macroscopic transparent layers. This final equilibrium situation, however, cannot be reached in many cases, especially when polymers are involved. A viscous opaque liquid system is obtained in which the collapse of the initially formed small domains is almost stopped at a rather small domain size.

Another difference between a polymer and a low molecular mass substance is the pronounced asymmetry that is introduced in the phase diagram. The maximum of the demixing curve is shifted to the solvent side. Because of the polydispersity of the polymers normally used, the flocculation curve, representing the temperatures at which opalescence sets in, will no longer represent the concentration of the coexisting phases. But for the general discussion of the formation of thermoreversible gels, the use of the simplified model, based on the monodisperse systems, will be sufficient. The flocculation curve, observed by optical means or by calorimetry, will represent, in a first approximation, the concentrations of the coexisting phases. Therefore, the term 'binodal' is used below for this flocculation curve.

8.3.1.2 Calorimetric observations

When L–L demixing occurs, heat is exchanged. This can be observed as an endothermic or exothermic signal in a dynamic experiment, and is illustrated by a well-known example that can be found in every book on physical chemistry.

Triethylamine and water are mutually soluble at low temperature but demix on heating. This system has a lower critical solution temperature at 292 K. The corresponding phase diagram is shown in Fig. 8.18. A mixture with the critical composition was heated in a Setaram Calvet Calorimeter BT 200 at a scanning rate of 0.3 K/min. When the system reached the critical temperature, an endothermic signal started. The extent of the demixing decreased as the temperature increased and the curve returned slowly to the base line, see Fig. 8.19 [9].

8.3.2 L–L demixing in polymer solutions

The detection of L–L demixing in polymer solutions by calorimetric techniques has been clearly illustrated with various polymers. A typical example is atactic polystyrene (aPS) in decalin [10, 11]. This system is characterized by an upper critical solution temperature. In Fig. 8.20, the exotherm observed on cooling is shown, together with the endotherm observed on heating. The temperature at the onset of the exotherm corresponds very well to the temperature at which the first traces of opalescence are observed. These temperatures are plotted in Fig. 8.21 as a function of concentration. A phase diagram characteristic of an L–L demixing is obtained.

223

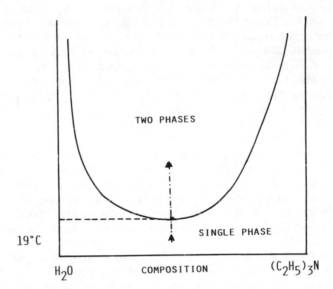

Fig. 8.18. *T*–concentration phase diagram for the system triethylamine–water

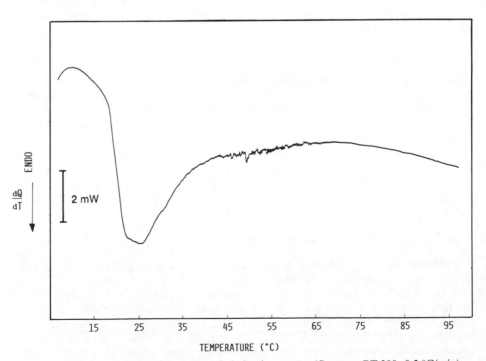

Fig. 8.19. Heating curve for the system triethylamine–water (Setaram BT 200, 0.3 °C/min)

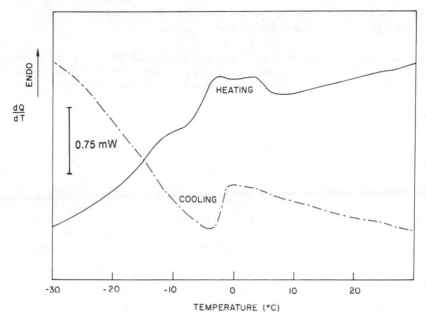

Fig. 8.20. Cooling and heating curve for the system aPS–decalin (DSC-2 Perkin-Elmer, 5 °C/min) (mass fraction polymer = 0.39)

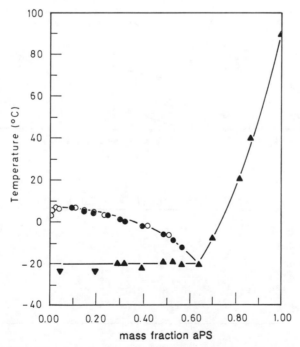

Fig. 8.21. *T*–concentration diagram for the system aPS–decalin: ○ L–L demixing by optical observation; ● L–L demixing by calorimetry; ▲ T_g by calorimetry; ▼ T_g by dilatometry; from cooling experiments

225

8.4 The combination of different thermal transitions and its importance in the study of thermoreversible gelation

A complex gelation mechanism is to be expected when different thermal transitions interfere. The solidification of the concentrated domains, formed during L–L demixing, can be at the origin of the solidification of the solution as a whole.

8.4.1 Combination of L–S and L–L demixing

The occurrence of L–L demixing in the vicinity of L–S demixing has a profound influence on the T_m–concentration relation of the dissolved substance [12]: its curvature is altered and tends to a plateau value in the lower concentration regions. This is observed even when the two domains do not really intersect. When a polymer is involved, various situations are possible. The melting line and crystallization line can be situated above the binodal. The other extreme situation is the intersection of both the melting line and the crystallization line with the binodal. A third possibility is a melting line situated above the binodal, while the crystallization line intersects the binodal. This last situation is shown schematically in Fig. 8.22. Cooling a polymer solution results in L–L demixing with the formation of diluted and concentrated domains. When the temperature at the intersection of the binodal and the crystallization line is reached, crystallization starts in the most concentrated phases. Further

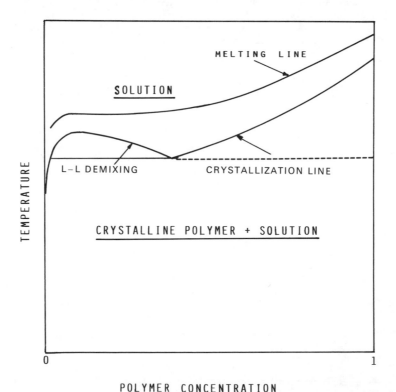

Fig. 8.22. Schematic representation of the interference of L–S and L–L demixing

226

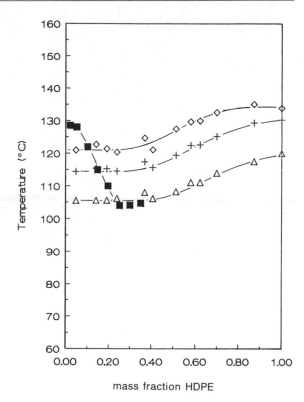

Fig. 8.23. T–concentration phase diagram for the system HDPE–amylacetate: $\diamond$ T_m^{end}; $+$ T_m^{max}; $\triangle$ T_c; $\blacksquare$ *T* of onset of L–L demixing (optical observations)

cooling can of course result in a crystallization in the more dilute phase, but this will depend on the shape of the crystallization line in this low-concentration region. Because crystallization sets in after this L–L demixing, the temperature at the onset of crystallization will be independent of the initial concentration, provided that this concentration is within the demixing region. This temperature corresponds to the temperature of onset of the solidification or gelation T_{gel}. Because L–L demixing sets in simultaneously at different places, a multiphase structure is obtained with mechanical characteristics that depend on the kind of material continuity realized throughout the solution. This depends on the molecular mass of the polymer, its concentration, and the thermal treatment of the solution. Crystalline porous materials with interesting properties can be prepared [13–17]. Within the demixing domain, the melting point remains constant. This type of phase behavior is shown in Fig. 8.23 for the system HDPE–amylacetate.

8.4.2 Combination of L–L demixing and vitrification

The addition of a solvent to an amorphous polymer results in a decrease in its glass transition temperature T_g. This T_g–concentration line can also intersect the demixing line, as shown schematically in Fig. 8.24. When a solution with a concentration within the demixing region is cooled to low temperatures, the concentrated domains, formed by the L–L demixing, will vitrify. If material continuity is realized throughout the solution by the concentrated domains, a solidification of the solution as a whole will take place. This mechanism is analogous to the one discussed in section 8.4.1. In view

227

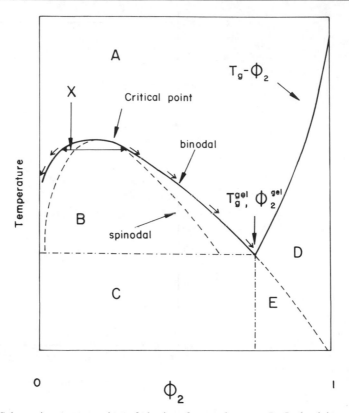

Fig. 8.24. Schematic representation of the interference between L–L demixing and the T_g–concentration curve. A: one phase solution; B: two phase solution; C: two phases—solution and glass; D: one phase glass; E: one phase glass or two phases—solution and glass

of the general definition given above, this amorphous glassy system is also called a thermoreversible gel. Its mechanical properties, however, are very different from those of the 'classic' gels obtained with biopolymers such as gelatin of carrageenans. The temperature at the intersection of both curves is called the gelation temperature T_{gel}. It corresponds to the T_g of the concentrated domains, and is independent of the initial polymer concentration if in this temperature region the flocculation curve and the coexisting lines coincide. Consequently, the glass transition of the solution will be independent of the overall polymer concentration for polymer concentrations between the pure solvent axis and the intersection point [8]. At higher concentrations, the solution will turn into a transparent glass without L–L phase demixing. A typical example is shown in Fig. 8.21 for aPS in decalin [10, 11]. The L–L demixing was investigated by optical and calorimetric techniques; T_g was measured by calorimetry and dilatometry. Gelation sets in at $-20\ °C$ and a polymer mass fraction of 0.60 in the concentrated domains.

Interesting morphologies can be obtained on removing the solvent from these glassy gels. The type of morphology depends on the overall polymer concentration, the molecular mass of the polymer, the cooling procedure and the mechanism of demixing. This is an easy and efficient technique for the preparation of membranes [18, 19]. This mechanism of structure formation is quite general, and can be observed with any polymer that has its T_g in the vicinity of the demixing region. The gelation temperature and the concentration in the concentrated vitrified domains depend on the exact location of the intersection point of the L–L demixing and the T_g–concentration line.

8.5 Conclusions

The use of a dynamic calorimeter can be of great benefit in the study of the thermoreversible gelation of polymer solutions. It reveals the occurrence of L–S or L–L demixing, the combination of these transitions and their possible combination with a glass transition. Great care is necessary, however, in the exact interpretation of the experimental data. The location of a transition temperature must be carried out keeping in mind the thermodynamic principles at the origin of these transitions. Furthermore, the dynamic character of the experiments has to be considered, together with the influence of the polymeric nature of the systems studied.

References

1 *Berghmans, H., Stoks, W.*, in: Integration of Fundamental Science and Technology. *Lemstra, P. J., Kleintjens, L. A.* (Eds.), Vol. 1, p. 218, Elsevier Applied Science, London, New York, 1986
2 *Berghmans H.*, in: Integration of Fundamental Science and Technology. *Lemstra, P. J., Kleintjens, L. A.* (Eds.), Vol. 2, p. 286, Elsevier Applied Science, London, New York, 1988
3 *Keller, A.*, in: Structure–Property Relationships in Polymeric Solids. *Hiltner, A.* (Ed.), Plenum Press, Amsterdam, 1983
4 *Russo, P. S.*, in: Reversible Polymeric Gels and Related Systems. ACS, Symposium Series, Vol. 350, p. 1, American Chemical Society, Washington D.C., 1987
5 *Wunderlich B.*: Macromolecular Physics. Vol. 1, 1973; Vol. 2, 1976; Vol. 3, 1980; Academic Press, New York
6 *Aerts, L., Berghmans, H.*: Bull. Soc. Chim. Belg. 99, p. 931 (1990)
7 *Gryte, C. C., Berghmans, H., Smets, G.*: J. Polym. Sci. Polym. Phys. Ed. 17, p. 1295 (1979)
8 *Aerts, L., Berghmans, H.*: unpublished results
9 *Vandeweerdt, P., Berghmans, H.*: unpublished results
10 *Arnauts, J., Berghmans, H.*: Polymer 28, p. 97 (1987)
11 *Arnauts, J., Berghmans, H.*: Physical Networks and Gels. *Burchard, W., Ross, Murphy, S. B.*, p. 35, Elsevier Applied Science, London, New York, 1990
12 *Stoks, W., Berghmans, H.*: J. Polym. Sci. Polym. Phys. Ed. 29, p. 609 (1991)
13 *Castro, A. J.*: US Patent 4,247,498, 1981
14 *Aubert, J. H., Clough, R. L.*: Polymer 26, p. 2047 (1985)
15 *Aubert, J. H.*: Macromol. 21, p. 3468 (1988)
16 *Young, A. T.*: J. Vac. Sci. Technol. 4(3), p. 1128 (1986)
17 *Young, A. T.*: J. Cell. Plast. 23, p. 55 (1987)
18 *Vandeweerdt, P., Berghmans, H., Tervoort, Y.*: Macromol. 24, p. 3547 (1991)
19 *Hikmet, R. M., Callister, S., Keller, A.*: Polymer 29, p. 1378 (1988)

Nomenclature

HDPE	high-density polyethylene
LCST	lower critical solution temperature
L–L demixing	liquid–liquid demixing
L–S demixing	liquid–solid demixing or crystallization from melt or solution
PEO	poly(ethylene oxide)
PPO	poly(2,6-dimethyl-1,4-phenylene oxide)
R	ideal gas constant

t	time
T	temperature
T_c	crystallization temperature
T_{eu}	eutectic melting temperature
T_g	glass transition temperature
T_{gel}	temperature of onset of gelation
T_m	melting point
T_m^o	equilibrium melting point of polymer
T_m^{end}	temperature at the end of the melting endotherm
T_m^{max}	temperature at the maximum of the melting endotherm
UCST	upper critical solution temperature
X_A	mole fraction of component A
ΔH	melting enthalpy

Chapter 9

The Crystallization and Melting Region

Vincent B. F. Mathot, DSM Research, P.O. Box 18, NL-6160 MD Geleen, The Netherlands

Contents

9.1 The melt as a reference state

As mentioned in chapter 5, it is important to ensure that there are stable isotherms at the beginning and end of the DSC measuring range, especially in the case of a c_p measurement.

On the low-temperature side, it obviously makes sense to start and end measurements well below the glass transition temperature T_g. There, conditions can be found under which processes such as aging and crystallization do not occur to any measurable extent, which means there are no measurable excess phenomena (see chapter 5). An added advantage in that case is that below T_g the measuring signal is well-defined, since in that region, as discussed in chapter 5, crystallinity and morphology have little influence on $c_p(T)$ (in other words, in that region $c_p(T)$ is hardly discriminating), and thus on the measuring signal, so that $c_p(T) \approx c_{p_s}(T) \approx c_{p_g}(T) \approx c_{p_c}(T)$. The same holds for a 'pseudo-c_p' measurement (see section 5.2.6). For practical reasons, however, in many cases the starting temperature is chosen—irrespective of T_g—in such a way that the signal does not change during the isothermal stay. Admittedly, this implies that one possibility for checking with literature data or with one's own data is lost.

On the high-temperature side, in the case of thermoplastics the measurements should preferably start and end in the melt. In the study of crystallization processes and subsequent melting [1–10] one should, however, reckon with the possibility of history effects [2]. Structures in the melt, for example catalyst residues, impurities and crystalline additives, can persist up to temperatures far above the highest polymer melting temperature. Crystallization, and thus the resulting morphology, can be influenced to a considerable degree by such structures [2, 11–15]. Being very suitable for the study of the crystallization process, DSC represents a powerful tool for the screening of nucleating agents in terms of their activity [16, 17]. Throughout this chapter we shall assume that an isotropic melt is started from, and that crosslinks, a liquid crystalline phase, orientation, constraints and other influences are absent. In this connection it should be noted that macromolecules may themselves give rise to history effects, cases in point being polyethylene [15, 18–23] and polyamides [24, 25].

In particular in DSC studies, one should be aware of the possibility of self-seeding [2, 26–29]: during cooling, incompletely molten macromolecular crystal residues can cause the nucleation process to start at higher temperatures than would normally (i.e. in the absence of self-seeding; isotropic melt) be the case. The crystals resulting from self-seeding are usually rather stable, since they were formed at a lower degree of supercooling than normal, which means that a melting point measured during the heating cycle will be higher than a melting point one would normally measure.

Figure 9.1 shows to what point melts of nylon 6, nylon 6.6 and nylon 4.6 should be heated to avoid self-seeding. Each sample was heated in a DSC-2 to 315 °C and, after 5 min, cooled to 200 °C at a rate of 5 °C/min to give it a well-defined thermal history. Next, the sample was heated to a variable temperature in the melt, the maximum heating temperature, and after a waiting time of 5 min it was cooled down to room temperature, the scanning rate being always 5 °C/min. The cooling curve was recorded. The maximum heating temperature was varied between the melting peak temperature T_m, determined in the absence of self-seeding, and 315 °C.

Figure 9.1 shows that, under the chosen conditions, up to about 30° above the highest melting temperature the DSC crystallization behavior of nylon 6 is influenced by the maximum heating temperature chosen. For nylon 6.6 this is about 15°, and for nylon 4.6 a few degrees.

232

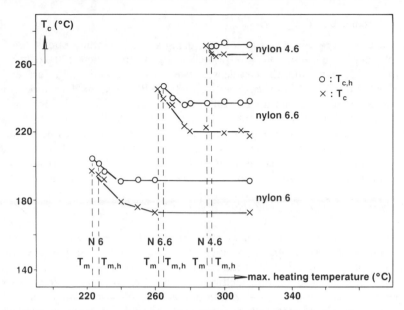

Fig. 9.1. Influence of the maximum heating temperature on crystallization for various nylons:
T_c and $T_{c,h}$ are the peak temperature and highest temperature for crystallization respectively;
T_m and $T_{m,h}$ are the peak temperature and the highest temperature for melting respectively

9.2 Linear and branched polymers

Unlike low molar mass substances, polymers do not have a single crystallization or
melting temperature. For various reasons, they always show a crystallization or
melting temperature distribution. When peak temperatures like T_c and T_m are
reported for polymers, this is mainly to characterize the distribution broadly. To a
first approximation, the polymers discussed in this section can still be characterized in
this simple manner, but for multiple-peaked DSC curves, as discussed in section 9.3,
for example, this simple characterization alone no longer suffices.

Sample masses of the polymers discussed here were kept as low as
0.800 ± 0.025 mg to minimize thermal lag and thus distortion of the DSC curve shape
while still allowing the peak areas to be determined quantitatively. That this is
possible was concluded from measurements on NBS SRM 1484 (see chapter 5). For
this linear polyethylene it was found that still lower sample masses [30, 31] (down to
15 µg in our investigations) no longer influence the curve shape due to the thermal lag
in the sample, so that 0.8 mg is the optimum sample mass. At a scanning rate of
$5\,^\circ$C/min the low sample mass leads to a crystallization peak temperature of
$T_c = 118.6\,^\circ$C and a melting peak temperature of $T_m = 133.1\,^\circ$C.

Serving as a reference sample in the study, the NBS SRM 1484 sample was
measured each time essential changes occurred in the DSC performance, for example
changes in curvature of the empty-pan signal as a function of temperature. The eight
measurements thus obtained (all performed on fresh samples) nevertheless exhibited
good reproducibility for the peak area upon cooling due to crystallization A_{cc} and the
crystallization peak temperature T_c:

$$A_{cc,average} = 157.9 \pm 1.5 \text{ J/g} \tag{9.1}$$

233

$$T_{c,average} = 118.6 \pm 0.2 \, °C \tag{9.2}$$

The measurement procedure was as follows: using a fixed scanning rate of 5 °C/min, the sample, kept in a continuously refreshed nitrogen atmosphere (20 cm³/min) was heated to 150 °C; after 5 min the sample was cooled to 50 °C and after a further 5 min it was reheated to 150 °C. In other experiments for NBS SRM 1484 a temperature of 180 °C was used. However, hardly any difference was noticeable, and therefore it was assumed that the procedure used suffices for removing crystal residues and obtaining reproducible measurements. The peak area during crystallization was determined as being the area enclosed by the DSC curve and a line drawn from that curve at the onset of crystallization (defined as being the first visible deviation from the melt signal) to an intersection with the curve 20° lower (see, however, section 9.2.1.2).

9.2.1 Influence of molar mass and thermal history

The molar mass occupies a prominent place among the many factors that have an influence on crystallization from the melt, and thus on the morphology and on thermal and other properties. Literature data [2, 3, 32–40] on the influence of the molar mass on crystallization and melting behavior are to a large extent based on experiments with isothermal crystallization of fractions of narrow molar mass distribution. This type of experiment offers certain advantages, for instance in theoretical interpretations [41, 42].

The influence of molar mass distribution and the role of the molar mass under more realistic (i.e. non-isothermal) conditions [43–46] has been studied less intensively. It is on these topics that the research reported here was focussed. To increase the relevance of the measurements to day-to-day practice, customary scanning rates were chosen, in most cases 5 °C/min.

9.2.1.1 Transition temperatures

Figure 9.2 shows the results of an investigation of T_c as a function of molar mass at various scanning rates during cooling S_c (0.31, 5 and 40 °C/min), and of T_m as a function of molar mass at a standard scanning rate during heating S_h after cooling at the same rate (5 °C/min), the samples being linear polyethylene (LPE) fractions [47]. These fractions were obtained from Société Nationale des Pétroles d'Aquitane [48] (now called Elf d'Aquitaine Production) and the U.S. Dept. of Comm., Nat. Bureau of Standards (now called the National Institute of Standards and Technology), where they were prepared by means of preparative size exclusion chromatography, as well as by in-house cross-fractionations. M_w/M_n for the fractions is around 1.5; in any case lower than 2. $[\eta]_{dec}^{135}$ and/or M_v (the viscosity–average molar mass) were known; for conversion purposes the relationship of these two was taken to be $[\eta] = KM_v^a$, based on the Mark–Houwink equation with constants K and a for a linear polyethylene chain [49]. The results of measurements on $C_{58}H_{118}$ (prepared in-house) and literature values [50] for $C_{64}H_{130}$ to $C_{140}H_{282}$ have been included.

Table 9.1 gives the molecular data of the fractions studied and the crystallization and melting characteristics. A rough peak evaluation was obtained by using two different lines, one in the case of a DSC cooling curve, covering 20°C (from the highest crystallization temperature $T_{c,h}$ to an intersection with the curve 20° lower) and the other in the case of a DSC heating curve, covering 30 °C (from the curve at the highest melting temperature $T_{m,h}$ to an intersection with the curve 30°

234

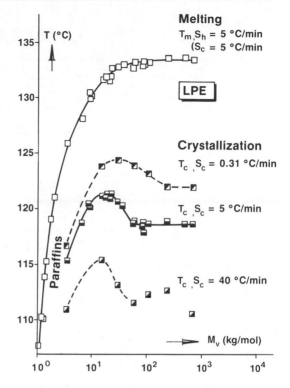

Fig. 9.2. Crystallization and melting peak temperatures T_c and T_m respectively, for LPE fractions with narrow molar mass distributions, obtained from DSC-2 cooling and heating curves (sample masses of 0.800 ± 0.025 mg), and T_m for paraffins, obtained from the literature, as functions of molar mass: influence of cooling rate on T_c

lower). The peak temperatures for crystallization and melting are denoted by T_c and T_m, respectively; the peak height from the base line by dq/dT and the width at half the height by $w_{1/2}$. The peak area enclosed by the DSC curve and the line obtained as indicated before is represented by A_{cc} for the cooling curve and by A_{hc} for the heating curve.

The melting temperatures increase with increasing molar mass as reported in several publications [50–54]. To describe the molar mass dependence of T_m for the data reported here, we first used the formula $T_m = A(M - B)/(M + C)$. This expression showed a good mathematical fit for paraffins and for paraffins plus LPEs [50]. Using multiple linear regression, the constants A, B and C were calculated for the 22 polyethylenes from Table 9.1, supplemented with 11 paraffins, viz. $C_{64}H_{130}$ up to and including $C_{140}H_{282}$, according to the data in Table 1 of ref. 50. The result is

$$T_m(M_v) = 406.6 \frac{M_v - 0.0588}{M_v + 0.0094} \quad (K); \qquad M_v \text{ in kg/mol} \tag{9.3}$$

with a standard deviation of 0.7 K absolute and a correlation coefficient of 0.9981.

Since the constant 0.0094 is almost negligible, the more practical formula $T_m(M_v) = A + B/M_v$ was also expected to yield a good fit. This proved correct:

$$T_m(M) = 133.4 - \frac{27.1}{M_v} \quad (°C); \qquad M_v \text{ in kg/mol} \tag{9.4}$$

with, surprisingly, a slightly higher correlation coefficient of 0.9983.

For linear chains the relationship between the number of end-groups per 1000 C, B_e, and M_n in kg/mol is expressed by $B_e = 28/M_n$. If, on account of the low dispersity, we assume $M_v \approx M_n$, we get the practical and empirical relation:

235

Table 9.1. Molecular structure data, crystallization and melting characteristics of LPE fractions with narrow molar mass distributions. Scanning rates of 5 °C/min, samples masses of 0.800 ± 0.025 mg

Sample	$[\eta]^{135}_{dec}$ dl/g	M_v kg/mol	Crystallization					Melting				
			T_c °C	$T_{c,h}$ °C	$\frac{dq}{dT}$ J/(g°C)	$w_{1/2}$ °C	A_{cc} J/g	T_m °C	$T_{m,h}$ °C	$\frac{dq}{dT}$ J/(g°C)	$w_{1/2}$ °C	A_{hc} J/g
F(K)-16-4	0.15	3.3	115.3	117	219	0.8	222	125.8	129.1	74	2.4	228
L 30-78-5-4	0.25	6.5	118.7	121.5	156	1.0	212	128.0	132.1	77	2.1	232
F4	0.31	8.6	120.4	122.8	171	1.0	227	130.3	133.5	93	1.7	235
L 30-78-5-6	0.32	9.0	120.1	122.5	155	1.0	224	129.8	133.1	86	1.8	232
F(V)-22-(K)	0.33	9.4	120.2	122.9	156	0.9	217	130.0	132.9	85	1.9	228
SNPA 1B-I	0.44[1]	13.6[2]	121.1	123.5	157	1.1	224	131.5	134.2	92	1.8	231
SNPA 1A	0.55	18.5	121.3	124.0	145	1.1	221	131.7	135.1	87	1.8	225
F6	0.57[1]	20[2]	120.9	123.8	156	1.0	218	131.4	134.1	92	1.7	225
F6 H-P	0.57[1]	20[2]	121.0	123.7	130	1.2	215	131.8	134.2	85	1.9	225
SNPA1	0.61	21.2	120.8	123.8	125	1.3	215	131.8	134.2	88	1.8	231
F(K)-15-6	0.63	22.1	121.3	124.5	130	1.3	216	132.6	134.5	86	1.8	210
SNPA 2B-I	0.84[1]	32.2[2]	120.6	124.3	95	1.8	202	132.7	135.0	75	2.0	210
SNPA 2A	0.98	39.9	120.2	123.2	104	1.5	200	132.9	135.0	62	2.0	209
SNPA 3A	1.33	60.0	118.6	123.1	85	1.6	182	132.6	135.5	48	2.8	191
SNPA 3B-I	1.40[1]	65.0[2]	118.8	123.1	90	1.7	181	133.1	135.9	42	3.2	185
SNPA 4	1.67	81.3	118.3	123.5	81	1.6	172	132.8	136.0	40	3.1	177
SNPA 4A	1.85	93.1	117.8	122.2	96	1.3	166	132.9	136.4	35	3.4	181
SNPA 4B-I	1.90[1]	95.5[2]	118.5	122.0	97	1.4	168	133.2	136.0	36	3.3	174
NBS SRM 1484	2.34[3]	119.6[2]	118.6	121.7	98	1.2	158	131.1	136.1	30	3.5	167
SNPA 5A[4]	3.50	218	118.8	121.7	94	1.2	148	133.5	137.2	23	4.6	154
SNPA 6A[4]	6.28	475	118.6	123.6	61	1.3	124	133.5	137.7	16	6.6	132
SNPA 7A[4]	8.67	730	118.6	123.9	43	1.7	108	133.3	137.3	14	7.0	118

[1]From M_v according to $[\eta]^{135}_{dec} = 6.17 \times 10^{-2} M_v^{0.75}$, M_v in kg/mol. Recently, [49], $[\eta]^{135}_{dec} = 7.11 \times 10^{-2} M_v^{0.725}$ was proposed resulting in slightly different values
[2]According to supplier
[3]From $[\eta]^{130}_{TCB}$ according to supplier
[4]Melting curve has shoulder on high-temperature side.
$T(K) = T(°C) + 273.2$

$$T_m(B_e) = 133.4 - \frac{27.1}{28} B_e \quad (°C)$$

$$= 133.4 - 0.97 B_e \quad (°C) \tag{9.5}$$

As a good approximation, the melting point depression relative to 133.4 °C happens to be equal to B_e, the number of end-groups per 1000 C:

$$133.4 - T_m(B_e) \approx B_e \quad (°C) \tag{9.6}$$

This expression can be used as a rule of thumb for non-branched fractions with narrow molar mass distributions, for scanning rates of 5 °C/min and sample masses of less than 0.8 mg.

The crystallization behavior, as shown in Fig. 9.2, is remarkable. After an initial expected increase T_c decreases from $M_v \approx 20$ kg/mol to become more or less constant (118–119 °C) at $M_v \approx 60$ kg/mol. It should be noted here that the crystallization temperature of polyethylene, compared with, for example, polypropylene and nylon 6, is influenced to a lesser extent by nucleating agents. The M-dependence of T_c illustrates the increasing resistance to crystallization with increasing molar mass.

As well as the absolute temperature, the degree of supercooling ΔT is important, this being the difference between the equilibrium transition temperatures of crystal $\leftrightarrow$ melt T_m° at the relevant molar mass and the actual sample temperature. This is because the temperature dependence of the crystal growth rate is determined mainly by the factor $e^{-C/(T \Delta T)}$ [2, 33]. There is no direct experimental way of determining T_m° as a function of M_v (or as a function of the number of C atoms in the main chain X, with $M = (X \cdot 14.026 + 2)/1000$ kg/mol), leaving aside for the moment the influence of the polydispersity in molar mass on T_m°, while all estimates are affected by the morphological model adopted [50, 51, 55–59]. Even if it is assumed only that an extrapolation according to X of paraffin values can be used, several routes can be followed, giving rise to an uncertainty in the value of ΔT that cannot be ignored. This is illustrated in Table 9.2 for two LPE fractions.

The increase of ΔT_c with M_v by a factor of 1.6 (see Table 9.2) indicates that for fractions of high molar mass crystallization is considerably more difficult than for the fractions of low molar mass. It is, moreover (Fig. 9.2), strongly affected by the

Table 9.2. Supercooling at the melting peak temperature ΔT_m and at the crystallization peak temperature ΔT_c for two LPE fractions

LPE fraction M_v (kg/mol)	F(K)-16-4 3.3	SNPA 6A 475
$\Delta T_m = T_m^\circ - T_m$ (K) $\Delta T_c = T_m^\circ - T_c$ (K)	$4.1^{1)} \leftrightarrow 5.9^{2)}$ $14.6^{1)} \leftrightarrow 16.4^{2)}$	$7.5^{1)} \leftrightarrow 13^{2)}$ $22.4^{1)} \leftrightarrow 27.7^{2)}$

[1] via $T_m^\circ = 414.3 \dfrac{X - 1.5}{X + 5.0}$ (K)   (9.7)

This is formula (2) in ref. 50; experimental melting curve

[2] via $T_m^\circ = 419.7 \dfrac{X - 3.48}{X + \ln X - 0.45}$ (K)   (9.8)

This is formula (5) in [50], but with $T^0 (X = \infty) = 419.7$ K, the value according to [32]; theoretical melting curve. More recent information can be found in refs. 3, 42, 60, 61

cooling rate, which shows that the time scales of crystallization and experiment are of the same order.

It should also be noted (see Table 9.2) that over the same range ΔT_m increases with M_v by a factor of 2, which shows that the hindrance of crystallization is also reflected in T_m, albeit at first glance (in Fig. 9.2) less conspicuously than in T_c.

On the changes in $T_c(M)$ on dynamic crystallization, as discussed here, few literature data are available [62, 64]. Quantitative comparison with isothermal experiments is difficult, and therefore the discussion is restricted to a qualitative interpretation. It is known that the isothermal crystallization rate as a function of M shows a maximum [37–39, 65–69] which shifts from $M_v \approx 100$ kg/mol when the supercooling is only slight to $M_v \approx 10$ kg/mol when it is strong. In Fig. 9.2 an analogous tendency is seen for the maximum in $T_c(M)$ as a function of the cooling rate, which makes a correspondence to the isothermal case plausible. In an absolute sense, a change in the cooling rate of a factor of 100 for LPE fractions thus yields a change in T_c of 5–10 °C, depending on the molar mass.

The reported increase in supercooling and the decrease of T_c between $M_v \approx 20$ kg/mol and $M_v \approx 60$ kg/mol might be explained from the fact that the diffusion of chains (chain reptation [5, 42, 70, 71]) towards the crystal front that is required for crystal growth becomes more difficult as the chains become longer. How complex the situation actually is will become clear if one remembers that there is, with increasing molar mass under the conditions used, a transition from extended chain crystallization for the paraffins to folded chain crystallization for very low molar mass macromolecules [72–76]. The lowest molar mass fraction in Table 9.1, F(K)-16-4, is expected to crystallize in Regime I [2, 10, 32, 42, 71, 77] region, where a single surface nucleus leads to completion of the substrate (substrate completion dominates). At higher M the supercooling increases at constant cooling rate, as a result of the diffusion of chains becoming more difficult, so that multiple nucleation can occur (Regime II: numerous surface nuclei are involved in substrate completion: secondary nucleation and substrate completion rates are comparable) [2, 32, 39, 42, 71, 77]. This may lead to fixation of chain parts at various crystal locations, thus also enabling the formation of tie molecules [78–80]. At $M_v \approx 60$ kg/mol the situation has evidently become so extreme that chain parts will crystallize individually, so that T_c is no longer dependent on M, and crystallization according to Regime III (distance between surface nuclei approaches the width of a stem: secondary nucleation dominates) [21, 38, 39, 77, 81, 82] occurs.

This observation is in line with the prediction by the classical nucleation theory [2, 83, 84], which shows that the critical free enthalpy ΔG^* which must be overcome for nucleation to occur becomes independent of M_v from $M_v \approx 100$ kg/mol onwards. ΔG^* and the critical length of a chain part l^* decrease with increasing ΔT_c, so that for smaller chain parts nucleation will become progressively easy. The need for chain diffusion therefore decreases. This would seem to explain the constancy—as reported here—of ΔT_c above a certain, cooling-rate dependent, molar mass.

As M increases, going along the $T_c(M)$ curve the morphology of the supermolecular structure [85] will change from d rods (lamellae organized as thin rods or rod-like aggregates), through g rods (rods whose breadth is comparable to their length) and a-type spherulites to b-type spherulites (the deterioration of spherulitic order proceeds from a via b to c) [44, 86]. At a very high M fringed micelle-type crystallization might even occur. Optical microscopic and SALS (small angle light scattering) experiments—in which the cooling rate was also varied (including quenching)—on a selection (as regards M) of the LPE fractions discussed here

238

yielded results [87] that were in agreement with the morphological phenomena (axialites (generic name for various types of rods), spherulites, etc.) described in the literature [32, 34, 43, 44]. As far as morphology is concerned it can also be concluded that, in the case of PE, isothermal experiments yield useful predictions for experiments under dynamic conditions.

It is intriguing that the specific $T_c(M)$ variation does not have a parallel in $T_m(M)$. This illustrates that in DSC studies the use of melting curves only can lead to a serious narrowing of the scope of information. Originally we assumed that only reorganization phenomena during heating (such as crystal perfectioning [88] and lowering of surface–volume ratio [1, 2, 89, 90]) disturbed the $T_c \leftrightarrow T_m$ correlation, but ref. 22 led us to think that—depending on crystallization temperature and cooling rate—even during cooling thickening of lamellae can occur to a considerable degree. (A lamellar crystallite is the most common crystal form [1, 4, 7, 46, 91, 92], with, characteristically, a smallest dimension of a few tens of nanometers and a largest dimension of up to a few microns; a number of such entities constitute the aforementioned supermolecular structures.)

Figure 9.3 shows results of calculations of the Raman length l_R under dynamic conditions for LPE. These calculations are based on isothermal measurements [15, 22, 93] and were performed on the assumption that the influences of time and temperature on l_R are not interrelated. We hope that in this way a qualitatively correct picture is obtained. The Raman length is defined as the part of a chain in a crystallite with contiguous trans–trans conformations—which is given by the LAM (longitudinal acoustic mode) measurement [94–96] in Raman spectroscopy. Since the

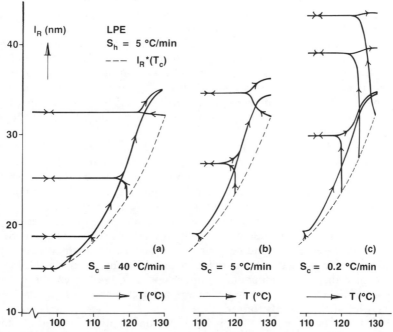

Fig. 9.3. The Raman length ($\rightarrow$) $l_R(T)$ for LPE as a function of the Raman length at the temperature at which the second stage of crystallization starts ($- - -$) $l_R^*(T_c)$; the cooling rate S_c and the subsequent heating rate S_h

chains are at an angle relative to the surface of a lamellar crystallite [46], the Raman length is greater than the crystallite thickness l [92].

For thermodynamic reasons, a melting temperature is directly related to a crystalline thickness (for example via the Gibbs–Thomson equation [3, 54, 97, 98]). This means that for LPE the relation of T_m and l can be examined via a determination of l_R (provided that the angle between l and l_R is known). Thus, the simulation of the influence of the experimental conditions on l_R gives at least qualitative information about T_m.

In the calculation it was assumed that any starting temperature indicated could be reached without crystallization occurring, for example by varying the molar mass, the nucleation density, etc. In particular, it was assumed that a starting temperature could be reached by quenching followed by further cooling at a constant rate. The initial l_R, the Raman length immediately after the formation of a crystallite nucleus at a starting temperature, indicated by $l_R^*(T_c)$ as a function of the temperature shows the well-known hyperbolic relationship [15]. Recent new experiments [99–101] show that the logarithmic increase in l_R^* to l_R is in fact already a second stage of crystallization. In the preceding first stage, a very rapid doubling, tripling and quadrupling of the long period appears to have taken place. This implies that the l_R^* given here will be a superposition of a genuinely initial l_R^* (which will therefore be smaller than the values given here) and values that are already higher. The observed phenomena can be explained by lamellar thickening, possibly via selective melting of the thinnest lamellae [102]. The calculation shows how l_R develops when a certain starting temperature is reached and nucleation and growth occur. In the case of LPE, the occurrence of nucleation and growth depends on the molar mass, the thermal history, the presence of nucleating agents, etc., as discussed earlier.

In the case of cooling at a rate of 40 °C/min, the calculation shows that at starting temperatures of about 120 °C or lower l_R hardly increases relative to $l_R^*(T_c)$, if at all; but on subsequent heating at a rate of 5 °C/min l_R increases continuously and may become twice as high. Apparently, after such a thermal history the relations of lamellar thicknesses determined at room temperature and melting temperatures (the use of the Gibbs–Thomson equation!), between T_c and T_m, etc. need not be unambiguous at all: original differences in lamellar thickness can be cancelled out completely during subsequent heating (see Fig. 9.3), so that the melting temperatures are ultimately the same.

If cooling takes place at a rate of 0.2 °C/min, l_R increases strongly starting at temperatures of about 120 °C and higher. In case of cooling at a rate of 0.2 °C/min starting at temperatures of about 120 °C and lower, a strong increase in l_R occurs on subsequent heating. This means, among other things, that in this example, with $S_c = 0.2$ °C/min and $S_h = 5$ °C/min, it will be difficult to establish a relation between the lamellar thickness at room temperature and T_c for starting temperatures of 125 °C and higher, whereas it will be relatively simple to establish such a relation with T_m.

Whether an increase in l_R and hence in crystal thickness actually occurs depends on several factors, in particular the mobility of the chains involved [46, 73, 103–112] (see for example the chain sliding diffusion concept [113, 114]) in combination with the topology of the chains in the crystalline and amorphous regions and the size and number [17, 46] of the crystals formed.

It should now be clear why there is no one-to-one relation between T_m and T_c: the relation depends on the time–temperature history. However, the fact that T_m is always higher than T_c is not exclusively due to the reorganization effects mentioned above, which cause the lamellar thickness to increase ($l > l^*$ and hence $T_m > T_c$).

After all, immediately after nucleation at T_c, as well as the longitudinal dimension of the developing crystallite the lateral dimension usually increases (and usually to a much higher extent, e.g. from nm to μm), which is one of the main causes of an increase in T_m. Suppose a tetragonal nucleus formed, with surface free enthalpies σ_{en} (end-surface) and σ_{sn} (side-surface [115]) and minimum stable dimensions $2a*$ and $l*$, grows into a lamellar crystal with relatively large lateral dimensions $a \gg a*$ at the same lamellar thickness $l = l*$. If we assume that for $\Delta g(T)$ the CP0 approximation can be taken (see section 5.2.3) and that for the end-surface free enthalpy of the lamella $\sigma_e \approx \sigma_{en}$, then we get a well-known relation of T_m and T_c [3, 98, 116, 117]: $T_m = (T_c + T_m^\circ)/2 = T_c + (T_m^\circ - T_c)/2$. The situation described here is intended as an illustration; it is just one of the many possibilities [58].

The calculation as shown in Fig. 9.3 gives an impression of the situations that can occur in a DSC experiment. It is clear that the choice of a certain combination of cooling and heating rates is crucial. Sometimes, for example in the case of a microcalorimeter, only low rates can be used. In some cases, when one would like to cool much faster, for example in order to approximate process conditions, and then perform heating at an even faster rate [118–121] in order to prevent reorganization of the crystal, the limitations of the available equipment make this impossible.

All in all, it has been shown that, should a situation like the one depicted in Fig. 9.3 indeed be the case, then not only the $T_c \leftrightarrow T_m$ correlations but also the correlations of T_c and T_m with the results of morphology studies at room temperature— such as lamellar thickness estimation [43, 46, 92, 122–124]—would become difficult or impossible to interpret if crystal reorganization were not taken into account.

9.2.1.2 Peak areas

The dependence on the molar mass of the DSC peak area for melting during heating A_{hc} of PE fractions is shown in Fig. 9.4. The DSC peak area for crystallization during cooling A_{cc} changes analogously. For most LPE fractions the phase transition seemed virtually completed in the limited temperature ranges cited for this (20° and 30° for

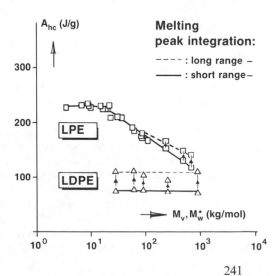

Fig. 9.4. Peak areas calculated from DSC-2 heating curves A_{hc} as functions of M_v and M_w^* for, respectively, LPE (see Table 9.1) and LDPE preparative SEC fractions (as in Fig. 9.5) with narrow molar mass distributions: heating rate 5 °C/min after cooling at a rate of 5 °C/min (sample masses of 0.800 ± 0.025 mg); influence of the length of the integration interval on A_{hc}

crystallization and melting respectively). In Fig. 9.4 this type of peak area determination is denoted by 'short-range peak integration'. Certainly for the samples of high molar mass the melting range is larger and the peak area is underestimated; the results after the required correction (in the direction indicated by the arrows) are shown in Fig. 9.4 as 'long-range peak integration'.

Figure 9.4 shows that, while initially its value is constant, the peak area for the LPE fractions decreases from $M_v \approx 20$ kg/mol onwards. The following relations describe the dependencies on M_v of the crystallization and melting peak areas during cooling and heating, A_{cc} and A_{hc} respectively

$$A_{cc} = 308 - 70 \log M_v \quad (J/g); \qquad M_v \text{ in kg/mol; integration interval } 20\,°C \qquad (9.9)$$

$$A_{hc} = 312 - 68 \log M_v \quad (J/g); \qquad M_v \text{ in kg/mol; integration interval } 30\,°C \qquad (9.10)$$

The decrease in the melting peak area with M has long been known from the literature [52, 53, 66, 67, 125, 126] and is related to the decrease in crystallinity with increasing molar mass [2, 45, 46, 51, 64, 127–131].

Calculation of the heat of crystallization, the heat of fusion and the crystallinity on the basis of the peak area determination as outlined here (short or long integration interval) is rather arbitrary, as is explained in more detail in section 5.2.5. LPE NBS SRM 1484 may serve as an example. Correct analysis of the crystallinity as a function of temperature (see Fig. 5.14) yields $\sim 66\,\%$ for the crystallinity at room temperature. The much-used but erroneous calculation that involves dividing A_{cc} and A_{hc} by the heat of fusion, $\Delta h(T_m^\circ) = 293$ J/g, would yield values of $54\,\%$ and $57\,\%$ respectively. Moreover, these results are rather arbitrary since they depend on the shape of the curve, the distance between the low-temperature intersection point of the line drawn under the DSC curve and room temperature, etc. Sometimes errors cancel each other so that a more or less correct value is accidentally obtained. It is clear that the procedure described above can serve only for comparison of the fractions.

The decrease in the peak area A with molar mass is assumed [129, 133] to be related directly to a combination of decrease in size and number of crystallites and associated changes in end-surface free enthalpy. These in turn are caused by changes in the manner of crystallization: see the discussion on $T_c(M)$ in the previous section.

For the lowest molar masses the LPE chains are present in lamellar crystals in extended form, or they are folded a few times with adjacent re-entry, so that $X_c \lesssim X$, where X_c is the number of units of a main chain in the crystal and X is the total number of units in the main chain. This situation, which is reminiscent of paraffins, though the end-groups are 'unpeeled', yields high values of A. The end-groups and the chain length distribution create transitional layers between the lamellar crystals. For $M > 20$ kg/mol X_c does increase somewhat, but the ratio X_c/X decreases rapidly, viz. $X_c \ll X$ for 20 kg/mol $< M_v < 100$ kg/mol and $X_c \lll X$ for $M_v > 100$ kg/mol.

With increasing molar mass, the chains are less and less situated completely inside crystals, and there is a two-phase structure: a crystalline phase and an amorphous phase. This implies that a dissipation problem arises in the transitional layer of crystallites with large lateral dimensions [134–145], since the amorphous phase has a lower density than the crystalline phase. To avoid overcrowding of the transitional layer, a partial re-entry of the chains into the crystal must occur. Depending on the nucleation and growth conditions, and in general also on chain stiffness, chain structure, etc., a certain ratio between adjacent and non-adjacent re-entry is established. Besides a transitional region, where the mobility of the chains is restricted by

242

the crystalline region (see the discussion about 'rigid' amorphous material in section 5.3.4), with increasing molar mass an interlamellar region or 'interzone' with a melt character as regards chain mobility ('mobile' amorphous fraction) is increasingly formed between the transitional regions. Molecular mobility determinations from broad-line proton NMR results for LPEs with M_v varying from 1.850 up to and including 3400 kg/mol were interpreted in this way using a three-component analysis of the spectrum [133, 146, 147].

The unit cell dimensions prove to be independent of the macroscopic density [148] (varied from 920 to 990 kg/m^3), so that this cannot be the cause of the depression of A with increasing M.

9.2.2 Influence of molar mass and chain branching

It is a well-known fact that a copolymer, in which the comonomer disturbs the crystallization, crystallizes and melts at lower temperatures than the corresponding homopolymer [149–151]. In section 9.3 various types of such ethylene copolymers are discussed.

There are also many polymers which, for various reasons, possess non-linear chains in the sense that short- and/or long-chain branching is present. In some cases this is inherent in the type of process or catalyst etc., but it is preferred, and increasingly attempted, to introduce a specific form of chain branching in order to control the processability and the ultimate properties. In specific cases the properties may even be said to be influenced directly by the polymerization process (including the catalyst used).

In this context it is important to know how branching affects the crystallization behavior, the morphology and the thermal properties. If one goes into the subject more deeply, one soon finds that subtle differences between apparently simple polymers are in fact of decisive importance for the properties of these polymers. For example, what is initially considered an almost linear polymer, without many possible variations, may be found to have a certain branching pattern after all. Further study then reveals that there are unmistakable differences between types/grades. Such studies are obviously impossible without dedicated analytical tools.

Fractionation is increasingly becoming an indispensable tool for such structure unravelling. There was a time, decades ago, when various fractionation techniques were widely employed. Then came a long period in which they were used by some specialist laboratories only, but nowadays companies and universities are rediscovering fractionation. Their main motive is often the increasing complexity of polyolefins. In particular, the development of new polyethylene and polypropylene copolymer types makes it absolutely necessary to gain detailed insight into the molecular structure. In the following sections, fractionation is therefore the recurrent theme; section 9.5 deals explicitly with some of its aspects.

9.2.2.1 High-density polyethylene

High-density polyethylene (HDPE) forms our point of departure. Contrary to what is often intuitively assumed, HDPEs are to some extent short-chain branched and should therefore be distinguished from LPEs. An illustration of what has been said in the preceding section is the statement in the literature [152–154] that the manner in which short-chain branching (SCB) is distributed across the molar mass distribution

(MMD) is of vital importance to the environmental stress cracking resistance (ESCR) [155] of polyethylene molded objects and to the long-term performance of polyethylene gas pipes. Preferably, SCB should be present in the long chains. Some catalysts achieve this automatically; some do so only under specific polymerization conditions, and some never do. Polymer chemists will, therefore, first want to know how the SCB relates to the MMD for a specific catalyst system. Only when this has been answered can efforts be made to work towards a predetermined goal. By way of illustration of this mode of thinking, a number of HDPEs are discussed below [47].

Hizex 7000 F, an HDPE with 1-butene as comonomer, contains 4 $CH_3/1000$ C. To study the SCB distribution across the MMD, various fractionations were performed. The nature of the problem to be addressed calls for a separation in terms of molar mass and optionally, if need be, a further separation (cross-fractionation) according to SCB. A crystallization fractionation, to begin with, would not serve its purpose here, for the chance of a rigorous separation according to a molecular structure parameter being achieved is small and cannot be properly estimated in advance (see the discussion on Hizex 7000 F in section 9.2.3).

In HDPE a separation according to molar mass is quite feasible, for HDPE is known to possess (practically) no long chain branching (LCB), only SCB being found. To meet the requirements for unambiguous results and labor intensiveness, two fractionation techniques are eligible for this polymer: size exclusion chromatography (SEC) and direct extraction (DE).

For a separation according to M, the first technique to be considered is 'analytical' SEC—as this is a commonly applied technique—used in a preparative way. Laboratories studying polyolefins need a high-temperature version in order to be able to dissolve the semi-crystalline samples. The influence of SCB on the SEC separation mechanism is known [49, 156, 157] and because of the low degree to which SCB is present in HDPE, its influence here is negligible, and the separation according to M using SEC is a rigorous one. This separation was performed on Hizex 7000 F, and the solvent free fractions (see section 9.2.3) were subjected to DSC measurements to determine the peak temperatures during crystallization and melting.

DE (see section 9.2.3) also appears to be an excellent preparative fractionation technique, with the advantage that it can be performed at such a level that IR and even NMR analyses of fractions are possible as well as DSC and morphological studies. Figure 9.5 shows the results of DSC measurements on Hizex 7000 F fractions obtained using the two fractionation techniques [47]. $T_c(M)$ and $T_m(M)$ of the fractions obtained by the two techniques exhibit analogous changes. Apparently, for HDPE the DE technique, just like SEC, separates according to molar mass, so that it provides an interesting possibility for preparative HDPE fractionation.

For comparison, the figures also include the lines for LPE (see Fig. 9.2). From the $T_m(M)$ plot it can readily be concluded where branching has been effected in the molecules, i.e. where the lines for LPE and Hizex do not coincide. As in this case the $T_c(M)$ values of the Hizex fractions are reasonably close to the $T_c(M)$ values of the LPE fractions, it may be assumed that the influence of molar mass for this HDPE is the same as that observed for LPE. The difference $T_c(M)_{LPE} - T_c(M)_{HDPE}$ is then to be attributed mainly to the influence of SCB, in accordance with IR data. The same of course holds for the $T_m(M)$ data. The importance of LPE lines, obtained and measured in a comparable manner, is clear. Comparison with a theoretical melting curve (Eqs. 9.7, 9.8) in this case would not easily have led to a correct answer. Without the LPE line, the changes in $T_c(M)$ would have given rise to a nonsensical

244

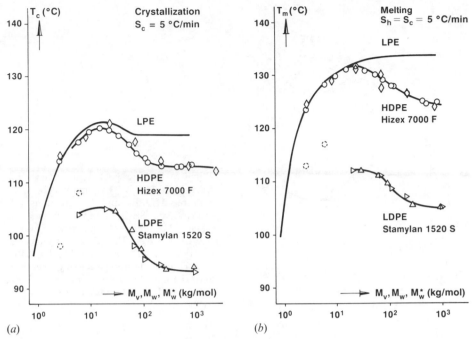

Fig. 9.5. (*a*) Crystallization peak temperatures obtained in cooling at a rate of 5 °C/min by DSC-2 for HDPE fractions ($\bigcirc$ direct extraction, sample masses of 0.800 ± 0.025 mg; $\diamond$ 'analytical' SEC) and LDPE fractions ($\triangle$ preparative SEC, sample masses of 0.800 ± 0.025 mg; $\triangleright$ 'analytical' SEC) as functions of M_v, M_w and M_w^* for LPE, HDPE and LDPE, respectively. LPE curve according to Fig. 9.2. A smaller peak, compared to the main peak ($\bigcirc$), in the DSC curve is indicated by $\because$. (*b*) melting peak temperatures in subsequent heating at a rate of 5 °C/min, otherwise as in (*a*)

interpretation because the retardation of crystallization, and ultimately the crystallization of parts of chains, with increasing M would have gone unrecognized.

For the HDPE shown here it follows from the $T_m - M_v$ profile that molecules with a mass of 20 kg/mol upwards are increasingly branched, the branches being caused by incorporated 1-butene comonomer. The amount of SCB of about 4 $CH_3/1000$ C measured for Hizex 7000 F is in fact an average value of an SCB distribution. The MMD in combination with the SCB distribution gives rise to a melting point distribution.

Figure 9.6 gives more information on the possibilities of the analytical SEC–DSC combination [158]. Three HDPEs are reported here, one of density 962 kg/m³ and two of density 949 kg/m³ (densities after compression molding). The HDPE with the higher density was found to have smaller melting point depressions relative to $T_m(M)$ for LPE than the HDPEs with the lower density. It is interesting to note that the two HDPEs with the same (overall) density of 949 kg/m³ exhibit different changes in $T_m(M)$, which suggests that the SCB distributions across the MMD are not the same. The HDPE denoted by O already exhibits a melting point depression at a low molar mass, and therefore probably has a more uniform distribution of SCB across the molecules.

245

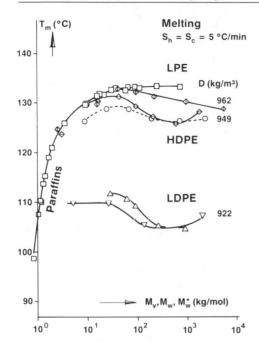

Fig. 9.6. Melting peak temperatures for fractions (Φ, $\bigcirc$, $\ominus$ 'analytical' SEC) of three HDPEs and for fractions ($\triangle$ preparative SEC, sample masses of 0.800 ± 0.025 mg; ∇: 'analytical' SEC) of two LDPEs from DSC-2 heating curves as functions of M_v, M_w and M_w^* for LPE, HDPE and LDPE respectively (densities and thermal history as indicated; LPE curve based on paraffins and a selection of LPEs)

Despite the very good agreement of various fractionation methods (see section 9.3.2) and the plausible interpretation of the differences between LPE and HDPE as regards $T_c(M)$ and $T_m(M)$ in terms of SCB, a critical note is called for. The $T_m(M)$ values for the HDPE with a density of 962 kg/m³ in Fig. 9.6 are below the LPE line. However, the $T_c(M)$ values—not reported here—in the M range from 10 to 200 kg/ mol are significantly above the LPE line, with a maximum of 2 °C around ~50 kg/ m³. Although this deviation also represents the maximum uncertainty as regards reproducibility (see section 9.2.3), it is remarkable that this deviating behavior is not reflected in $T_m(M)$.

A possible explanation may be diversity in amount and activity of (heterogeneous) nucleation centers (additives, nucleating agents, etc.) between the LPE and HDPE fractions, which might be caused for instance by the fractionation method and subsequent working-up. Differences in activity may affect the moment at which crystal nuclei are formed and start to grow: whether the crystals formed can increase in dimension and become more perfect will then depend on the possibilities for reorganization and annealing (in time and depending on the temperature). As discussed earlier, in particular the increase in lateral and longitudinal dimensions leads to an increase in melting temperature relative to the crystallization temperature. So it is plausible that the crystallization behavior, more than the melting behavior, reflects nucleation effects, the underlying assumption being that reorganization and annealing may more or less wipe out original differences.

The DSC curves for the HDPE fractions are single-peaked and extend across normal temperature ranges. No further (cross-)fractionation, for instance according to SCB, was carried out. For other fractionations of Hizex 7000 F, see section 9.2.3.

246

9.2.2.2 Low-density polyethylene

LDPE is a polyethylene type that has been produced in high-pressure processes for more than 50 years. Nevertheless, much is still unclear with respect to its structure and the relation between the structure and the crystallization and melting behavior [35, 44, 47, 159–164].

LDPE has a complex structure, as different types of SCB may be present [165] which, owing to the limited possibilities of characterization by means of IR and NMR, can be quantified unambiguously by those methods only up to a length of five C atoms. Longer side branches cannot be distinguished from one another, or from 'really long' side branches [165–167]. Long chain branching (LCB), which is so characteristic of LDPE, covers side chains whose length is of the same order of magnitude as that of the 'backbone chain'. The presence of LCB influences molecular dimensions [168] to such an extent that molecules with a (true) molar mass M have, in SEC, the same elution volume as linear molecules with a molar mass M^*, with $M^* < M$. The SEC (apparent) molar mass values are therefore distinguished further on by means of an asterisk, for instance M_w^*. SCB does not affect the molecular dimensions to a measurable extent and can therefore be accounted for very easily [49, 156]. It turns out that a linear branch must be at least about 12 carbons long in order to be measured as a 'long' branch by SEC [167].

Stamylan 1520 S is an LDPE with a density of 922 kg/m³ (after compression molding); it contains about 20 CH_3/1000 C. In view of the presence of one long chain branch per 1000 C, molar masses determined for this polymer are distinguished by an asterisk. In the figures, M (for HDPE) and M^* (for LDPE) are plotted on the same molar mass axis.

Stamylan 1520 S was fractionated by means of 'analytical' (see section 9.2.2.1) and preparative SEC, yielding fractions with M_w^*/M_n^* and M_z^*/M_w^* ratios of about 1.1 and 2 respectively. As can be seen in Fig. 9.5, the $T_c(M^*)$ and $T_m(M^*)$ results obtained by the two methods are in excellent agreement [47].

In the absence of knowledge of the $T_c(M)$ variation for LPE fractions, one would be inclined to ascribe the sharp drop of T_c (12 °C!) with M^* in Fig. 9.5(a) for Stamylan 1520 S to an increase in the degree of SCB. IR analysis shows, however, that the degree of SCB decreases slightly. Table 9.3 gives some relevant information.

The DSC curves are almost all single-peaked, if we ignore the small crystallization peak whose area is very small compared to that of the main peak. This low-temperature peak is always found in LDPE and often in ethylene-containing polymers (see for instance Figs. 5.6 and 9.11). There also appears to be a relation with the density: the higher the density, the higher is the position of the peak on the temperature axis. For the LDPE discussed here, with density 922 kg/m³ after compression molding, the peak lies at about 60 °C. Apart from this, there are all sorts of differences. For instance, the shapes of the curves differ greatly, while the crystallization rate for fractions having a high M_w^* is *higher* than for fractions with a low M_w^*.

Figure 9.5(b) shows a substantial decrease (7 °C) with M^* also for T_m, see also Fig. 9.6. Evidently, in this type of LDPE the decrease of T_c with M^* is so great that reorganization cannot prevent it from being clearly reflected in $T_m(M^*)$.

It is to be expected that a constant SCB across the MMD will yield an analogous and possibly even larger decrease of T_c and T_m with M^*. This is in accordance with earlier investigations [169–173], which, strangely, have hardly attracted attention. Since we observed the changes in $T_c(M^*)$ and $T_m(M^*)$ not only for the LDPE

247

Table 9.3. Molecular structure and DSC data on Stamylan 1520 S: preparative (P) and 'analytical' (A) SEC fractions

	M_w^* kg/mol	$B^{1)}$	$B_e^{2)}$	T_c K	T_m K
Whole polymer	100			372.0	383.0
A	6			(~ 377)	n.d.
A	20			378.0	385.0
P	28	23.0	2.8	377.5	385.0
P[3)]	60	20.4	1.0	374.0	384.0
A	66			371.0	384.0
P	90	18.7	0.4	370.5	382.5
A	105			368.5	381.5
A	206			367.5	380.0
P	258	17.9	0.1	366.5	378.5
P	885	19.1	0.0	367.0	378.0
A	940			366.0	378.0

[1)]B: total amount (end- and side-chain) groups/1000 C by IR
[2)]B_e: end-groups/1000 C based on M_n measurement
[3)]SEC curve has shoulder on the low M side.
Temperature values are rounded to the nearest half degree (to 0.5 or 0.0)

described here but also for other LDPEs with different structures which were also investigated using PSEC, we think these changes are a characteristic phenomenon in LDPEs [168]. To find an explanation, it is desirable to start with cross-fractionation so as to obtain detailed insight into how the molecular structure (M^* and SCB, LCB) relates to the crystallization and melting behavior. 'Iso-branching lines' for $T_c(M)$ and $T_m(M)$ could be a useful result; a possible next step might be further differentiation in terms of the SCB distribution in the chain. In this connection, the role of LCB and in particular the hindrance of the crystallization process by LCB should be considered.

It is interesting to note that the polyethylene types considered here show a T_c decrease above about the same values of M^* and M (~ 20 kg/mol) for LDPE and LPE/HDPE respectively. This could give the impression that molar mass (degree of polymerization, extended chain length) plays a smaller part than molecular dimensions in the melt in the crystallization process. After all, M^* and M, being deduced from SEC values, are determinative of the molecular dimensions in solution, which, in turn, are related to the molecular dimensions in the melt. However, we think this impression is wrong. Since for LDPE $M > M^*$, due to LCB, the decrease in T_c due to reduced chain mobility hindering crystallization occurs at a higher M than in the case of LPE and HDPE. We think this can be explained as follows: although the chain as a whole becomes practically immobile due to the LCB points (on the analogy of crosslinks), the mobility of chain *parts* remains at a reasonable level even at a higher M due to the larger number of free ends. Apparently, when M is corrected to M^*, LDPE scales to LPE/HDPE again. In this sense, as well as M, $[\eta]$ and M^* are suitable quantities for evaluating relations between molecular structure and crystallization and melting behavior. Their advantage is that they are easy to determine.

By implication, for LDPE, too, the molar mass represents an important parameter as well as the SCB. The fact that at densities of ~ 920 kg/m^3 the molar mass is generally still an important parameter in the crystallization process also appears from results obtained with ethylene-1-hexene copolymers of narrow molar mass and composition distributions [174], and the same is true for hydrogenated polybutadienes [174–176]. These polybutadienes [174, 175], which are ethyl-branched (density after compression molding 915 kg/m³) without being long-chain-branched, exhibit a depression of T_m in an M_w range that is comparable to the M_w^* range for LDPEs. Recent relevant information can be found in ref. 280.

As far as the melting peak area and, more generally, the crystallinity are concerned, the molar mass has less influence than SCB (see Fig. 9.4 and refs. 44, 150; however, see also ref. 176).

9.2.2.3 Conclusions and consequences for some relations

The results discussed here clearly illustrate once more that overall quantities such as $CH_3/1000\,C$, density, heat of fusion and T_c and T_m, determined on the whole polymer, do not necessarily point to one definite molecular structure. Samples showing the same values of the overall quantities may, for instance, differ in SCB distribution across the MMD, as is shown above for HDPE. This may result in differences in, for example, thermal and mechanical properties. Fractionation according to molar mass is an effective tool for revealing such differences.

From the above, two important molecular causes of the fact that polymers almost always have a melting point distribution can be deduced. The polyethylenes shown can be regarded as blends of molecules, the molecules differing not only in length (and hence in molar mass) but also in degree of branching. In addition, in view of the blend character the possibility of mutual influencing of chains and chain parts has to be taken into account. Interaction of this kind could result in effects such as segregation in the melt, retarded/accelerated crystallization and co-crystallization. In this chapter no attention is paid to the blend character and its consequences (except in section 9.5.2.5).

Important differences between the fractions as regards the shape of the DSC curve, for example the differences occurring in the case of the LDPE PSEC fractions (not reported here), no doubt point to additional differences in SCB distribution. This alone makes evaluation of the shape of the DSC curve a necessary refinement, which, however, is very difficult to perform. In such cases more information will have to be obtained on the SCB—and more specifically on the nature, length and distribution of the branches—at certain molar masses, and (cross-)fractionation according to SCB and further analysis of the fractions will be needed. In view of the above, one cannot expect simple relations to exist between crystallization parameters such as T_c, T_m and transition enthalpies on the one hand and molecular structure parameters such as SCB, LCB and M on the other.

For instance, the differences in T_c for LPE and HDPE fractions can be ascribed to SCB in a qualitative sense, but for the quantification of these differences one must take into account also the value of M and in some cases the influence of seed nuclei, as well as the fact that PE was dissolved during fractionation, the processing route, including solvent removal, etc.

The influence of various parameters as sketched above often causes a spread in correlations. In the case of LDPE fractions with narrow molar mass distributions, for example, this is the case for the relation of T_m and SCB. Only for fractions with a low molar mass does an expression analogous to Eq. 9.6 appear to hold if instead of B_e the total amount of branching per 1000 C, B, is used. Nor is there a good correlation between T_m and T_c for LPE, HDPE, LDPE and LLDPE (linear low density polyethylene, see section 9.3) fractions with narrow molar mass distributions, for example:

$$6 \leq T_m - T_c \leq 16 \quad (^\circ C) \tag{9.11}$$

However, if one considers the fractions with $M_w^* \approx 100$ kg/mol separately, the band width is much narrower:

$$T_m - T_c \approx 12 \pm 1 \quad (^\circ C) \tag{9.12}$$

9.2.3 Fractionation

As emerged in the preceding sections, a good separation technique based on the molar mass is of crucial importance, for it allows us to substantially deepen our insight into the molecular structure and the polymerization underlying this structure. In addition, such a technique will yield further information on the crystallization and melting behavior, while making it easier to study the relations with properties. It has also become clear that, besides molar mass, the nature, amount and distribution of the branches (particularly SCB) play a major role, so that a combination of separations according to M and SCB is very useful indeed. In the following sections, supplementary information and comments on fractionations are presented together with experimental data.

9.2.3.1 Fractionation according to molecular dimension

There are several methods to achieve separation according to, or mainly according to, M [47, 150, 151, 177–193]. SEC and DE are referred to above and discussed below. A distinction is sometimes made between 'analytical' and 'preparative' separations, depending on whether or not fractions can be collected in sufficient quantities to allow analysis by other techniques. The success of collection, working-up and further analysis is determined to a large extent by the sensitivity of the measuring equipment and the skills of the experimenter. This implies that the qualifications given will change with time, and this makes them rather arbitrary. The term 'preparative' is used mostly in cases where individual fractions are actually isolated and characterized [187, 190].

As stated in section 9.2.2, in SEC separation takes place according to molecular dimensions in solution, and the influence of SCB and LCB is known [49, 156, 157, 194]. The HDPE and LDPE 'analytical' SEC fractionations reported here were carried out with a Waters GPC-200, using 1,2,4-trichlorobenzene (TCB) at 140 °C. M_w^*/M_n^* and M_z^*/M_w^* amounted to about 1.1.

As on-line spectroscopy for quantitative determination of SCB was not possible in 'analytical' SEC at the time the research reported here was performed [195], an alternative method was sought that would give qualitative knowledge of SCB during or after separation according to M (or M^* in the case of LCB). DSC was chosen, because the crystallization and melting behavior is very sensitive to the presence of SCB, and because DSC was expected to be sensitive enough for measurement of the extremely small heat effects (due to the very low sample masses). It is evident that without calibration DSC measurements can yield only qualitative statements with respect to SCB.

'Analytical' SEC fractions of masses varying between about 10 μg and 1 mg were collected and precipitated on a filter [196]. The filter was then placed upside down in a DSC pan to promote contact between sample and pan bottom. A pan with an unloaded filter was used for reference purposes. Peak temperatures were determined for the fractions [47]. Since the sample mass cannot be determined exactly, normalization of the peak area on the basis of sample mass is not feasible and the heats of crystallization and fusion for the fractions cannot be meaningfully compared. As remarked in section 9.2 [31], below about 1 mg the peak temperatures are hardly affected by the sample mass, making comparison of fractions possible.

Direct extraction is a fractionating technique [47, 63, 154, 197, 202] that uses the differences in solubility caused by differences in molecular structure. From the sample to be fractionated, a gel is prepared with the aid of a solvent/non-solvent mixture. The ratio of solvent to non-solvent is varied at a fixed temperature, and the fractions are isolated through vibration (necessary for rapid diffusion from the gel). The direct extraction experiments on the HDPE Hizex 7000 F were performed by Chemische Werke Hüls AG, with p-xylene/ethylene glycol monoethyl ether mixtures at 119 °C; the fractions obtained varied in mass from about 80 mg to 0.8 g. 2 g/l of BKF was used as stabilizer. For DE on copolymers, see section 9.5.

High-temperature PSEC on LDPE Stamylan 1520 S [47] was performed at the National Physical Laboratory using a Waters Anaprep apparatus, with TCB at 135 °C. Yields per fraction varied between 30 mg and 3 g. For PSEC on a copolymer, see section 9.5.

In view of the dissimilarity of DE and SEC with regard to the separation mechanism, the experimental conditions, the polydispersity of the fractions and the working-up procedures, it would not be surprising if there were many differences between the three above-mentioned methods. However, as became apparent above, the reported thermal behavior of the fractions is in very good agreement.

This proves the accuracy of the analytical SEC–DSC method. Previously [158] it had already been established that the reproducibility for $T_c(M)$ and $T_m(M)$ is good: generally within 1 °C, with 2 °C as the maximum difference. Comparison with the PSEC results in order to determine the accuracy (Fig. 9.5) shows that the fractions differ at most by 1 °C in terms of peak temperatures, which is better than expected. It should be noted, though, that the SEC M-values are most accurate because the fractions were treated at about the same time and the same calibration curve could be used.

It follows from the above that DSC can apparently lead to quantitative results for T_c and T_m with 'analytical' SEC fractions from which the solvent has been removed, which means that 'analytical' SEC–DSC is a powerful analysis combination for polymers. Thus, for polymers, it is possible to obtain reliable information on crystallization and melting behavior as a function of molecular dimensions in an unequivocal way.

9.2.3.2 Fractionation according to crystallizability/dissolvability

As noted in section 9.2.2, it may be necessary to effect further separation according to another structural parameter after separation according to molar mass. Certainly in the case of polymers where chain branching has been introduced, there may be a need for further information with regard to nature, length and distribution of the branches. Cross-fractionation according to SCB and further analysis is then the appropriate route.

It should be realized, though, that many of the current SCB fractionation methods are based on crystallization [181, 185, 203] and dissolving [47, 193, 201, 202]. A case in point is temperature rising elution fractionation (TREF) [180, 182,

192, 204–227]. In carrying out TREF and other crystallization fractionations, and in the interpretation of the results obtained, the phenomena discussed in this chapter should be taken into account.

In general, and for HDPE in particular, a crystallization fractionation need not lead to rigorous separation according to a single molecular parameter. More specifically, it should not be assumed that separation takes place only according to SCB. Often this is implicitly assumed, as it is known that SCB has a major influence on a polymer's dissolution behavior and it is wrongly assumed that the influence of the molar mass distribution can be ignored. As stated above, this need not always be the case. Being overall quantities, peak temperatures, too, are influenced by both SCB and M. As demonstrated, these peak temperatures should be considered the resultant of crystallization and melting point distributions that are related to nucleation, and growth and melting, respectively, of a size distribution of crystals. The causes of this distribution range from the molar mass distribution, the type of SCB and the distribution of SCB along the molecules to non-molecular factors such as nucleating agents, thermal history, processing conditions, etc. The molecules, moreover, influence one another and should be considered to be components of a blend, with all the associated possible variations, from completely miscible to highly immiscible. In addition, the crystallites' size distribution is not fixed owing to their metastable character, while the perfection of the crystals and the state of the amorphous phase, be it mobile or rigid, are not fixed either.

The conclusion must be that in case of cross-fractionation the sequence in which the fractionation techniques are applied can be important [47]. Separation according to M is fairly independent of SCB and in the case of LCB at least a separation according to M^*, linked with the molecular dimensions in solution, is achieved. After this, separation according to SCB of a fraction with a narrow M or M^* distribution using a crystallization/dissolution method may be successful if the molecules differ in SCB distribution (SCBD).

If, on the other hand, there are differences with respect to the distribution of short chain branches only within each molecule and not between molecules, then of course no separation can be effected. In such a case, a sufficiently high rise in temperature above the crystallization temperature will first cause the molecule parts with the smallest crystallized sequences to dissolve. Parts of the same molecule that are still stable in crystals because these contain longer crystallized sequences will dissolve only at higher temperatures. Upon a further increase in temperature the longest crystallized sequences will ultimately become unstable in the crystals, and the entire molecule will dissolve. Only then can a change in concentration in solution be measured, for instance by means of the concentration detector in a TREF set-up, provided of course that the required diffusion from crystal to solution is possible. The signal obtained represents the entire molecule, but at a temperature that is characteristic of the longest crystallized sequences in that molecule. A useful complementary tool in such a case is the microcalorimeter (see Fig. 9.10), since at each temperature at which molecule parts dissolve there may be a measurable heat of solution. In principle, therefore, all sequences capable of crystallization are detected, unless the sensitivity is too low.

Separation according to SCB using a crystallization/dissolution method and subsequent separation according to M [180, 210, 212, 216, 219, 221, 223, 224, 226] yields results that are much less rigorous, and is to be recommended only if the system is already roughly known, so that the results can be interpreted in a straightforward way.

For the sake of argument, let us take the melting behavior of the HDPEs in Figs. 9.5 and 9.6 as a starting point. Small, linear or slightly branched molecules may have the same melting point as large, heavily branched molecules. In the first step of a crystallization/dissolution method the polymer is crystallized in solution. In the second step in the analytical method continuous heating is applied and a detection method, for instance IR, is used to determine the concentration of the molecules that have dissolved. By means of a calibration curve the dissolution temperature is translated into an SCB content [208, 209, 220, 221, 224, 228, 229], and this of course is where the danger lies, since the large, heavily branched molecules may be eluted at the same temperature as the small, linear or slightly branched HDPE ones. After all, there is no reason for the behavior in solution to differ essentially from that in the melt: the dissolution peak temperature as a function of M, $T_d(M)$, may develop in roughly the same way as shown in Figs. 9.5 and 9.6, though of course at lower temperatures. Since in solution, in contrast to the melt, the crystallization behavior of the longer chains will not be, or will be to a much lesser extent, hindered by a lack of mobility, it seems probable that the crystallization behavior in solution will nevertheless be different: it is to be expected that a decrease like the one observed for LPE, see $T_c(M)$ in Fig. 9.2, will be absent or will occur to a lesser extent. This, incidentally, will cause the $T_d(M)$ curve for LPE [209] and for HDPE to be steeper than the $T_m(M)$ curve. Whether there are still large molecules with such an amount of SCB that the dissolution temperature is the same as that of the small linear or slightly branched molecules will depend on the specific SCB distribution. A fraction eluted at this dissolution temperature would in that case have a bimodal molar mass distribution. As explained below, the above applies to the HDPE Hizex 7000 F discussed earlier [47].

To gain more insight into this matter, besides the direct extraction reported above (Fig. 9.5), a preparative crystallization/dissolution method was used. The sample [2 (m/m) %] was dissolved at 134 °C in p-xylene and then crystallized on a stirrer covered with hemp [230], by cooling to 50 °C at the rate of 8 °C/h. Following this, fractions were drained off at various increasing solution temperatures, the yields being about 0.1–2 g. The stabilized (1 g/l Topanol + 1 g/l Irgafos) solution was kept under nitrogen during the whole procedure. Subsequently the fractions were subjected to several characterization methods, including DSC and SEC.

Figure 9.7 shows DSC curves for two fractions obtained by the direct extraction method and by the crystallization/dissolution method. As expected, the DSC curves show a single peak.

Figure 9.8 shows the corresponding SEC curves. The direct extraction fractions (fr. 3 and fr. 11) are unimodal. Dissolution in the crystallization fractionation process at temperatures between 85 and 87.5 °C (fr. VI) does indeed result in a bimodal SEC curve. A fraction obtained after dissolution at a temperature between 90 and 92.5 °C also still shows a bimodal curve. However, if dissolution takes place between 95 and 100 °C (fr. X in Fig. 9.8), the curve is unimodal but broad; this could correspond to a maximum in the $T_d(M)$ curve, analogous to the maximum in the $T_m(M)$ curve (see Fig. 9.5).

This is summarized in Fig. 9.9 [231]. Assuming that $T_d(M)$ develops as indicated, the various cross-sections perpendicular to the M-axis and the T_d-axis, which can be realized by fractionation, yield hypothetical DSC and SEC curves. These curves are in agreement with the experimental results reported above.

It should be noted that a fraction obtained by elution in a small temperature range ΔT_d in the crystallization/dissolution method will yield a different temperature

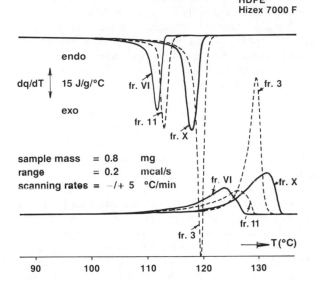

HDPE
Hizex 7000 F

endo

dq/dT | 15 J/g/°C

exo

fr. VI

fr. 11

fr. X

fr. 3

sample mass = 0.8 mg
range = 0.2 mcal/s
scanning rates = –/+ 5 °C/min

fr. VI

fr. X

fr. 3

fr. 11

T (°C)

90   100   110   120   130

Fig. 9.7. DSC-2 cooling and heating curves for some Hizex 7000 F fractions: fractions 3 and 11 obtained by direct extraction at 119 °C with *p*-xylene/ethylene glycol monoethyl ether mixtures of 160 ml/140 ml and 195 ml/ 105 ml respectively; fractions VI and X were obtained by crystallization/dissolution fractionation, with dissolution taking place at temperatures between 85 °C and 87.5 °C and between 95 °C and 100 °C respectively

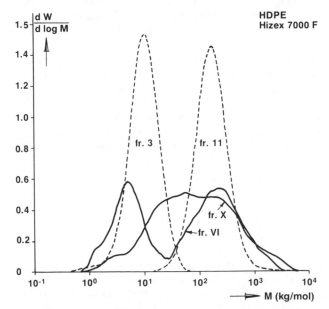

$\dfrac{dW}{d\log M}$

1.5

1.4

1.2

1.0

0.8

0.6

0.4

0.2

0

fr. 3

fr. 11

fr. X

fr. VI

HDPE
Hizex 7000 F

10⁻¹   10⁰   10¹   10²   10³   10⁴

M (kg/mol)

Fig. 9.8. Normalized SEC curves for the Hizex 7000 F fractions given in Fig. 9.7: the direct extraction fractions 3 and 11 are unimodal with M_n, M_w, M_z values 7.4, 12, 17 and 130, 210, 440 kg/mol respectively; the crystallization/dissolution fraction VI is bimodal with M_n, M_w and M_z values of 9, 210 and 820 kg/mol respectively; the crystallization/dissolution fraction X is unimodal with M_n, M_w and M_z values of 21, 280 and 1380 kg/mol respectively

range ΔT_m in the DSC measurement. After all, the DSC measurements were made on solvent-free fractions, and the crystallization from the melt and the subsequent melting process are represented in Fig. 9.9. Additional effects such as nucleation and mobility restrictions may then play a role.

In general, it is to be expected that a calorimetric measurement involving crystallization and dissolution of the sample in the measuring cell in a comparable

254

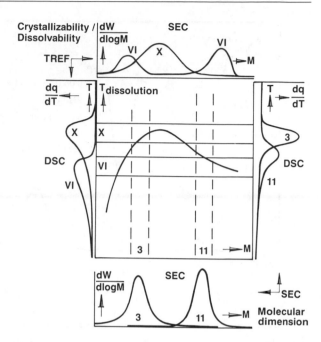

Fig. 9.9. Explanation of the DSC and SEC results (see Figs. 9.7 and 9.8) for Hizex 7000 F: dissolution peak temperature vs. molar mass curve $T_d(M)$ for fractions with narrow molar mass distributions of a hypothetical polymer assumed to be a blend varying from small molecules containing little or no SCB to large molecules containing much SCB; compare with $T_m(M)$ for some HDPEs in Fig. 9.6 (fractionation according to molar mass or crystallizability/dissolvability would yield the DSC and SEC curves as indicated)

solution and under optimally comparable conditions will yield results that correspond much more unequivocally to those of fractionation (see the discussion in section 9.5.2) than a calorimetric measurement involving crystallization and melting of the sample as such. In view of the low concentration [typically 0.05 (m/m) %] in the crystallization/dissolution method and the very low scanning rates, the chances of the corresponding heat effects being detected are slim when using DSC equipment. In principle, a microcalorimeter like the Setaram C 80 (see chapter 10) offers a better chance of success.

To illustrate this, the following experiment was performed. In a concentration of 0.05 (m/m) % in o-dichlorobenzene, LPE SNPA 4A (a fraction with a narrow molar mass distribution; for further data see Table 9.1) with DBPC added as anti-oxidant, was dissolved in a nitrogen atmosphere in the C 80, being stirred for 7 h at 120 °C (for TREF typically 7 h at 150 °C). Subsequently, without stirring, cooling was effected at a rate of 1.5 °C/h down to 30 °C, followed by heating to 130 °C. Figure 9.10 shows the cooling and heating curves. In spite of the combination of extremely unfavorable measuring conditions in the form of a low concentration and low scanning rates, the microcalorimeter is quite capable of supplying useful information. Moreover, no use was made in this case of further optimization possibilities, such as correction on the basis of empty-cell measurements.

From the picture outlined above it follows that in the case of the Hizex 7000 F sample, and thus more generally for comparable HDPEs, the crystallization/dissolution method used here, and analogously TREF, etc., do not always lead to a rigorous separation according to a molecular parameter; it is therefore better to use crystallization/dissolution or TREF for cross-fractionation according to SCB following a fractionation according to M (see section 9.5.3) than the other way round (TREF followed by SEC).

255

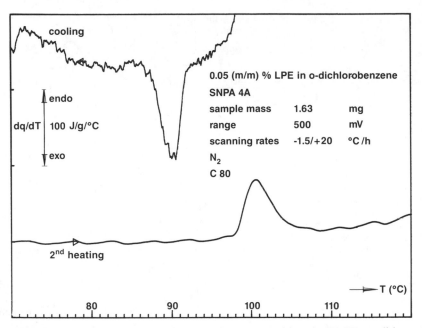

Fig. 9.10. Microcalorimeter results under conditions approaching the TREF conditions: C 80 cooling and heating curves of a 0.05 (m/m) % LPE solution (cooled at 1.5 °C/h after dissolution with stirring for 7 h at 120 °C, and subsequently heated at 20 °C/h)

9.3 Copolymerization

The preceding sections showed that, as well as thermal history and molar mass, the SCB content plays a major role. Fractionation experiments and analysis, based on these experiments, of the molecular structure in relation to the thermal behavior indicated that the SCBD across the molecules may vary and that this SCBD is an important parameter. As will become clear, the nature of the SCB is also important.

The following sections discuss copolymers whose comonomer content is so high that it determines most or even all of the crystallization and melting behavior. The nature of the SCB and the way it is distributed within the chains (the intramolecular distribution) and between the chains (the intermolecular distribution) will prove to be of overriding importance.

In uncovering the molecular structure, fractionation techniques are again indispensable. For the copolymers discussed here, the molecular information will prove to be linked so strongly to the thermal behavior that in this specific case DSC can almost be used as a 'molecular fingerprinting' technique.

9.3.1 Single-peaked and multiple-peaked DSC curves

Figure 9.11 shows a typical picture of the thermal behavior of the types of copolymer discussed in the following sections [217].

For the DSC curves of the two ethylene copolymers, an EP (ethylene–propylene) copolymer and a 1-octene LLDPE (ethylene-1-octene LLDPE) [232], the crystallization and melting processes extend across large temperature ranges. The exact extent of these ranges is discussed later; reference curves will be used for their determination (chapter 5).

Another point worth noting is that crystallization and melting of the EP copolymer yields single peaks (apart from the little exothermic peak in the cooling curve just above 60 °C, which is often found at low temperatures in polymers with ethylene crystallinity (see also the discussion on LDPE)). The LLDPE, on the other hand, yields two peaks on crystallization (again ignoring the exothermic peak just above 60 °C) with maxima around 97 °C and around 108 °C, and three peaks upon melting, with one maximum around 106 °C and two maxima around 122 °C [233]. Since research (not reported here) has shown the two peaks around 122 °C to relate mainly to recrystallization behavior [234], the LLDPE may be said to have a 'low-temperature peak' and a 'high-temperature peak'.

Leaving aside for the moment the differences between the two copolymers, which are discussed extensively in the following sections, Fig. 9.11 once again illustrates that density and peak enthalpy are 'overall' quantities. After all, the two copolymers have roughly the same density (at room temperature), which is characteristic of an important class of commercial polymer grades (analogous to the LDPEs shown in Fig. 9.6; incidentally, LDPEs resemble the EP copolymer discussed here as far as their DSC curves are concerned). The fact that the peak areas in Fig. 9.11 are almost identical is in keeping with the fact that the copolymers have the same density. For

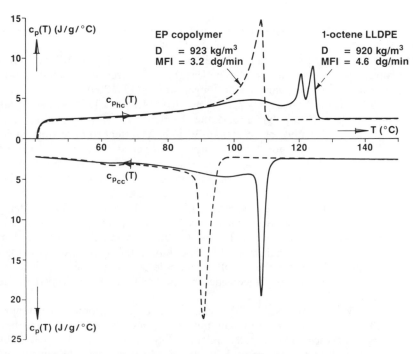

Fig. 9.11. DSC-2 continuous specific heat capacity curves for cooling $c_{p_{cc}}(T)$ at a rate of 5 °C/min, from 200 °C to 40 °C and subsequent heating $c_{p_{hc}}(T)$ at a rate of 5 °C/min: isothermal stays of 5 min, sample masses of 4.870 mg (EP) and 4.976 mg (LLDPE)

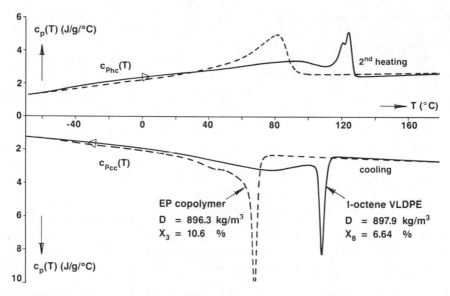

Fig. 9.12. DSC-2 continuous specific heat capacity curves for cooling $c_{p_{cc}}(T)$ at a rate of 10 °C/min from 180 °C to −70 °C and subsequent heating $c_{p_{hc}}(T)$ at a rate of 10 °C/min for an EP copolymer and a VLDPE: isothermal stays of 5 min

both types of copolymer, therefore, a good density–peak area relation can be established.

The fact that the crystallization and melting behavior of the LLDPE is character-ized by several peak temperatures makes it clear that it is necessary to study the shape of the DSC curves; overall quantities no longer suffice [151, 200].

Figure 9.12 shows two copolymers with lower densities [235]. Crystallization and melting of the EP copolymer and the 1-octene VLDPE (very low density polyethylenes have densities of less than 915 kg/m³ after compression molding [202, 218, 227, 235–238]) do not differ essentially from crystallization and melting of the copolymers of Fig. 9.11. Again, the densities and peak areas are comparable, while the crystallization and melting temperature distributions are totally different. Compared to the Fig. 9.11, peak broadening (see peak widths) and smaller heat effects (see also peak heights) are observed in Fig. 9.12. To obtain stable low-temper-ature isotherms for these copolymers it is necessary to extend the measurements to much lower temperatures, in this case to about −70 °C, whereas for the copolymers in Fig. 9.11 about 40 °C was sufficient [202]. The data in the figure also show that, to obtain the same density as for the VLDPE, the mole percentage of propylene in the EP copolymer discussed should be considerably higher than the mole percentage of octene in the VLDPE.

In Fig. 9.13, copolymers with the same type of comonomer (1-octene) are presented [202, 220], but this time the amount of comonomer has been kept approx-imately the same. Once more, totally different DSC curves are obtained, but in contrast to the previous figures the peak areas, too, differ clearly. The density of the ethylene-1-octene (EO) copolymer will therefore be substantially lower than that of the VLDPE.

258

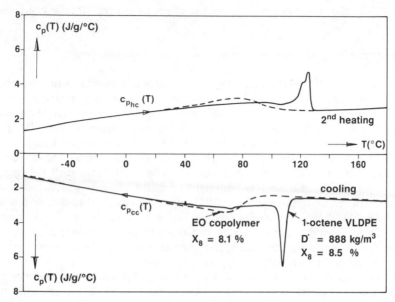

Fig. 9.13. DSC-2 continuous specific heat capacity curves for cooling $c_{p_{cc}}(T)$ at a rate of 10 °C/min from 180 °C to -70 °C and subsequent heating $c_{p_{hc}}(T)$ at a rater of 10 °C/min, for an EO copolymer and a VLDPE: isothermal stays of 5 min, sample masses of 9.259 mg (EO) and 10.005 mg (VLDPE)

The examples cited involve two types of copolymers with strongly differing crystallization and melting temperature distributions. One could, roughly speaking, distinguish between single-peaked DSC curves and multiple-peaked DSC curves. Apparently, the crystallization is hindered in markedly different ways in the two types of copolymer.

At the same amounts of 1-octene comonomer (see Fig. 9.13) the peak areas (and densities) of the two types of copolymer differ strongly in the sense that the 'single-peaked copolymer' has a much smaller peak area than the 'multiple-peaked copolymer'. The amount of comonomer to be incorporated in the molecules to achieve a certain peak area (density) is much lower for a 'single-peaked copolymer' than for a 'multiple-peaked copolymer'; one might say that the 'effectivity' of density and peak area reduction through comonomer incorporation differ strongly for the two types of copolymer [184, 239].

It should, however, be noted that this holds only if the same comonomer is used. Figure 9.12 shows that it need not hold if different comonomers are applied, for instance when propylene is used instead of 1-octene. The mole percentage of propylene required to achieve a certain density is not lower, but higher, than the required mole percentage of 1-octene.

The above illustrates clearly that the nature of the comonomer is important [3, 150, 151, 207, 220, 222, 240, 241]. For the copolymers discussed here, it appears that the occurrence of single-peaked and multiple-peaked DSC curves is to be attributed to the molecular structure, notably the distribution of the comonomer units within and between the molecules [184, 242–244]. Moreover, because of the vast temperature range across which the phenomena extend, the crystallization mechanisms may differ. For instance, on cooling, different crystallization regimes may be traversed (see

259

section 9.2.1), and a transition from folded chain to fringed micelle crystallization may occur.

9.3.2 Homogeneity and heterogeneity of comonomer incorporation

For HDPE (see section 9.2.2 and Fig. 9.6) a relation was established between the amount of SCB and the crystallization and melting temperature at a given molar mass and combination of cooling/heating rate. It was concluded that, depending on the type of HDPE, the SCB distribution among the molecules may vary. HDPEs were discussed in which the smaller molecules are not branched or hardly branched, while the larger molecules are branched. One HDPE was discussed in which, as well as the larger molecules, the smaller molecules are branched. To distinguish between these situations, the SCBD in these HDPEs might be called more or less 'heterogeneous'. By contrast, the term 'homogeneous' SCBD might then be used to denote a situation in which the SCB in all molecules is the same as regards nature, amount and distribution. It is evident that there are many possible situations that could be termed 'heterogeneous', while the definition of 'homogeneous' can be refined further, for instance by linking it to a polymerization mechanism.

On the basis of the DSC curves alone it cannot be established whether HDPE possesses a heterogeneous or a homogeneous SCBD; the curves are all single-peaked and have roughly the same shape. What one does find, though, are subtle differences in peak temperatures as a function of the branching content, peak area or density, which no doubt originate in part in the heterogeneity of the SCBD. On closer study, the shapes of the DSC curves, too, appear to differ. Of course, when interpreting the curve shape for these polyethylenes one should pay due attention to the experimental conditions, sample conditions (see chapter 4) and instrumental effects (see chapter 3).

The DSC curves of the various copolymers discussed in the preceding section are so different that one may hope that they do yield some links with the SCBD. It makes sense to assume that copolymers with multiple-peaked DSC curves exhibit substantial heterogeneity. Copolymers with a single-peaked DSC curve, on the other hand, might be homogeneous, though they need not be, since, as in the case of HDPE, it is conceivable that more or less the same crystallization and melting temperatures are found for different combinations of molar mass and SCB, resulting in a single-peaked DSC curve. Likewise, a single-peaked DSC curve may be the result of various SCBDs that are close together. It should be borne in mind that, in contrast to LPE and HDPE, in the copolymers considered here considerable peak broadening has taken place.

For copolymers with a low density ($D \lesssim 920$ kg/m^3 after compression molding), as reported here, 'homogeneous' and 'heterogenous' are defined as follows with respect to comonomer incorporation [200] (see also ref. 184).

We speak of a *homogeneous copolymer* when the way in which the comonomer is incorporated during polymerization can be described by a single set of chain propagation probabilities of (co)monomer incorporation in the chain (P set), or alternatively by the combination of a single set of reactivity ratios (r set) and a single monomer–comonomer ratio. Statistically there are no differences within and between the molecules. All other copolymers are *heterogeneous copolymers*. Two special cases are distinguished, viz. 'intramolecular heterogeneity' of comonomer incorporation when the heterogeneity is manifested within the molecules, and 'intermolecular heterogeneity' when the heterogeneity is between molecules.

The criterion given for homogeneity is a very narrow one and is hard to meet. It excludes HDPE and LDPE, even though the DSC curves of these polyethylenes are single-peaked and bear much resemblance to the DSC curves of homogeneous copolymers of the same density (see remark below).

It will be clear that the definition of heterogeneity is very comprehensive and that all kinds of combinations [210] of intramolecular and intermolecular heterogeneity are possible. Nevertheless, this simple classification already allows the formulation of a working hypothesis for the description of the crystallization and melting behavior of the copolymers discussed here. This working hypothesis is based on the following considerations.

In copolymers where the predominant monomer unit is capable of crystallization, as in this case the ethylene units, the comonomer units that are incapable of crystallization form hindering factors during crystallization. Several forms of hindrance [1, 3] are conceivable.

It is known that a propylene unit may be included to a certain extent as a defect [127, 150, 241, 245–248] in the ethylene crystal lattice at interstitial positions (inclusion). When crystallization proceeds under 'equilibrium' conditions [249], for example at low cooling rates, most [250, 251] of the 1-butene units are excluded from the crystal (exclusion). On rapid cooling, however, more of the 1-butene units do get included in the crystal as amorphous defects; it can be said that in such a case the kinetics of the crystallization process inhibit the realization of a thermodynamically more stable situation.

The longer or the bulkier the side groups originating from the comonomer units incorporated in the chain, for instance 1-hexene, 4-methyl-1-pentene, 1-octene, the greater is the probability that the comonomer is excluded from the crystals [150, 245, 252–254]. There are possibly still substitution defects in the lattice due to inclusion [241, 246, 247, 255]. Crystallization of such a copolymer is determined to a major extent, if not completely, by the comonomer distribution among the chains. In other words, the comonomer distribution in a chain, and more generally the SCBD in a chain, defines a sequence length distribution of the crystallizable monomer units [188, 256]. These sequences form the building blocks of crystals.

Sequences are qualified below using the terms '(very) short' and '(very) long'. These qualifications will remain rather vague, because the sequence length will be linked to the dimension of the crystallite eventually formed in the chain (sequence) direction, a dimension which is to a large extent kinetically determined (see Fig. 9.3).

When the sequences are 'very long', they will be capable of forming crystals of a thickness comparable to their length only under special conditions (such as high pressure). Under normal conditions there is, at a low degree of supercooling, a nucleation barrier which is too high to be overcome by the driving force of crystallization. If the temperature is sufficiently lowered, so that supercooling and thus the driving force are sufficiently increased to overcome the nucleation barrier, then nucleation and subsequent growth will be possible. However, during such supercooling the 'very long' sequences will undoubtedly be much longer than the probable initial crystal thickness at that temperature $l^*(T_c)$. It is in this sense that we call these sequences 'very long'. This is a favorable condition for folding to occur, and the way in which this happens also fixes the ultimate morphology with respect to the degree of adjacent re-entry, the number of intercrystalline links and tie-molecules [78–80, 257], etc. (see also the regime theories, section 9.2.1).

261

When the sequences are 'long', it depends on the temperature of nucleation and crystal growth whether there are sequences whose length exceeds the probable initial crystal thickness at that temperature. If there are such sequences, folding may again take place. In that case, although the sequence length distribution affects the crystallization behavior, it is not directly reflected in the crystal dimension distribution. After folding, sequences of varying lengths may be incorporated into the same crystal of a certain thickness, provided they are longer than $l^*(T_c)$, and all will melt at the same temperature. The finite sequence length may give rise to defects [252, 258, 259, 260–265] and even hindering of crystal growth in the lateral direction [264]. Under optimal conditions for equilibrium crystallization, in particular when chain mobility is increased, an unambiguous relation between sequence length and crystal dimension may be expected. In connection with this, crystallization at an increased temperature and pressure is an interesting option [4, 265–268].

When the sequences are 'short', the system will yield initial crystal thicknesses of the same order of magnitude as the length of the sequences. When the comonomer units cannot be included in the crystal, these units really inhibit the growth of the crystal in the chain direction and the crystal dimension in that direction will probably be at most equal to the length of the sequence concerned. In this case, there is a maximum chance that the crystal dimension distribution reflects, or at any rate is closely related to, the distribution of the crystallizable sequences [72, 108, 164, 269]. Lateral dimensions will be restricted.

When the sequences are 'very short', fringed micelle-like nucleation and growth become likely, especially because the crystallizability will have been reduced to such an extent that crystallization will take place far below T_m°, already in the neighborhood of T_g. In such a situation the mobility of chains and chain parts, and hence the folding capacity and the likelihood that sequences of approximately the same length find one another, will have been strongly reduced. For this reason the dimensions of the crystallites will probably be greatly limited in lateral directions also. Such a situation may also occur during heating-up from the glassy state, in which case we speak of 'cold crystallization' [270]. Longer side chains, for example C_6 or longer, originating from the comonomer, will then also play a role in the crystallization process [3].

As mentioned above, terms like 'short' and 'long' as used here are only qualitative. They are more or less defined by the description given above.

From the above discussion it will be clear that in practice, due to kinetic causes, we are far removed from thermodynamic equilibrium, in contrast to what is supposed in, for example, Flory's copolymer crystallization theory [55, 271–273].

When the sequence length distribution of crystallizable units in the chains includes both short and long sequences (in relation to $l^*(T_c)$), the presence of both folded-chain and fringed-micelle crystals is conceivable. In the literature little or no attention has been paid to such situations, in the first place because such extremely heterogeneous copolymer systems have not (yet) attracted any great interest or have only recently reached the commercial phase, and in the second place because crystallization of systems with 'short' sequences is intrinsically difficult (minor heat effects, which may be spread out over large temperature ranges), so that high experimental demands are to be met. In such cases, great caution is to be exercised in uncovering the molecular structure using crystallization fractionation. After all, a transition from one crystallization mechanism to another (e.g. folded chain to fringed micelle, possibly resulting in different crystal populations) during fractionation, for instance when the sequences are of the same magnitude as $l^*(T_c)$, may be mistaken for a difference in molecular structure.

262

The VLDPEs referred to in the preceding section possess short and long sequences, while the EP copolymers discussed in the next section contain short sequences. Besides copolymers, one may think more generally of polymers in which a sequence length distribution of crystallizable units is present because a molecular structure factor hinders the crystallization capability, as does tacticity in the case of PVC [274, 275] and many other polymers.

Precisely because the phenomenon of crystallization being hindered by 'molecular defects' in the chain is so widespread in macromolecules, studies on the homogeneity/heterogeneity of the presence of such defects in the molecules represent an important research field. As discussed above, the relation between the defect-related sequence length distribution(s) of the crystallizable units in the chains on the one hand and the crystallite size distributions on the other is of vital importance for gaining the understanding needed in the translation of molecular structure to thermal properties and vice versa. Because of its general nature, it can be supposed that what is discussed here for polymers with ethylene crystallinity can also be used for other polymers.

Figure 9.14 shows a rough classification of a number of commercial polymers with ethylene crystallinity. The crystallization capacity varies from minimum (EPDM rubber) to maximum (LPE), and the density varies accordingly. The densities given in Fig. 9.14 are no more than rough indications; LPE, for example, can approximate a density of 1000 kg/m^3 under special conditions (e.g. high pressure). All non-linear polyethylenes are grouped together under short-chain-branched polyethylenes but, on the basis of the above, a distinction is made between polyethylenes with single-peaked and those with multiple-peaked DSC curves, and likewise between polyethylenes with

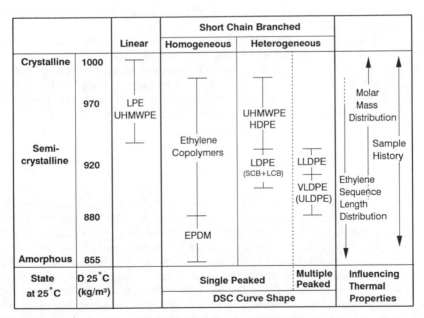

Fig. 9.14. Classification of a number of linear and branched polymers with ethylene crystallinity according to density, DSC curve shape and homogeneity/heterogeneity of incorporation of non-ethylene units (abbreviations explained in text)

homogeneous and those with heterogeneous branching distribution or comonomer incorporation.

In the case of heterogeneous ethylene polymers the nature and the origin of the SCB may vary considerably, as mentioned above when the term 'heterogeneous' was introduced. It should once again be pointed out that the homogeneous/heterogeneous distinction does not coincide with the single-peaked/multiple-peaked DSC curves distinction. This is largely due to the strict definition of homogeneity. For example, DSC curves of HDPEs and LDPEs are single-peaked but, on account of their molecular structure, these polyethylenes must be classified as heterogeneous ethylene polymers. Chapter 5 gives some examples of homogeneous and heterogeneous co-polymers of the same density ($888-920$ kg/m^3) after compression molding. If we compare the DSC curves of LLDPEs and LDPEs of the same density, the (single-peaked) DSC curves of the LDPEs suggest that they are homogeneous ethylene copolymers, while according to the definition given above LLDPE and LDPE are both heterogeneous.

UHMWPE stands for ultra-high molecular weight polyethylene, which may be produced in either a linear or a branched form. The density of the linear version may also vary; see Fig. 9.4 showing the peak area as a function of M, and ref. 276. ULDPE (ultra-low density polyethylene) is another term for VLDPE, although it has also been used for homogeneous copolymers of very low density [277].

As far as the thermal properties are concerned, Fig. 9.14 indicates by arrows the polymers for which the molar mass distribution (MMD) is an important parameter and those for which the ethylene sequence length distribution (ESLD) is important. It also gives a rough indication of overlaps in density ranges. 'Sample history' is a collective term for other important parameters that can be operative in the entire density range, for example the effects of crystallization during polymerization, crystal-lization upon stretching, the structure of the melt, nucleating agents, processing conditions, thermal history, etc.

For LPE in particular, the cooling rate is very important for the density at room temperature. For LPE NBS SRM 1484, for example (see Table 9.1), cooling at $0.2\,°C/min$, $5\,°C/min$, $50\,°C/min$ and quenching results in a density at room tempera-ture of 959, 950, 945 (compare ref. 278) and 934 kg/m^3 respectively. For LPE F(V)-22-(K) (see Table 9.1) these densities are 986, 976, 973 and 964 kg/m^3 respec-tively. The differences in density between these LPEs at the same cooling rate are due to the influence of molar mass (see Fig. 9.4).

9.4 Homogeneous copolymers

As defined above, the comonomer distribution will be considered homogeneous if the kinetics of the polymerization process can be described by a single set of chain propagation probabilities of (co)monomer incorporation in the chain (a P set) or with a single set of reactivity ratios (an r set), in combination with a constant monomer–comonomer ratio in the reactor. In the case of copolymers it implies a single value for the product of the reactivity ratios $r_m r_c$, where m stands for monomer and c for comonomer. The chain will then show a single-peak comonomer sequence length distribution and, correspondingly, a single-peak ethylene sequence length distribution. The term 'homogeneity' is further understood to imply that there are no differences between individual molecules as far as chain statistics are concerned (see Fig. 9.15 [200]).

264

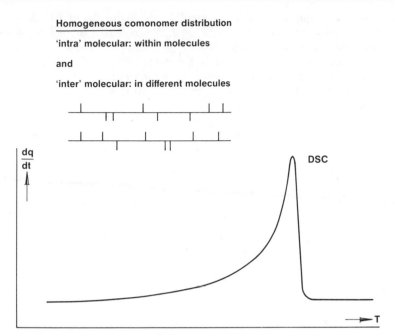

Homogeneous comonomer distribution

'intra' molecular: within molecules

and

'inter' molecular: in different molecules

Fig. 9.15. Schematically represented chain structure of a homogeneous copolymer whose homogeneity is both 'intramolecular' and 'intermolecular' and the relation between the chain structure and the shape of the DSC curve

If the comonomer units hinder the crystallization process, the distributions mentioned above can lead to a crystallite size distribution on cooling. If the crystallization is hindered to a great extent (see the comments on 'short sequences' in section 9.3.2) it is most likely that there will be a one-to-one correspondence between the distribution of crystallizable sequences on the one hand and the crystallization temperature distribution (nucleation and crystal growth and the resultant DSC cooling curve), crystallite size distribution and melting point distribution (DSC heating curve) on the other. In this way the DSC curve can reflect the sequence length distribution, and if the latter is single-peaked then the result may be a single-peaked DSC heating curve.

If, on the other hand, the DSC curve is not single-peaked, this may be an indication that the sequence length distribution is not single-peaked, provided that other possible causes such as transition to a different nucleation mechanism or reorganization/recrystallization can be excluded.

It is possible to produce homogeneous copolymers with the aid of certain catalysts [174, 184, 207, 239, 242, 279, 280], for instance those used in the production of EPDM rubber. A good example of the homogeneous copolymers that can be produced in this manner is homogeneous ethylene–propylene copolymers, which are obtained via polymerization with the aid of a (promoted) catalyst system consisting of an aluminum alkyl combined with a vanadium component [49, 239, 281].

It has recently proved possible to determine the chain structure of such copolymers with the aid of a specially developed polymerization model [282–284] based on the results of ^{13}C-NMR measurements [165, 239, 285, 286]. This determination was complicated by the fact that the propylene is incorporated into the chains in two different ways, viz. 'normally' and 'invertedly' (see definition in Fig. 9.16), which meant that a terpolymer model had to be used. The resultant first-order Markovian terpolymer model is not only suitable for EP copolymers of the kind described above, but is more generally applicable to catalyst systems with which, in the synthesis of homogeneous copolymers, one of the two units is incorporated in the chain normally as well as invertedly. Of course, copolymerization without inversion is described by the model as a special case.

The final model was successfully applied to a set of 19 EP copolymers [282, 287], yielding the required information on the microstructure of the copolymer molecules. Figure 9.16(a) gives an impression of such a microstructure.

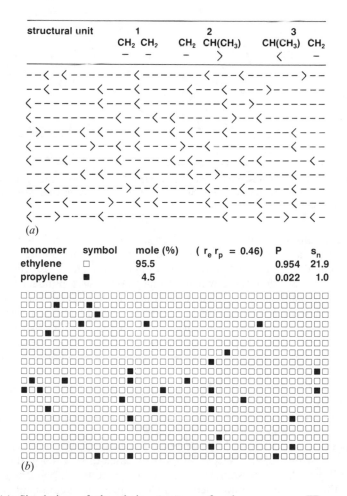

Fig. 9.16. (a) Simulation of the chain structure of a homogeneous EP copolymer with $X_e = 62\%$. Terpolymer ($--$ ethylene; $-<-$ normal propylene; $->-$ inverted propylene) presentation: the chain is constructed by linking the end of each line to the beginning of the next line. (b) Simulation of the homogeneous EP copolymer, $X_e = 95.5\%$, shown in Fig. 9.11. Copolymer presentation (ethylene and propylene): $P_{ee} = 0.954$ and $P_{pp} = 0.022$ are chain propagation probabilities; s_n is the number-average sequence length (chain construction as in (a))

Characterization of the ethylene–propylene copolymer EJ 157 yielded an ethylene mole percentage X_e of 62 %, an $[\eta]_{dec}^{135} = 2.90$ dl/g and an $M_n = 88$ kg/mol. The chain structure of the EP copolymer has been simulated with the aid of a computer via a Monte Carlo process. The chain growth follows from the generation of random numbers between 0 and 1 and comparison of these with the chain propagation probabilities (P set). These probabilities P_{ij} relate to the addition of a j unit to a chain ending in an i unit, with the possibility of each of units i and j being 1, 2 or 3 (see Fig. 9.16(a)). P_{ij} is defined by r_{ij} and the ratio of the concentrations of i and j during polymerization. $P_{23} = 0$ because with vanadium catalyst systems the 23 sequence never occurs in practice. As only a limited number of units can be represented in Fig. 9.16(a), qualitative use only is illustrated. The results of the evaluation of the sequence lengths in the 'chain' after generation of a sufficiently large number of 'monomeric' units always agree completely with those obtained analytically. Although the number of simulated units is small, it is clear that the methylene sequences formed by linked CH_2 units are short. Less than 5 % of the sequences are longer than 10 units; less than 1 % are longer than 20 units.

If the molecular microstructure has been analyzed and the r set has been determined in the manner described above, we can study specific situations. In Fig. 9.16(b) the EP copolymer of Fig. 9.11, where $X_e = 95.5$ %, is represented as a copolymer, which means that a propylene unit shown in the figure may represent either a normal or an inverted propylene. A 'pseudo'-$r_e r_p$ can also be calculated for a specific sample, by pretending that copolymerization of ethylene and propylene units takes place, ignoring the fine structure that is the result of inversion of the propylene unit [288]. In this case $r_e r_p = 0.46$ was calculated.

Figure 9.16 shows that for a certain polymerization system, in this case homogeneous polymerization characterized by a single r set, a tremendously wide range of chain microstructures can be obtained by varying the monomer–comonomer ratio of the feed. It goes without saying that the effect of this variation on the crystallization behavior and various properties will already be substantial in the case of the incorporation of only a small amount of comonomer. In the case of a large amount of comonomer it will have an overriding effect.

Figure 9.17 shows the ethylene and propylene sequence length distributions $w_{s:e}$ and $w_{s:p}$ respectively for EJ 157. s:e and s:p stand for ethylene and propylene

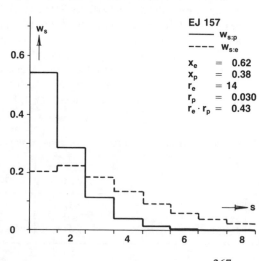

EJ 157

$w_{s:p}$ (solid line)
$w_{s:e}$ (dashed line)

$X_e = 0.62$
$X_p = 0.38$
$r_e = 14$
$r_p = 0.030$
$r_e \cdot r_p = 0.43$

Fig. 9.17. Ethylene $w_{s:e}$ and propylene $w_{s:p}$ sequence length distributions for the EP copolymer of Fig. 9.16(a)

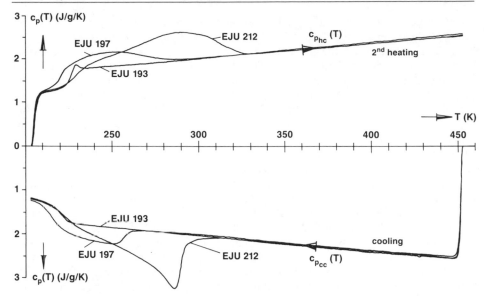

Fig. 9.18. DSC-2 continuous specific heat capacity cooling curves $c_{p_{cc}}(T)$ and subsequent heating curves $c_{p_{hc}}(T)$ for three homogeneous EP copolymers with various ethylene mole fractions: crystallization, (de)vitrification and melting phenomena; scanning rates of $-/+10$ K/min between 203 K and 453 K (isothermal stays of 5 min, sample masses of between 8 mg and 17 mg)

sequences of length s, respectively; $w_{s:e}$ and $w_{s:p}$ stand for the mass fractions of s:e and s:p. Also indicated are the ethylene mole fraction x_e, the propylene mole fraction x_p, the reactivity ratios r_e and r_p and the product $r_e r_p$.

It may be hard to imagine that a copolymer like EJ 157 can still crystallize, with its methylene and ethylene sequences being so short [150, 164, 289]. Figures 9.18 and 9.19 show that the crystallization capability of EP copolymers with an ethylene content of $\sim 60\%$, polymerized with the aid of this specific catalyst system, is indeed minimal [281, 290, 291]. Sample EJU 193 (see Fig. 9.18) is amorphous and shows a glass transition only; EJU 197 and EJU 212 crystallize and melt gradually, and these processes partly coincide with the (de)vitrification. Possibly, part of the sequences capable of crystallization in these samples cannot crystallize due to vitrification.

Table 9.4 gives the ethylene content, the intrinsic viscosities and the average molar masses of the three ethylene–propylene copolymers mentioned [49, 284].

What has been said in section 5.2.5 on the determination of the transition enthalpy from melt heat capacities is very aptly illustrated here. We see that the melt

Table 9.4. Molecular structure data of some ethylene–propylene copolymers

	X_e %	$[\eta]_{dec}^{135}$ dl/g	M_n^* kg/mol	M_w^* kg/mol	M_z^* kg/mol
EJU 193	45	0.37	5.7	15.4	26
EJU 197	65	2.40	78	190	350
EJU 212	75	1.48	41	91	160

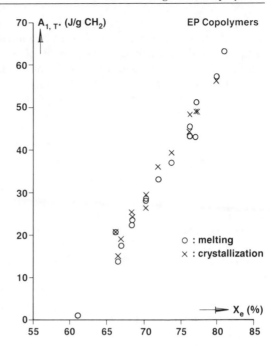

Fig. 9.19. Peak areas for crystallization and melting obtained from DSC-2 cooling and heating curves at scanning rates of $-/+5$ K/min, A_{l,cc,T^*} and A_{l,hc,T^*}, respectively, for homogeneous EP copolymers. The effect of the ethylene mole percentage. Sample masses varied between 19 and 28 mg

heat capacities $c_{p_a}(T)$ can be excellently described by a straight line with a positive temperature coefficient, as is usually the case with polymers (see section 5.1.4), and that the heat capacities of the amorphous EJU 193 can be described by a straight line over more than 250 K. In the case of the EP copolymers mentioned above, the $c_{p_a}(T)$ (in J/g per K) is not significantly dependent of the ethylene–propylene ratio. The example shows that, as such, extrapolation from the melt, giving $c_{p_l,extr \to T}(T)$, is a suitable method for estimating $c_{p_a}(T)$ (see section 5.2.5).

The crystallization and melting peak areas were determined via extrapolation from the melt. In view of the fact that it is not possible to distinguish the glass transition regions from the crystallization and melting region of the samples in question, the extrapolation was each time performed to the point T^* at which the straight line according to the extrapolation from the melt intersects the DSC curve. In this way the peak areas A_{l,T^*} (see section 5.2.5) were determined, with T^* differing for different copolymers. Figure 9.19 shows the result for a series of homogeneous EP copolymers that were produced—using the same catalyst—under slightly different polymerization conditions from the copolymers discussed above. The peak areas, expressed in joules per gram of methylene units (from ethylenes and propylenes), show a simple correlation with X_e. Crystallization and melting lead to the same results, if calculated in the manner described above. In the case of ethylene contents of less than ~ 60 mol.%, vitrification will prevent crystallization, if the ethylene sequences concerned have any crystallization capability at all [278, 289].

It is not easy to translate the molecular microstructure quantitatively into crystallization temperature distribution, crystallite size distribution and melting point distribution. So far, analytical models [2, 3, 55, 117, 164, 270–273, 292–308] have yielded useful information. Another promising way of tackling this problem is via computer simulation with the aid of Monte Carlo methods [309–311].

269

The molecular microstructure determined with the aid of the polymerization model described above can be used as a point of departure for simulations. As an example, we shall briefly describe a model that is currently being developed [312] for simulation of the crystallization of homogeneous ethylene–propylene copolymers with 'short' (see the previous section) crystallizable ethylene sequences.

First, it is assumed that only the ethylene units in the chains can take part in the crystallization process and that propylene units form defects in the lattice in interstitional places [241]. On the basis of an energy balance criterion on a monomer-unit scale, a Monte Carlo procedure decides whether or not an ethylene unit crystallizes or melts: an on-event or off-event respectively. This principle ensures that the system moves towards a local thermodynamic equilibrium. The timescale on which on–off events take place is assumed to be large compared to that of individual chain-unit movements in the melt.

The parts of the chains that are in the amorphous state are phantom, i.e. they do not occupy lattice sites. Whenever the status of an ethylene unit is altered from amorphous to crystalline its position is fixed at a certain lattice coordinate. If in this way the ethylene unit becomes attached to an already existing crystallized ethylene unit, its orientation is the same as that of the already existing unit. Each crystallized ethylene unit has an orientation along one of the three principal axes. A propylene unit is always phantom. By the same reasoning, two crystallized ethylene sequences can merge crystallographically only if they end up with the same orientation.

The free enthalpy difference between the crystalline state and the melt (which is the reference state) consists of three components. The first component is the driving force of melting, $\Delta g(T) = g_a(T) - g_c(T)$. This free enthalpy differential function is extensively discussed in section 5.2.2 and is in fact a volume term: in the case of completely and perfectly crystallizing systems it is assumed that there are no surface terms.

In the case of real systems, however, allowance must be made for surface effects. The second component is hence an end-surface (longitudinal; in the direction of the chain) and a side-surface (lateral; across the chain [115]) term.

The third component is due to the conformational entropy of the part of the chain that is in the amorphous state and is attached to the crystallized part of the chain [2, 97, 258, 313–316]. This important component, which is often ignored, is accounted for analytically by assuming that the chain parts in the amorphous state behave like unperturbed flexible (Gaussian) chains. Since the crystallization timescale is much shorter than that of full relaxation of long chains, it is assumed that the contribution of chain parts to the entropy levels at a certain part length l_{max}.

Figure 9.20 gives an impression of how the crystallized ethylene units are arranged spatially. The simulation, with $l_{max} = 10$, was performed on a $(20 \times 20 \times 20)$ lattice with periodic boundaries for a system of 20 polymer chains, each consisting of 400 monomer units with an ethylene content of 65 %. The crystallized units of one of the polymer chains are highlighted, and the connecting amorphous (phantom) chain parts are represented by straight lines just to indicate their order in the polymer chain. Obviously, the connecting amorphous chain parts can be much longer than the straight lines shown.

Further research will have to be done to refine and quantify the model (for instance, the final percentage crystallized ethylene units, about 60 %, in Fig. 9.20 and the percentage of the maximum attainable lowering of the energy reached, about 35 %, are unrealistically high values) and to enable translation of the model parameters into

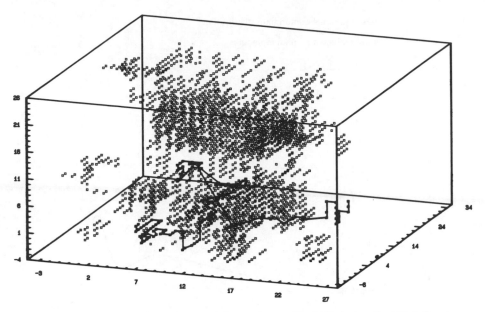

Fig. 9.20. An example of a spatial arrangement of crystallized ethylene units (■) belonging to 20 EP copolymer chains, each consisting of 400 monomer units and having an ethylene content of 65 %, obtained via Monte Carlo simulation of the crystallization process: a single EP copolymer chain is highlighted, the phantom amorphous chain parts are represented by straight lines

experimental quantities such as temperature and c_p. The model does not include vitrification processes.

The model can also be made suitable for copolymers with different monomer units or for homopolymers with different chain microstructures, for example differences in tacticity. In principle, it should also be possible to model heterogeneous copolymers, provided that the crystallizable sequences are 'short'.

9.5 Heterogeneous copolymers

This section discusses ethylene copolymers with a more complex ethylene sequence length distribution than those dealt with above. A multiple-peaked DSC curve of the kind illustrated in Fig. 9.21 may be an indication of heterogeneous comonomer incorporation in the molecules [184, 188, 200, 207, 233, 317–319] if the peaks are not too close to one another and if these peaks are not caused by reorganization/recrystallization, etc. [233, 320, 321]. The curve bears a close resemblance to LLDPE and VLDPE curves (see Figs. 9.11–9.13); we shall use these ethylene copolymers to show how fractionating techniques and DSC can lead to a clarification of molecular structures and an understanding of thermal behavior.

Two special cases of heterogeneity have already been mentioned, viz. 'intramolecular' and 'intermolecular' heterogeneity. Figure 9.21 schematically represents the corresponding molecule parts and suggests a link with the DSC curve. Having established the relation between the structure of a molecule and crystallization

271

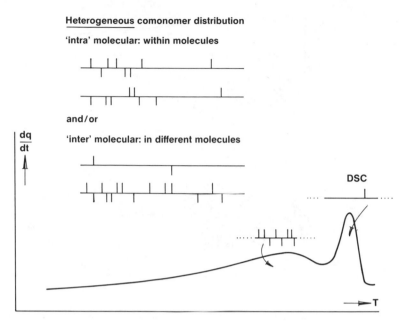

Fig. 9.21. Two examples of heterogeneity of comonomer distribution and a possible explanation for the shape of the DSC curve

behavior (see section 9.3.2), we can now set up a working hypothesis based on the assumption that the two DSC peaks (two melting point distributions) are the result of two different crystallite size distributions, which are attributable to the different crystallization behavior of molecule parts relating to (at least) a bimodal ethylene sequence length distribution. This may be the result of, for example, a polymerization in which use is made of a catalyst with (at least) two reactive sites [151, 184, 197, 216, 228, 288, 322–335]. However, other causes are also conceivable, including causes associable with specific polymerization conditions [210]: a variable monomer–comonomer ratio, the use of several reactor units, etc. Obviously, a crystallization model should allow for interactions between ethylene sequences originating from different parts of the distribution.

In Fig. 9.21, the 'low-' and 'high-' temperature DSC peaks could indicate, for instance, molecule parts with '(very) short' and '(very) long' sequences respectively (see section 9.3.2), resulting from the incorporation of '(very) many' and '(very) few' comonomer units respectively. The division into 'intramolecular' and 'intermolecular' heterogeneity of comonomer distribution leaves room for a wide variety of mixed forms and for differences in distribution between short and long molecules.

In the case of 'intermolecular' heterogeneity, the obvious approach is to try to separate strongly branched molecules from weakly branched molecules.

In order to throw light on the structure of some heterogeneous copolymers, we now show some results obtained with fractionation methods that are rather different in character, viz. direct extraction (DE); preparative size exclusion fractionation (PSEC); crystallization/dissolution fractionation and temperature rising elution fractionation (TREF).

9.5.1 Fractionation according to molecular dimension

Figure 9.22 shows SEC curves for a 1-octene LLDPE and some of its fractions obtained via DE [200]. A good separation according to molecular dimension was achieved, which offered the possibility of studying the comonomer content of LLDPEs as a function of this parameter. In the absence of LCB, a separation according to molecular dimension implies a unique separation according to molar mass; see section 9.2.

Figure 9.23 shows the results of such a study [200, 201]. For a 1-butene LLDPE (gas-phase process) and a 1-octene LLDPE (solution process) having the same density after compression molding and melt flow index [Fig. 9.23(a)], the mole percentages of incorporated comonomer were determined by means of infrared

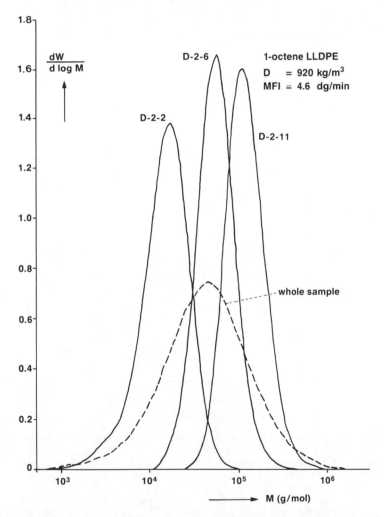

Fig. 9.22. Normalized SEC curves for an 1-octene LLDPE (same as in Fig. 9.11) and fractions thereof obtained via direct extraction

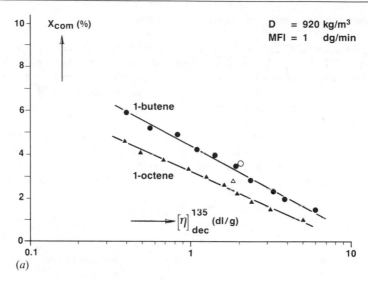

(a)

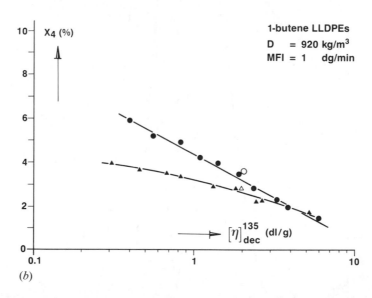

(b)

Fig. 9.23. (a) Mole percentages of comonomer and intrinsic viscosities of a gas-phase-process 1-butene LLDPE ($\bigcirc$) and a solution-process 1-octene LLDPE ($\triangle$) and direct-extraction fractions thereof ($\bullet$ and $\blacktriangle$ respectively). For the 1-butene LLDPE, $[\eta]_{dec}^{135}$ (dl/g) and X_4 (%) are 2.00 and 3.56 respectively for the whole polymer and 2.04 and 3.63 respectively for the sum of fractions. For the 1-octene LLDPE, $[\eta]_{dec}^{135}$ (dl/g) and X_8 (%) are 1.90 and 2.84 respectively for the whole polymer and 1.84 and 2.80 for the sum of fractions respectively. (b) As in (a), for the same gas-phase-process 1-butene LLDPE ($\bigcirc$, $\bullet$) and a high-pressure-process 1-butene LLDPE ($\triangle$, $\blacktriangle$). For the high-pressure process 1-butene LLDPE $[\eta]_{dec}^{135}$ (dl/g) and X_4 are 1.96 and 2.61 respectively for the whole polymer and 1.85 and 2.84 for the sum of fractions

274

measurements on whole (i.e. unfractionated) samples and on the fractions obtained via DE.

The mole percentages are shown here as a function of the intrinsic viscosity $[\eta]_{\text{dec}}^{135}$, not of the molar mass. The two quantities are related by the Mark–Houwink equation, taking into account the comonomer content and the width of the molar mass distribution [49, 156, 167, 194].

As can be seen in Fig. 9.23(a), the comonomer content of the fractions increases considerably as $[\eta]$, or the molar mass, decreases [195, 197, 198, 225]. The fractions of the 1-octene LLDPE contain less comonomer over the whole of the molar mass range than the fractions of the 1-butene LLDPE. The same can be said of the whole samples, although they have the same density. It would be too simple, however, to ascribe this phenomenon exclusively to differences in crystallization behavior caused by differences in the side-chain lengths.

That there are also other causes is seen in Fig. 9.23(b), which compares the same gas-phase-process 1-butene LLDPE with a high-pressure-process 1-butene LLDPE. These products also have the same density, but different 1-butene contents. This is partly attributable to a great difference in comonomer distribution over the molar masses. At the same time there may be differences with respect to chain statistics [213].

These results suggest that we already know whether the heterogeneity observed is intramolecular and/or intermolecular, since the low-temperature peak in the DSC heating curve (see Fig. 9.24) could be attributable [200] to strongly branched material of low molar mass and the high-temperature peak to weakly branched material of high molar mass. The DSC curves of such fractions would consequently possess only one peak, like, for example, the curves shown in Fig. 9.24.

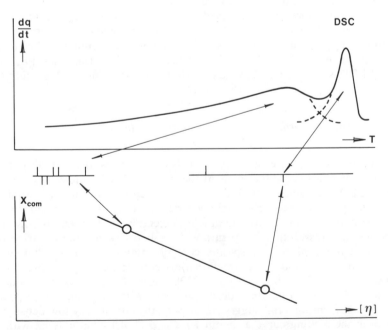

Fig. 9.24. An (incorrect) explanation for the shape of the DSC curve

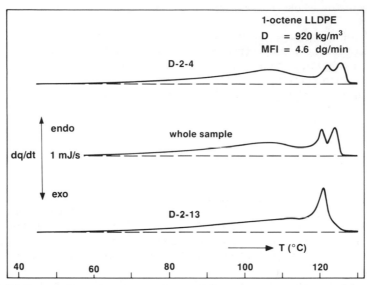

Fig. 9.25. DSC curves for an 1-octene LLDPE (as in Figs. 9.11 and 9.22) and two direct-extraction fractions thereof for heating at a rate of 5 °C/min after cooling from 150 °C at a rate of 5 °C/min: sample mass 0.8 mg

However, this reasoning is not borne out by the facts. Figure 9.25 [200] shows that a low molar mass fraction (D-2-4) and a high molar mass fraction (D-2-13) of a 1-octene LLDPE still possess the characteristic long melting ranges and produce DSC curves with more than one peak. The DSC curves suggest that for each direct-extraction fraction in Fig. 9.23, the comonomer content is still an average of a distribution, which means that there is heterogeneity of comonomer distribution over the entire molar mass range [151, 192, 200, 222]. This also means that in comparing different LLDPEs (as in Fig. 9.23) the heterogeneity at each molecular length must be taken into account.

9.5.2 Fractionation according to crystallizability/dissolvability

One might try to separate strongly branched molecules from weakly branched molecules, leaving aside for the moment the lengths of these molecules. Crystallization of an LLDPE in solution followed by drainage of dissolved material at various increasing temperatures gives rise to a good relation (Fig. 9.26) between the dissolution temperature T_d and the average mole percentage of 1-butene for the fraction obtained at that dissolution temperature. The linearity or non-linearity of such relations is determined by the specific crystallization/dissolution behavior of the polymer in question under the fractionating conditions chosen [192, 208, 209, 211–213, 215, 218, 219, 221, 224–226, 229]. Figure 9.27 shows diagrammatically [200] that such a fractionation indeed leads to fractions giving single-peak DSC curves, varying from material suggestive of polyethylene of very low density, with a maximum in the melting curve at about 80 °C, to HDPE-like material with a peak temperature of 132 °C. The fact that it is possible to obtain fractions producing such

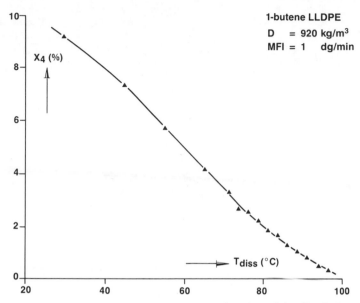

Fig. 9.26. Average mole percentages of 1-butene as a function of the dissolution temperature determined for fractions of a gas-phase-process 1-butene LLDPE (same as in Fig. 9.23) obtained via a crystallization/dissolution method

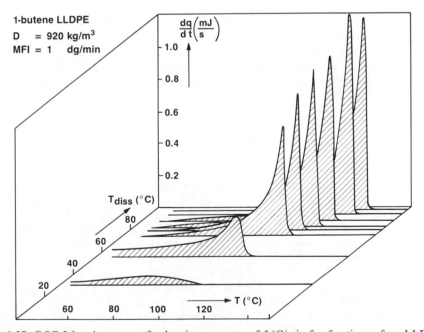

Fig. 9.27. DSC-2 heating curves for heating at a rate of 5 °C/min for fractions of an LLDPE (same as in Fig. 9.26) obtained via a crystallization/dissolution method

277

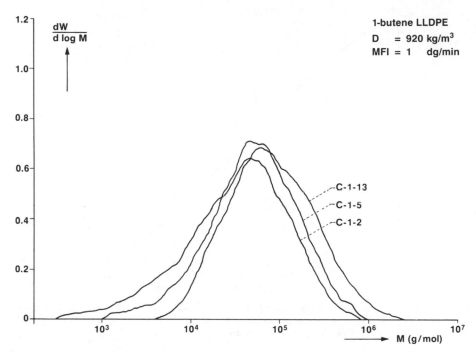

Fig. 9.28. Normalized SEC curves for some fractions of the LLDPE shown in Fig. 9.27

DSC curves proves that the heterogeneity of the comonomer distribution is of an intermolecular nature [200, 222, 227].

Figure 9.28 gives the SEC curves for a number of fractions [200]. The molar mass distributions are broad and there is evidence that fractionation according to molar mass has also taken place. However, this was to be expected considering the type of method used—crystallization on a stirrer [185] covered with hemp and subsequent dissolution [230].

9.5.2.1 Heat capacity, enthalpy and crystallinity

With the methods described in chapter 5, the copolymers can be analyzed further. This is illustrated below using the results of the heat capacity measurements on the 1-octene VLDPE with density 888 kg/m³ (see Fig. 9.13) as a point of departure [235].

On the analogy of the procedure described in section 5.2.4, the measured specific heat capacities $c_{p_{cc}}(T)$ and $c_{p_{hc}}(T)$ can be compared with the reference curves for completely crystalline polyethylene $c_{p_c}(T)$ and completely amorphous polyethylene $c_{p_a}(T)$. In the melt, $c_{p_{cc}}(T)$ and $c_{p_{hc}}(T)$ closely resemble $c_{p_a}(T)$; only at the lowest temperatures do these curves lie between the $c_{p_c}(T)$ curve and the $c_{p_a}(T)$ curve, which means that, on the assumption that the two-phase model applies here, virtually no crystallization or melting takes place at these temperatures. Obviously, crystallization and melting take place over extremely wide temperature ranges: more precisely, between about 120 °C and about −60 °C and between about −60 °C and about 130 °C respectively.

The assumption that the two-phase model applies here is based on the results of an evaluation of a series of VLDPEs [202]. The results of the DSC measurements and the analysis of these results with the aid of the two-phase model were internally consistent, which meant that there was no clear need to assume a third phase, for instance a 'rigid amorphous' phase (see section 5.3.4). The crystalline phase may, incidentally, also be regarded as well-defined, since virtually all the 1-octene groups will be outside the crystals.

The $c_{p_{cc}}(T)$ and $c_{p_{hc}}(T)$ can be integrated to $h_{cc}(T)$ and $h_{hc}(T)$ curves. Figure 9.29 shows the $h_{hc}(T)$ curve for the VLDPE after numerical integration of $c_{p_{hc}}(T)$ and calibration at 141.4 °C, 779 J/g [336, 337]. Also given are the reference curves $h_c(T)$ and $h_a(T)$, which were obtained via integration of $c_{p_c}(T)$ and $c_{p_a}(T)$ respectively.

These data can be used to calculate the enthalpy-based mass crystallinity heating curve $W_{hc}^c(T)$, according to Eq. 5.43

$$W_{hc}^c(T) = \frac{h_a(T) - h_{hc}(T)}{h_a(T) - h_c(T)} \cdot 100\%$$

The $W_{hc}^c(T)$ curve (see right-hand axis) shows a continuous decrease in $W_{hc}^c(T)$ from −60 °C onwards. Its value of about 25 % at 23 °C is considerably lower than its value of about 37 % at −60 °C. The highest melting temperature is about 130 °C, which means that the entire melting range covers no less than about 200 °C. The crystallization and melting ranges are hence extremely wide. When carrying out measurements or sample manipulations in the corresponding temperature range (for example X-ray

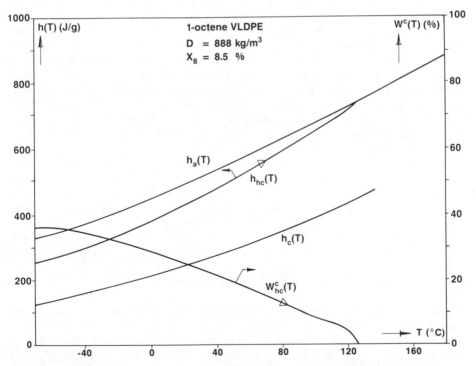

Fig. 9.29. Specific enthalpy heating curve $h_{hc}(T)$, reference specific enthalpy curves $h_a(T)$ and $h_c(T)$ and the enthalpy-based mass crystallinity heating curve $W_{hc}^c(T)$ for a VLDPE

279

measurements at room temperature or interpretation of electron micrographs after fixation/staining at the same temperature or a higher temperature) it must not be forgotten that there is a non-crystalline fraction at every temperature, which may yet crystallize on cooling. Another conclusion that can be drawn from the above is that for a complete thermal evaluation of such a VLDPE, it is necessary to choose isotherms below $-60\,°C$ and above $130\,°C$.

9.5.2.2 Crystallinities as a function of temperature *vs.* one-point measurements

A comparison [202] of the enthalpy-based mass crystallinities of VLDPEs with their volume-based mass crystallinities at $23\,°C$ would suggest that they agree, at least within the measuring accuracy of the two analytical methods. The absolute maximum deviation for seven VLDPEs and one LLDPE, whose density varied, at $23\,°C$, from $888\,kg/m^3$ to $919\,kg/m^3$, was $3\,\%$. For the VLDPE considered here these values are $26\,\%$ and $25\,\%$ respectively at $23\,°C$. This could mean that the samples in question show no appreciable deviations from the two-phase model. Nevertheless, we must be very careful in drawing such conclusions.

There are a number of preconditions that must be met before the results obtained according to the calorimetric method mentioned above [257, 292, 338–342] may be compared with those obtained according to other methods [298, 343] based on density [3, 128, 150, 151, 174, 344–346], IR [347], NMR [345, 348], Raman [150, 159, 174, 345, 346], X-ray [151, 160, 207, 248, 269, 345, 349–351, 372] and other measurements. First, the thermal history must of course be the same. That is not the case here: the compression-molding procedure employed in preparation for the density measurement is not the same as the DSC cooling procedure. Furthermore, a comparison with a one-point measurement (as in this case, at $23\,°C$) is of course very risky and extremely unsatisfactory, particularly in the case of copolymers that still contain a substantial fraction of crystallizable matter at the measuring temperature.

For a reliable comparison, within the two-phase model, of results obtained with different analytical methods it is necessary to determine 'corresponding' crystallinities as a function of the temperature [159, 160, 163, 349, 352–354]. This requires much know-how and, unfortunately, also involves a lot of work, but it is not possible to draw reliable conclusions (sometimes even qualitatively) from similarities and dissimilarities without carrying out such a determination.

What has been stated above for the comparison of results obtained with different methods also holds for the comparison of DSC results obtained from different samples and/or from samples with different thermal histories. The same procedure as indicated above must be followed in this case too. It is essential to use reference states (see section 5.2 for an explanation). It is also very important to perform a total analysis (see section 5.3.4) of $c_p(T)$ as well as of $h(T)$ and $W^c(T)$ in order to be able to check the internal consistency of the results obtained with the calorimetric method.

9.5.2.3 Heat capacities and crystallinities of fractions

As mentioned above, the multiple peaks of the DSC specific heat capacity curves suggest that the VLDPEs are heterogeneous copolymers. If the heterogeneity is predominantly intermolecular, as appeared to be the case with LLDPE, then fractionation via crystallization/dissolution is again a suitable method of research [202, 237].

5 g of the VLDPE sample were dissolved, with stirring, in 1.5 l of xylene at $120\,°C$. After 4 h the stirring was stopped and the material was crystallized from the

280

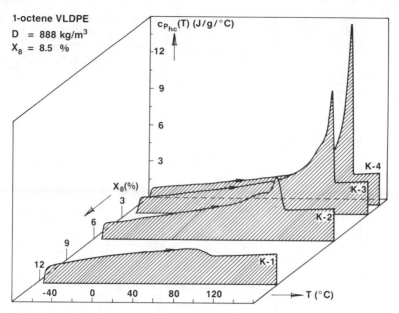

Fig. 9.30. DSC-2 specific heat capacity heating curves as a function of the mole percentage of 1-octene for fractions of a VLDPE obtained via a crystallization/dissolution method

solution by cooling to room temperature at a rate of less than 1 °C/min. Next, the fractions were drained off, first at room temperature (K-1) and then, after heating with stirring, at 68 °C (K-2), 91 °C (K-3) and finally 110 °C (K-4). The yields ranged from 0.3 g to 3 g. DBPC was used as an antioxidant.

Figure 9.30 includes the specific heat capacity curves $c_{Phc}(T)$ of the fractions, as obtained from the DSC heating curves, plotted for the mole percentages of 1-octene X_8 (%) in question. The curves have been plotted up to 160 °C, although the measurements were carried out up to 180 °C. The curves of the fractions are virtually all single-peaked, unlike the curve obtained for the whole sample (see Fig. 9.13). In the case of the first fraction obtained, K-1, the highest melting temperature is ~ 85 °C and the material in the whole polymer that causes the high-temperature peak is absent, whereas in the last fraction obtained, K-4, with a highest melting temperature of more than 130 °C, the low-temperature peak is absent. The behavior of the other fractions lies between these two extremes. As was also established in the case of LLDPE (see Fig. 9.26) there is an unambiguous connection between the dissolution temperature, the highest melting temperature and the average comonomer content (although this connection differs from that for the LLDPE).

The results are characteristic of intermolecular heterogeneity. Since, in addition, the first fraction represents no less than 66 (m)% of the material, it follows that the heterogeneity of VLDPE is largely, if not wholly, intermolecular. This means that VLDPE, like LLDPE, should be regarded as a blend of ethylene copolymer molecules with widely different comonomer incorporation percentages, the blend having been synthesized in the reactor. In fact, a range of molecules, from HDPE-like molecules to material with a very low crystallinity or even amorphous material, seems to be present. Further work will have to be done to demonstrate intermolecular heterogeneity also in the material melting at low temperatures, in particular by means of

281

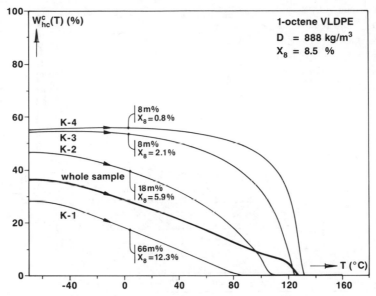

Fig. 9.31. Enthalpy-based mass crystallinity heating curves for the VLDPE shown in Fig. 9.30, based on the heat capacity curves shown there

low-temperature fractionation of this portion of the material, which, especially in terms of mass percentage and comonomer content (more than 12 mol-%), is an important portion.

The crystallinity curves [237] shown in Fig. 9.31 for the VLDPE and its fractions again clearly illustrate that the fractions differ greatly in melting behavior and crystallinity. At room temperature, about a quarter of the material (in terms of mass) is crystalline.

9.5.2.4 Morphology

In view of the relation of crystallite size distribution and melting point distribution, qualitative conclusions regarding the morphology can often be drawn from a DSC curve. Multiple-peaked DSC curves often imply a multiple-peaked crystallite size distribution, provided that the peaks do not lie too close to one another. This is a well-known phenomenon in the case of LLDPE; the contrast between the morphology of this heterogeneous copolymer [355] and that of homogeneous ones [264] is illustrative.

The blend character of VLDPEs is reflected in the morphology in a peculiar way, as revealed by transmission electron microscopy. Some VLDPEs examined by electron microscopy [236, 237, 356] showed a morphology in which regions with greatly different crystallinities were observed. In the case of a VLDPE with a density of 903 kg/m^3 after compression molding, dispersed regions of very low to zero crystallinity were visible [357, 358], and the material concerned could be extracted [358]. In the VLDPE with a density after compression molding of 888 kg/m^3 these regions have formed a continuous phase. The complementary phase consists of nodular entities (compact semi-crystalline domains or CSDs) which are interconnected and thus form a continuous phase as well. Figure 9.32 shows thick (13–14 nm) and relatively long (0.5–1 μm) lamellae located within CSDs, connecting CSDs and located in the surrounding low-crystalline phase. They also occur as single lamellae and are thought to be related to the HDPE type of material in VLDPE. On these lamellae, small lamellar crystallites can be observed.

The (co-continuous) phase outside the CSDs and lamellae seems to be structureless. However, it should be borne in mind that at the temperature at which the fixation/staining treatment [258, 359, 360] was performed (45 °C), an appreciable portion of the material was in the molten state. Moreover, crystals smaller than 3–4 nm cannot be made visible with the fixation/staining technique used. This means the phase that seems to be structureless may consist of uncrystallizable to poorly crystallizable material.

The morphology of the VLDPE under consideration is thus characterized by distinct co-continuous regions of highly different crystallinity. Internally, the CSDs show a mixture of the above-mentioned long lamellae and shorter ones (about 0.15 μm), which are also slightly thinner (down to 6–7 nm). Besides segregation via crystallization, thermodynamic demixing in the melt is being studied as a possible cause of the occurrence of CSDs, on the basis of evidence obtained for blends of molecules differing in branching [260, 356, 361–364].

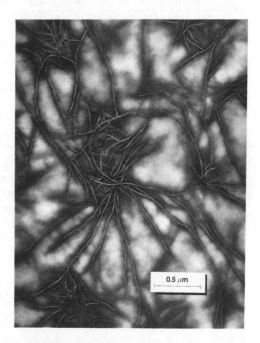

Fig. 9.32. Morphology as revealed by TEM of the VLDPE of Figs. 9.29–9.31; after staining/fixation via intensive chloro-sulphonation: compact semi-crystalline domains (CSDs) with lamellae inside and between them

9.5.2.5 Molecule interactions

The crystallization curve of the fraction with the lowest comonomer content (K-4) starts at a higher temperature than that of the whole sample. The corresponding melting curve has a single peak, and melting also ends at a higher temperature than in the case of the whole sample (see Fig. 9.31). This increase in crystallization and melting temperature [151, 225] is not surprising, given the fact that the whole sample should be regarded as a blend [365]. After all, in the reverse situation (blending), the

283

least branched component will show a lower crystallization and melting temperature in the blend than in isolation [126, 363, 364, 366–373].

These phenomena reflect interactions between molecules. Of course, they complicate the interpretation of DSC curves describing crystallization from the melt and the subsequent melting of these non-fractionated copolymers which are to be regarded as multicomponent blends consisting of molecules that differ in terms of length as well as in branching content and branching distribution. The multicomponent nature and the molecule interactions are reflected in the crystallization behavior, the resulting crystallite size distribution and the final melting point distribution.

Fractionation according to molecular dimension is a prerequisite in this case. This procedure, which is not difficult to perform using SEC (or DE in specific cases), can also be regarded as a useful way of preventing the occurrence of some of the aforementioned intermolecular interactions. The obvious next step, cross-fractionation according to crystallizability/dissolvability to determine the branching content and distribution, is surrounded by much more uncertainty, as the results of the measurements can be difficult to interpret. An advantage of fractionation by crystallization/dissolution is that the experimental conditions (quality of the solvent, concentration, time–temperature scheme, etc.) can be chosen so that the molecules can operate relatively independently of one another, which also minimizes intermolecular and intramolecular interactions [218, 224].

The character of the blend, and hence the interactions that are likely to take place between molecules and molecule parts, can be inferred from melting point depressions but also from the results of calorimetric measurements of crystallization in solution. Compared with crystallization from the melt, in crystallization in solution branched molecules can be hindered much less by intramolecular and intermolecular interactions and, as a result, crystallize much better [374, 375]. In such a case, their heat effects will be greater when they are heated. An example is LDPE.

Figure 9.33 shows that in low-concentration solutions of heterogeneous copolymers, too, it is possible to measure crystallization and dissolution [376–378]. The cooling curve and a second heating curve shown in Fig. 9.33 were obtained for the VLDPE in xylene with the aid of a C 80. The results were not optimized by corrections based on empty-cell measurements. The concentration of the solution was about 1 %. The curves obtained in the measurements in solution are also multiple-peaked, as in the case of measurements in the melt (see Fig. 9.13). The dissolution curve shows the 'high-temperature' peak (from about 87 °C to about 102 °C) and the 'low-temperature' peak (up to about 87 °C). A striking feature is the substantial amount of recrystallization observable between about 87 °C and about 102 °C. The calorimeter used is not suitable for measurement at still lower temperatures, hence it was impossible to study the entire crystallization and dissolution process (see section 9.5.2.6).

9.5.2.6 Use of microcalorimetry in fractionation

The curves in Figs. 9.33 and 9.10 show that the microcalorimeter is capable of providing background information for fractionation. If the specific information shown in these figures is needed, especially under realistic fractionating conditions as in Fig. 9.10, the usual DTA [379] and DSC instruments are not sufficiently sensitive, because of the extremely slow cooling and heating rates involved, the extremely low concentrations of polymer in solution and the wide temperature range over which the heat effects take place.

284

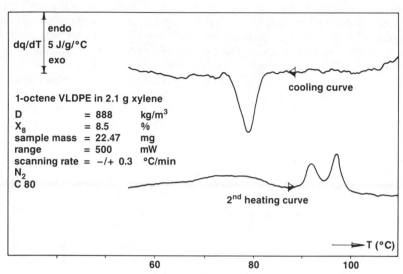

Fig. 9.33. C 80 cooling curve for cooling at a rate of 0.3 °C/min and subsequent heating curve for heating at the same rate, for a 1.1 (m/m) % solution in xylene of the VLDPE shown in the preceding figures

The curves in Figs. 9.10 and 9.33 are shown as they were measured (after division by sample mass and scanning rate, of course), i.e. no empty-cell measuring results were subtracted, which means that the result can be improved further. As was to be expected, the curves not only have a different shape but have also shifted roughly 30° to lower temperatures compared with the curves obtained in measuring from the melt; compare Fig. 9.33 with Fig. 9.13, and Fig. 9.10 with Table 9.1.

Dissolution effected after crystallization in solution is usually accompanied by reorganization and recrystallization [3, 126, 380–384], which is one of the reasons that extremely slow cooling rates are employed in fractionation. The microcalorimeter can prove most useful in this respect in helping to determine the optimum experimental conditions for fractionation, as can be seen in Fig. 9.33: if one were to fractionate under comparable conditions, one would have to take into account a split-up of the 'high-temperature' peak (this split peak corresponds to the crystallization peak between about 73 °C and about 84 °C); one would also have to investigate whether or not the phenomenon observed is indeed recrystallization and find a way of dealing with this in the fractionation procedure. Of course, if in a TREF experiment an entire chain dissolves and is carried off with the solvent, no recrystallization of that chain occurs, but in the case of partial dissolution (see section 9.2.3) recrystallization can occur.

The minimum temperature at which the C 80 microcalorimeter can be used is about 40 °C, which means that only a portion of the crystallization and dissolution region of the currently investigated VLDPE can be covered. If information is required on temperatures below 0 °C, use will have to be made of a different type of microcalorimeter, for example Setaram's BT 215 (see chapter 10). Such minimum measurement temperatures are a problem in fractionation, because solvents suitable for low temperatures are few and far between, and handling is also more difficult at low temperatures.

Microcalorimetry is thus expected to prove very useful for specialists that want to obtain a better understanding of what is happening under specific fractionating conditions: with the aid of a microcalorimeter, crystallization and dissolution measurements can be performed under these conditions. To illustrate this, let us assume an intramolecularly heterogeneous copolymer analyzed by TREF. The curve will probably be relatively narrow with a single peak. After all, the longer crystallized sequences in a molecule will have a dominant influence on the dissolution of the molecule and hence on the detection of this process (see discussion in section 9.2.3). A microcalorimeter, on the other hand, records the heat effect of dissolution of all crystallized sequences—including the shorter ones—and will therefore give a relatively wide curve, possibly with several peaks.

9.5.3 Cross-fractionation

Figure 9.34 shows the results obtained for LLDPE with the aid of the two fractionation methods discussed in sections 9.5.1 and 9.5.2 [200]. The conclusion that can be drawn from Fig. 9.34 is that for any length of chain there are molecules with many branches as well as molecules with few branches. This is confirmed by the fact that in the case of LLDPE the direct extraction method gives fractions with narrow molar mass distributions and multiple-peaked DSC curves with wide crystallization and melting ranges, whereas the crystallization/dissolution method gives fractions that yield single-peaked DSC curves and show broad molar mass distributions.

The obvious next step is to try to corroborate the conclusion drawn by means of cross-fractionation [150, 151, 182, 192, 198, 222, 385, 386]. To this end a 1-octene

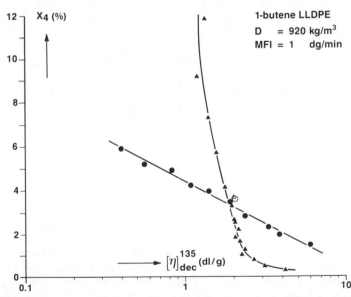

Fig. 9.34. Results for fractions obtained from the 1-butene LLDPE shown in Figs. 9.26 and 9.27 via the direct extraction method (●) and a crystallization/dissolution method (▲): ○ and △ refer to mathematically summed fractions ($[\eta]_{dec}^{135}$ (dl/g) and X_4 (%) are 2.00 and 3.56 respectively for the whole polymer, 2.04 and 3.63 for the sum of DE fractions and 1.99 and 3.69 for the sum of crystallization/dissolution fractions)

Table 9.5. Molecular structure data for some cross-fractions of a 1-octene LLDPE

Cross-fractions		m %	X_8 %	M_n^* kg/mol	M_w^* kg/mol	M_z^* kg/mol	M_w^*/M_n^*	M_z^*/M_w^*
D-1-2	C-1-1	38.7	7.4	8	18	32	2.3	1.7
	C-1-2	32.7	3.5	10	18	29	1.8	1.6
	C-1-3	13.6	n.d.[1]	n.d.[1]	n.d.[1]	n.d.[1]	—	—
	C-1-4	15.1	n.d.[1]	n.d.[1]	n.d.[1]	n.d.[1]	—	—
D-1-9	C-1-1	16.2	2.6	130	210	360	1.6	1.7
	C-1-2	22.7	2.2	130	200	310	1.5	1.6
	C-1-3	52.2	1.2	130	200	330	1.6	1.7
	C-1-4	9.0	1.2	n.d.	n.d.	n.d.	—	—

[1]n.d. = not determined

LLDPE was fractionated according to molar mass via the direct extraction method, after which a low molar mass fraction, D-1-2, and a high molar mass fraction, D-1-9, were further separated via a flocculation–crystallization/dissolution method [200]. The latter treatment no longer effected a separation according to molar mass: the cross fractions that were large enough to be characterized had the same molar mass distributions and the same $[\eta]$ values as the direct-extraction fractions from which they originated. Table 9.5 lists the molecular structure data in question.

Figure 9.35 shows the results of this cross-fractionation. The fractions have a narrow molar mass distribution and, in combination with the DSC curves, the corresponding IR results show that fractionation according to comonomer content has indeed taken place.

Figure 9.36 shows the results of cross-fractionation [231] of a 1-butene LLDPE, which was first fractionated according to molecular dimension with the aid of

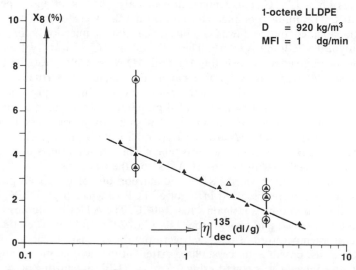

Fig. 9.35. Cross-fractionation (⬥) of an 1-octene LLDPE via a crystallization/dissolution method after direct extraction (▲): △ refers to mathematically summed fractions ($[\eta]_{dec}^{135}$ (dl/g) and X_8 (%) are 1.90 and 2.84 respectively for the whole polymer and 1.84 and 2.80 for the sum of DE fractions)

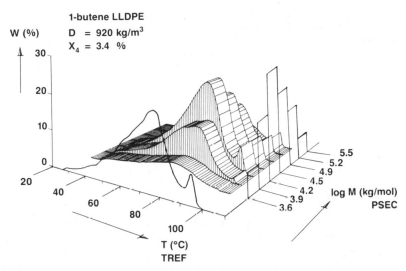

Fig. 9.36. Cross-fractionation of a 1-butene LLDPE via analytical temperature rising elution fractionation after preparative size exclusion chromatography: in addition, the mathematical sums of the separate values for the fractions are indicated in two directions (arbitrary scales)

preparative SEC (PSEC) and then according to crystallizability/dissolvability via analytical TREF (ATREF). In the case of LLDPE this means fractionation according to molar mass followed by fractionation according to SCBD, or according to ethylene sequence length distribution (ESLD), with which the latter is directly related.

For the PSEC 5 g of sample was accurately weighed and dissolved in 1.1 litres of 1,2,4-trichlorobenzene (TCB), together with 2 g of DBPC (antioxidant). The dissolution time was 4 h at 140 °C. This temperature was also the operating temperature of the apparatus. The initial polymer concentration was 0.3 (m/m) %. In two days, sufficient fractionated material was obtained, via 100 ml injections, for DSC, IR, ASEC (analytical SEC) and ATREF. The sums of the M_n^* and M_w^* values obtained for the fractions agreed well with the M_n^* and M_w^* values obtained for the whole sample (19 and 70 respectively), which means that there was virtually no loss or decomposition of material during fractionation and the further processing of the fractions. As the sample was first fractionated preparatively according to molar mass, the distribution along the log M axis in Fig. 9.36 is discontinuous. The first and the last values of log M at each 'M-slice' correspond to the beginning and end of the period in which the fraction concerned was taken. The retention times were calibrated on the basis of molar mass by means of ASEC on the fractions.

For cross-fractionation via ATREF, 150 mg portions of processed polymer, or as much as possible, were dissolved in 10 ml of TCB to which 1 g/l DBPC had been added. The dissolution time was again 4 h at 140 °C. The ATREF columns were filled with this solution, sealed and cooled from 140 °C to 20 °C in 80 h. Then the columns were stored at a temperature of 0 °C (refrigerator) for at least one night to allow as much material as possible to crystallize. After fitting a column in the elution installation, the polymer was eluted using 0.5 ml of TCB/min starting at 20 °C, while the temperature was increased by 20 °C/h. The polymer content of the eluent was continuously determined with the aid of an infrared detector, set at 3.41 μm (C–H stretch vibration).

The volume of each M-slice was normalized on the basis of the relative mass of each PSEC fraction. The final relative mass W is indicated on the z-axis. The number of PSEC fractions can of course be increased to obtain a more continuous distribution in the M direction. This would make a comparison of polymer types via a comparison of so-called contour plots (iso-W contours) more meaningful.

The overall picture thus obtained agrees with that obtained via direct extraction fractionation and fractionation according to crystallization/dissolution. The HDPE-like material ('high-temperature' TREF peak) is relatively prominent at higher molar masses, while its dissolution temperature is practically molar-mass-independent. The 'low-temperature' TREF peak shifts to lower dissolution temperatures with decreasing molar mass. Moreover, the TREF curves appear to become slightly narrower with increasing molar mass. Another striking fact (see the discussion of molecule interactions in the preceding section) is that the 'low-temperature' peak obtained in ATREF (which corresponds to the 'low-temperature' peak of the DSC curve) is found to represent much more material than the 'high-temperature' peak obtained in ATREF (which corresponds to the 'high-temperature' peak of the DSC curve), whereas from the dq/dTs of the DSC curves of the pure polymer the amounts seem at first sight to be more balanced.

Acknowledgments

The author wishes to express his appreciation to M. F. J. Pijpers, who performed most of the experiments, to W. H. P. Derks, J. van Ruiten and L. C. E. Struik for critical reading and stimulating discussions, to H. J. Rhebergen for the English translation of the original Dutch text and to G. Schuler for the drawings. Finally, thanks are due to DSM for giving their permission for publication.

References

1 *Wunderlich, B.:* Macromolecular Physics, Vol. 1: Crystal Structure, Morphology, Defects. Academic Press, New York, 1973
2 *Wunderlich, B.:* Macromolecular Physics, Vol. 2: Crystal Nucleation, Growth, Annealing. Academic Press, New York, 1976
3 *Wunderlich, B.:* Macromolecular Physics, Vol. 3: Crystal Melting. Academic Press, New York, 1980
4 *Bassett, D. C.:* Principles of Polymer Morphology. Cambridge University Press, 1981. *Vaughan, A. S., Bassett, D. C.,* in: Comprehensive Polymer Science, Vol. 2: Polymer Properties. *Booth, C., Price, C.* (Eds.), p. 415, Pergamon Press, Oxford, 1989
5 Disc. Faraday Soc. 68: Organization of Macromolecules in the Condensed Phase (1979)
6 *Keller, A.:* Rep. on Prog. Phys. 31, p. 623 (1968)
7 *Keller, A.,* in: Integration of Fundamental Polymer Science and Technology. *Kleintjens, L. A., Lemstra, P. J.* (Eds.), p. 425, Elsevier Applied Science, Barking, UK, 1986
8a *Dosière, M.,* in: Handbook of Polymer Science and Technology. *Cheremisinoff, N. P.* (Ed.), Vol. 2, p. 367, Marcel Dekker, New York, 1989
8b *Mandelkern, L.,* in: Comprehensive Polymer Science, Vol. 2: Polymer Properties. *Booth, C., Price, C.* (Eds.), p. 363, Pergamon Press, Oxford, 1989
9 *Phillips, P. J.:* Rep. Prog. Phys. 53, p. 549 (1990)
10 *Armitstead, K., Goldbeck-Wood, G.:* Adv. Polym. Sci. 100, p. 219 (1992)
11 *Koutsky, J. A., Walton, A. G., Baer, E.:* J. Polym. Sci. B5, p. 185 (1967)
12 *Boon, J., Challa, G., Van Krevelen, D. W.:* J. Polym. Sci., Part A-2 6, p. 1835 (1968)
13 *Binsbergen, F. L.:* Polymer 11, p. 253 (1970)
14 *Turturro, A., Olivero, L., Pedemonte, E., Alfonso, G. C.:* Br. Polym. J. 5, p. 129 (1973)

15 Barham, P. J., Chivers, R. A., Jarvis, D. A., Martinez-Salazar, J., Keller, A.: J. Polym. Sci.: Polym. Lett. Ed. 19(11), p. 539 (1981)
16 Beck, H. N.: J. Polym. Sci., Part A-2 4, p. 631 (1966)
17 Narh, K. A., Odell, J. A., Keller, A., Fraser, G. V.: J. Mater. Sci. 15(8), p. 2001 (1980)
18 Schultz, J. M.: J. Polym. Sci. A-2 7, p. 821 (1969)
19 Rault, J., Robelin, E.: Polym. Bull. 2(6), p. 373 (1980)
20 Rault, J., Sotton, M., Rabourdin, C., Robelin, E.: J. Physique 41(12), p. 1459 (1980)
21 Robelin, E., Rousseaux, F., Lemonnier, M., Rault, J.: J. Physique 41(12), p. 1469 (1980)
22 Chivers, R. A., Barham, P. J., Martinez-Salazar, J., Keller, A.: J. Polym. Sci.: Polym. Phys. Ed. 20, p. 1717 (1982)
23 Phuong-Nguyen, H., Delmas, G.: Macromol. 25, p. 408 (1992)
24 Khanna, Y. P., Reimschuessel, A. C.: J. Appl. Polym. Sci. 35, p. 2259 (1988)
25 Khanna, Y. P., Reimschuessel, A. C.: Polym. Eng. Sci. 28, p. 1600 (1988)
26 Banks, W., Gordon, M., Sharples, A.: Polymer 4, p. 289 (1963)
27 Banks, W., Sharples, A.: Makromol. Chem. 67, p. 42 (1963)
28 Blundell, D. J., Keller, A., Kovacs, A. J.: J. Polym. Sci. B4, p. 481 (1966)
29 Vidotto, G., Lévy, D., Kovacs, A. J.: Kolloid-Z. u. Z. Polym. 230, p. 289 (1969)
30 Mandelkern, L., Stack, G. M., Mathieu, P. J. M.: Anal. Calorim. 5, p. 223 (1984)
31 Hatakeyama, T., Watanabe, T., Hashimoto, T., Hatakeyama, H.: Thermochim. Acta 183, p. 167 (1991)
32 Hoffman, J. D., Frolen, L. J., Ross, G. S., Lauritzen, Jr., J. I.: J. Res. NBS 79A(6), p. 671 (1975)
33 Hoffman, J. D., Davis, G. T., Lauritzen, Jr., J. I., in: Treatise on Solid State Chemistry. Hannay, B. (Ed.), Vol. 3, Chapter 7, p. 497, Plenum Press, New York, 1976
34 Maxfield, J., Mandelkern, L.: Macromol. 10(5), p. 1141 (1977)
35 Mandelkern, L., Maxfield, J.: J. Polym. Sci.: Polym. Phys. Ed. 17, p. 1913 (1979)
36 Allen, R. C., Mandelkern, L.: J. Polym. Sci.: Polym. Phys. Ed. 20, p. 1465 (1982)
37 Magill, J. H., in: Polymer Handbook. Brandrup, J., Immergut, E. H. (Eds.), 3rd Edn., VI/279, Wiley, New York, 1989
38 Fatou, J. G., Marco, C., Mandelkern, L.: Polymer 31, p. 890 (1990)
39 Fatou, J. G., Marco, C., Mandelkern, L.: Polymer 31, p. 1685 (1990)
40 Fatou, J. G., in: Encycl. Polym. Sci. Eng., Suppl. Vol. Kroschwitz, J. I. (Ed.), p. 231, Wiley, New York, 1990
41 Hoffman, J. D., Guttman, C. M., DiMarzio, E. A.: Disc. Faraday Soc. 68, p. 177 (1979)
42 Hoffman, J. D., Miller, R. L.: Macromol. 21, p. 3038 (1988)
43 Voigt-Martin, I. G., Fischer, E. W., Hagedorn, W., Hendra, P., Mandelkern, L., Mehler, K., in: Proc. 26th Int. Symp. on Macromol., IUPAC Macro Mainz, p. 1250, 1979
44 Mandelkern, L., Glotin, M., Benson, R. A.: Macromol. 14, p. 22 (1981)
45 Mandelkern, L.: Polym. J., 17(1), p. 337 (1985)
46 Voigt-Martin, I. G.: Adv. Polym. Sci. 67, p. 194 (1985)
47 Mathot, V. B. F., Pijpers, M. F. J.: Polym. Bull. 11, p. 297 (1984)
48 Peyrouset, A., Prechner, R., Panaris, R., Benoit, H.: J. Appl. Polym. Sci. 19, p. 1363 (1975)
49 Scholte, Th. G., Meijerink, N. L. J., Schoffeleers, H. M., Brands, A. M. G.: J. Appl. Polym. Sci. 29, p. 3763 (1984)
50 Wunderlich, B., Czornyj, G.: Macromol. 10(5), p. 906 (1977)
51 Fatou, J. G., Mandelkern, L.: J. Phys. Chem. 69, p. 417 (1965)
52 Ford, R. W., Ilavsky, J. D., Scott, R. A., in: Analytical Calorimetry. Porter, R. S., Johnson, J. F. (Eds.), p. 41, Plenum Press, New York, 1968
53 Bair, H. E., Salovey, R.: J. Macromol. Sci.-Phys. B3(1), p. 3 (1969)
54 Wunderlich, B.: Thermal Analysis. Academic Press, San Diego, CA, 1990
55 Flory, P. J.: J. Chem. Phys. 17(3), p. 223 (1949)
56 Flory, P. J., Vrij, A.: J. Am. Chem. Soc. 85, p. 3548 (1963)
57 Broadhurst, M. G.: J. Res. NBS 70A(6), p. 481 (1966)
58 Gopalan, M., Mandelkern, L.: J. Phys. Chem. 71(12), p. 3833 (1967)
59 Lindenmeyer, P. H., Beumer, H., Hosemann, R.: Polym. Eng. Sci. 19(1), p. 51 (1979)
60 Mandelkern, L., Stack, G. M.: Macromol. 17(4), p. 871 (1984)
61 Mandelkern, L., Prasad, A., Alamo, R. G., Stack, G. M.: Macromol. 23, p. 3696 (1990)

62 *Lehtinen, A., Jääskeläinen, P.:* Kem.–Kemi 15(5), p. 472 (1988)
63 *Lehtinen, A., Jääskeläinen, P.,* in: Proc. 3rd Int. Symp. Polym. Anal. Charact. (ISPAC3), Brno, p. 135, 1990
64 *Jääskeläinen, P., Lehtinen, A.,* in: Proc. 3rd Int. Symp. Polym. Anal. Charact. (ISPAC3), Brno., p. 139, 1990
65 *Banks, W., Gordon, M., Roe, R. J. Sharples, A.:* Polymer 4, p. 61 (1963)
66 *Mandelkern, L., Fatou, J. G., Ohno, K.:* J. Polym. Sci. B 6, p. 615 (1968)
67 *Mandelkern, L., Allou, Jr., A. L., Gopalan, M.:* J. Phys. Chem. 72, p. 309 (1968)
68 *Barrales-Rienda, J. M., Fatou, J. M. G.:* Polymer 13, p. 407 (1972)
69 *Ergoz, E., Fatou, J. G., Mandelkern, L.:* Macromol. 5, p. 147 (1972)
70 *de Gennes, P. G.:* J. Chem. Phys. 55, p. 572 (1971)
71 *Hoffman, J. D.:* Polymer 23, p. 656 (1982)
72 *Hay, J. N., Wiles, M.:* J. Polym. Sci. Polym. Chem. Ed. 17, p. 2223 (1979)
73 *Ungar, G., Keller, A.:* Polymer 27, p. 1835 (1986)
74 *Ungar, G.,* in: Integration of Fundamental Polymer Science and Technology, *Lemstra, P. J., Kleintjens, L. A.* (Eds.), Vol. 2, p. 346, Elsevier Applied Science, Barking, UK, 1988
75 *Stack, G. M., Mandelkern, L.:* Macromol. 21, p. 510 (1988)
76 *Stack, G. M., Mandelkern, L., Kröhnke, C., Wegner, G.:* Macromol. 22, p. 4351 (1989)
77 *Hoffman, J. D., Miller, R. L.:* Macromol. 22, p. 3038 (1989)
78 *Keith, H. D., Padden, Jr., F. J., Vadimsky, R. G.:* J. Polym. Sci., Part A2 4, p. 267 (1966)
79 *Keith, H. D., Padden, F. J., Vadimsky, R. G.:* J. Appl. Phys. 42(12), p. 4585 (1971)
80 *Phillips, P. J., Edwards, B. C.:* Polym. Lett. Ed. 14, p. 449 (1976)
81 *Phillips, P. J.:* Polym. Prepr. (Am. Chem. Soc. Div. Polym. Chem.) 20(2), p. 483 (1979)
82 *Hoffman, J. D.:* Polymer, 24, p. 3 (1983)
83 *Mandelkern, L., Fatou, J. G., Howard, C.:* J. Phys. Chem. 68(11), p. 3386 (1964)
84 *Mandelkern, L., Fatou, J. G., Howard, C.:* J. Phys. Chem. 69(3), p. 956 (1965)
85 *Mandelkern, L.,* in: Proc. Golden Jubilee Conf. Polyethylenes, The Plastics and Rubber Institute, D1, Chameleon Press, London, 1983
86 *Chiu, G., Alamo, R. G., Mandelkern, L.:* J. Polym. Sci., Part B: Polym. Phys. 28, p. 1207 (1990)
87 *Mathot, V. B. F.:* unpublished results
88 *Kanig, G.:* Colloid Polym. Sci., 255, p. 1005 (1977)
89 *Sanchez, I. C., Colson, J. P., Eby, R. K.:* J. Appl. Phys. 44(10), p. 4332 (1973)
90 *Sanchez, I. C., Peterlin, A., Eby, R. K., McCrackin, F. L.:* J. Appl. Phys. 45(10), p. 4216 (1974)
91 *Geil, P. H.:* Polymer Single Crystals. Wiley Interscience, Chichester, UK, 1963
92 *Voigt-Martin, I. G., Mandelkern, L.:* J. Polym. Sci., Part B: Polym. Phys. 27, p. 967 (1989)
93 *Barham, P. J., Jarvis, D. A., Keller, A.:* J. Polym. Sci. Polym. Phys. Ed. 20, p. 1733 (1982)
94 *Snyder, R. G., Krause, S. J., Scherer, J. R.:* J. Polym. Sci. Polym. Phys. Ed. 16, p. 1593 (1978)
95 *Snyder, R. G., Scherer, J. R.:* J. Polym. Sci. Polym. Phys. Ed. 18, p. 1421 (1980)
96 *Glotin, M., Mandelkern, L.:* J. Polym. Sci. Polym. Phys. Ed. 21, p. 29 (1983)
97 *Zachmann, H. G.:* Kolloid-Z. u. Z. Polym. 231, p. 504 (1969)
98 *Lauritzen, Jr., J. I., Hoffman, J. D.:* J. Res. NBS 64A(1), p. 73 (1960)
99 *Martinez-Salazar, J., Barham, P. J., Keller, A.:* J. Mater. Sci. 20, p. 1616 (1985)
100 *Barham, P. J., Chivers, R. A., Keller, A., Martinez-Salazar, J., Organ, S. J.:* J. Mater. Sci. 20, p. 1625 (1985)
101 *Barham, P. J., Keller, A.:* J. Polym. Sci., Part B: Polym. Phys. 27, p. 1029 (1989)
102 *Vonk, C. G.:* J. Polym. Sci., Part B: Polym. Phys. 28, p. 1871 (1990)
103 *Yasuniwa, M.:* Polym. J. 4(5), p. 526 (1973)
104 *Takamizawa, K., Ohno, A., Urabe, Y.:* Polym. J. 7(3), p. 342 (1975)
105 *Reneker, D. H., Mazur, J.:* Polymer 24, p. 1387 (1983)
106 *Wunderlich, B., Grebowicz, J.:* Adv. Polym. Sci. 60/61, p. 1 (1984)
107 *Ichida, T., Tsuji, M., Murakami, S., Kawaguchi, A., Katayama, K.:* Colloid Polym. Sci. 263, p. 293 (1985)
108 *Vonk, C. G., Koga, Y.:* J. Polym. Sci. Polym. Phys. Ed. 23, p. 2539 (1985)
109 *Ungar, G., Keller, A.:* Polymer 28, p. 1899 (1987)

110 *Wunderlich, B., Möller, M., Grebowicz, J., Baur, H.:* Adv. Polym. Sci. 87 (1988)
111 *Sumpter, B. G., Noid, D. W., Wunderlich, B.:* J. Chem. Phys. 93(9), p. 6875 (1990)
112 *Noid, D. W., Sumpter, B. G., Wunderlich, B.:* Polym. Commun. 31, p. 304 (1990)
113 *Hikosaka, M.:* Polymer 28, p. 1257 (1987)
114 *Hikosaka, M.:* Polymer 31, p. 458 (1990)
115 *Hoffman, J. D., Miller, R. L., Marand, H., Roitman, D. B.:* Macromol. 25, p. 2221 (1992)
116 *Mandelkern, L.:* J. Polym. Sci. 47, p. 494 (1960)
117 *Mandelkern, L.:* Crystallization of Polymers. McGraw-Hill, New York, 1964
118 *Hager, Jr., N. E.:* Rev. Sci. Instrum. 35(5), p. 618 (1964)
119 *Hager, Jr., N. E.:* Rev. Sci. Instrum. 43(8), p. 1116 (1972)
120 *Wu, Z. Q., Dann, V. L., Cheng, S. Z. D., Wunderlich, B.:* J. Thermal Anal. 34, p. 105 (1998)
121 *Fröchte, B., Khan, Y., Kneller, E.:* Rev. Sci. Instrum. 61(7), p. 1954 (1990)
122 *Michler, G. H., Brauer, E.:* Acta Polymerica 34(9), p. 533 (1983)
123 *Voigt-Martin, I. G., Fischer, E. W., Mandelkern, L.:* J. Polym. Sci. Polym. Phys. Ed. 18, p. 2347 (1980)
124 *Voigt-Martin, I. G., Mandelkern, L.:* J. Polym. Sci. Polym. Phys. Ed. 19, p. 1769 (1981)
125 *Mandelkern, L., Fatou, J. G., Denison, R., Justin, J.:* Polym. Lett. 3, p. 803 (1965)
126 *Jackson, J. F., Mandelkern, L.,* in: Analytical Calorimetry. *Porter, R. S., Johnson, J. F.* (Eds.), p. 1, Plenum Press, New York, 1968
127 *Richards, R. B.:* J. Appl. Chem. 1, p. 370 (1951)
128 *Tung, L. H., Buckser, S.:* J. Phys. Chem. 62, p. 1530 (1958)
129 *Mandelkern, L.:* J. Polym. Sci. C 15, p. 129 (1966)
130 *Mandelkern, L.,* in: 20th Army Materials Research Conf., p. 369, Syracuse University Press, 1975
131 *Chu, C. C.:* Polym. Prepr. (Am. Chem. Soc. Div. Polym. Chem.) 19(2), p. 773 (1978)
132 *Hendus, H., Illers, K. H.:* Kunststoffe 57(3), p. 193 (1967)
133 *Kitamaru, R., Horii, F.:* Adv. Polym. Sci. 26, p. 139 (1978)
134 *Flory, P. J.:* J. Am. Chem. Soc. 84, p. 2857 (1962)
135 *Frank, F. C.:* Disc. Faraday Soc. 68, p. 7 (1979)
136 *Peterlin, A.:* Macromol. 13(4), p. 777 (1980)
137 *Mansfield, M. L.:* Macromol. 16, p. 914 (1983)
138 *Flory, P. J., Yoon, D. Y., Dill, K. A.:* Macromol. 17, p. 862 (1984)
139 *Yoon, D. Y., Flory, P. J.:* Macromol. 17, p. 868 (1984)
140 *Marqusee, J. A., Dill, K. A.:* Macromol. 19, p. 2420 (1986)
141 *Vonk, C. G.,* in: Integration of Fundamental Polymer Science and Technology. *Kleintjens, L. A., Lemstra, P. J.* (Eds.), p. 471, Elsevier Applied Science, Barking, UK, 1986
142 *Vonk, J.:* Polym. Sci., Part C: Polym. Lett. 24, p. 305 (1986)
143 *Marquse, J. A.:* Macromol. 22, p. 472 (1989) and references therein
144 *Kumar, S. K., Yoon, D. Y.:* Macromol. 22, p. 3458 (1989) and references therein
145 *Mandelkern, L.:* Acc. Chem. Res. 23, p. 380 (1990)
146 *Bergmann, K., Nawoti, K.:* Kolloid-Z. u. Z. Polymere 250, p. 1094 (1972)
147 *Kitamaru, R., Horii, F., Murayama, K.:* Macromol. 19, p. 636 (1986)
148 *Kitamaru, R., Mandelkern, L.:* J. Polym. Sci., Part A2 8, p. 2079 (1970)
149 *Jackson, J. F.:* J. Polym. Sci., Part A1, p. 2119 (1963)
150 *Alamo, R., Domszy, R., Mandelkern, L.:* J. Phys. Chem. 88, p. 6587 (1984)
151 *Hosoda, S.:* Polym. J. 20(5), p. 383 (1988)
152 *Ishikawa, N., Shimizu, T., Shimamura, Y., Gotoh, Y., Omori, K., Misaka, N.,* in: Proc. 10th Plast. Fuel Gas Pipe Symp., New Orleans, 1987, p. 175, American Gas Association, Arlington, 1987
153 *Huang, Y.-L., Brown, N.:* J. Polym. Sci., Part B: Polym. Phys. 29, p. 129 (1991)
154 *Scholten, F. L., Rijpkema, H. J. M.,* in: Proc. Plastics Pipes Symp. VIII, September 1992, Eindhoven. The Plastics and Rubber Institute, London, 1992
155 *Bubeck, R. A., Baker, H. M.:* Polymer 23, p. 1680 (1982)
156 *Scholte, Th. G.,* in: Developments in Polymer Characterization. *Dawkins, J. V.* (Ed.), Vol. 4, Applied Science Publishers, Barking, UK, 1983
157 *Cinquina, P., Gianotti, G., Borghi, D.:* Polymer 32(11), p. 2049 (1991)
158 *Booij, H. C., Meijerink, N., Pijpers, M. F. J.:* unpublished results

159 *Strobl, G. R., Hagedorn, W.:* J. Polym. Sci. Polym. Phys. Ed. 16, p. 1181 (1978)
160 *Strobl, G. R., Schneider, M. J., Voigt-Martin, I. G.:* J. Polym. Sci. Polym. Phys. Ed. 18, p. 1361 (1980)
161 *Glotin, M., Mandelkern, L.:* Macromol. 14(5), p. 1394 (1981)
162 *Michler, G. H., Gruber, K.:* Acta Polymerica 32(6), p. 323 (1981)
163 *Strobl, G. R., Engelke, T., Meier, H., Urban, G.:* Colloid Polym. Sci. 260(4), p. 394 (1982)
164 *Rosenberger, B., Asbach, G. I., Kilian, H.-G., Wilke, W.:* Makromol. Chem. 189, p. 2627 (1988)
165 *Randall, J. C.:* Rev. Macromol. Chem. Phys. C29(2, 3), p. 201 (1989)
166 *Bugada, D. C., Rudin, A.:* J. Appl. Polym. Sci. 33, p. 87 (1987)
167 *Grinshpun, V., Rudin, A., Russell, K. E., Scammell, M. V.:* J. Polym. Sci., Part B: Polym. Phys. 24, p. 1171 (1986)
168 *Kulin, L. I., Meijerink, N. L., Starck, P.:* Pure Appl. Chem. 60(9), p. 1403 (1988)
169 *Murata, K., Kobayashi, S.:* Kobunshi Kagaku 26, p. 536 (1969)
170 *Shirayama, K., Kita, S.:* Kobunshi Kagaku 28, p. 321 (1971)
171 *Hosoi, M., Naoi, T., Kawai, T., Kuriyama, I.:* Kobunshi Kagaku 29, p. 557 (1972); Eng. Edn. 1, p. 848 (1972)
172 *Otocka, E. P., Roe, R. J., Bair, H. E.:* J. Polym. Sci. Polym. Phys. Ed. 12, p. 1245 (1974)
173 *Gianotti, G., Cicutta, A., Romanini, D.:* Polymer 21, p. 1087 (1980)
174 *Alamo, R. G., Mandelkern, L.:* Macromol. 22, p. 1273 (1989); *Alamo, R. G., Chan, E. K., Mandelkern, L., Voigt-Martin, I. G.:* Macromol. 25, p. 6381 (1992)
175 *Hser, J.-H., Carr, S. H.:* Polym. Eng. Sci. 19, p. 436 (1979)
176 *Alamo, R. G., Chan, E. K., Mandelkern, L.:* Polym. Prepr. (Am. Chem. Soc. Div. Polym. Chem.) 30(2), p. 312 (1989)
177 *Nakajima, A.:* Chem. High Polymers, Jpn. (Kubunshi Kagaku) 7, p. 64 (1950)
178 *Tung, L. H.:* J. Polym. Sci. 20, p. 495 (1956)
179 *Francis, P. S., Cooke, Jr., R. C., Elliot, J. H.:* J. Polym. Sci. 31, p. 453 (1958)
180 *Wijga, P. W. O., van Schooten, J., Boerma, J.:* Makromol. Chem. 36, p. 115 (1960)
181 *Koningsveld, R., Pennings, A. J.:* Recueil 83, p. 552 (1964)
182 *Shirayama, K., Okada, T., Kita, S.-I.:* J. Polym. Sci., Part A 3, p. 907 (1965)
183 Polymer Fractionation, *Cantow, M. J. R.* (Ed.), Academic Press, New York, London, 1967
184 *Elston, C. T.:* Can. Pat. No. 984,213, 1967
185 *Pennings, A. J.:* J. Polym. Sci., Part C 16, p. 1799 (1967)
186 *Wild, L., Ranganath, R., Ryle, T.:* J. Polym. Sci., Part A-2 9, p. 2137 (1971)
187 *Bello, A., Barrales-Rienda, J. M.,* in: Polymer Handbook, 2nd Edn., *Brandrup, J., Immergut E. H.* (Eds.), IV/175, Wiley, New York, 1975
188 *Seeger, M., Cantow, H.-J., Marti, S.:* Z. Anal. Chem. 276, p. 267 (1975)
189 Fractionation of Synthetic Polymers. *Tung, L. H.* (Ed.), Marcel Dekker, New York, 1977
190 *Pollock, D. J., Kratz, R. F.,* in: Polymer Physics. Methods of Experimental Physics. Vol. 16, p. 13, *Fava, R. A.* (Ed.), Academic Press, New York, 1980
191 *Johnson, J. F., Cantow, M. J. R., Porter, R. S.,* in: Encyclopedia of Polymer Science and Technology (EPST), 1st Edn., Vol. 7, p. 231, 1987
192 *Schouterden, P., Groeninckx, G., Van der Heyden, B., Jansen, F.:* Polymer 28, p. 2099 (1987)
193 *Koningsveld, R., Kleintjens, L. A., Geerissen, H., Schützzeichel, P., Wolf, B. A.,* in: Comprehensive Polymer Science, Vol 1: Polymer Characterization. *Booth, C., Price, C.* (Eds.), p. 293, Pergamon Press, Oxford, 1989
194 *Schröder, E., Winkler, E.:* Plaste und Kautschuk 12, p. 910 (1974)
195 *Markovich, R. P., Hazlitt, L. G., Smith, L.:* Polym. Mater. Sci. Eng. 65, p. 98 (1991)
196 *Pijpers, M. F. J.:* personal communication
197 *Böhm, L. L.:* Makromol. Chem. 182, p. 3291 (1981)
198 *Springer, H., Hengse, A., Hinrichsen, G.:* J. Appl. Polym. Sci. 40, p. 2173 (1990)
199 *Holtrup, W.:* Makromol. Chem. 178, p. 2335 (1977)
200 *Mathot, V. B. F.,* in: Polycon '84 LLDPE, p. 1, The Plastics and Rubber Institute, London, 1984
201 *Mathot, V. B. F., Schoffeleers, H. M., Brands, A. M. G., Pijpers, M. F. J.,* in: Morphology of Polymers, *Sedlácek, B.* (Ed.), p. 363, Walter de Gruyter, Berlin, New York, 1986
202 *Mathot, V. B. F., Pijpers, M. F. J.:* J. Appl. Polym. Sci. 39(4), p. 979 (1990)

203 Allen, G., Booth, C., Jones, M. N.: Polymer 5, p. 257 (1964)
204 Desreux, v.: Recueil Trav. Chim. Pays-Bas 68, p. 789 (1949)
205 Desreux, V., Spiegels, M. C.: Bull. Soc. Chim. Belg. 59, p. 476 (1950)
206 Hawkins, S. W., Smith, H.: J. Polym. Sci. 28, p. 341 (1958)
207 Shirayama, K., Kita, S.-I., Watabe, H.: Makromol. Chem. 151, p. 97 (1972)
208 Wild, L., Ryle, T. R., Knobeloch, D. C., Peat, I. R.: Polym. Prepr. (Am. Chem. Soc. Div. Polym. Chem.) 18(2), p. 182 (1977)
209 Wild, L., Ryle, T. R., Knobeloch, D. C., Peat, I. R.: J. Polym. Sci. Polym. Phys. Ed. 20, p. 441 (1982)
210 Constantin, D., Hert, M., Machon, J.-P.: Makromol. Chem. 179, p. 1581 (1978)
211 Bergström, C., Avela, E.: J. Appl. Polym. Sci. 23, p. 163 (1979)
212 Nakano, S., Gotoh, Y.: J. Appl. Polym. Sci. 26, p. 4217 (1981)
213 Kimura, K., Shigemura, T., Yuasa, S.: J. Appl. Polym. Sci. 29, p. 3161 (1984)
214 Knobeloch, D. C., Wild, L.: SPE Polyolefins IV Conf. Prepr., p. 427, February, 1984
215 Cady, L. D., in: Broadening the Horizons of Linear Low Technology; Retec, p. 107, SPE, Akron, OH (Oct. 1, 2, 1985)
216 Usami, T., Gotoh, Y., Takayama, S.: Macromol. 19(11), p. 2722 (1986)
217 Mathot, V. B. F., Pijpers, M. F. J.: Thermochim. Acta 93, p. 3 (1985)
218 Kelusky, E. C., Elston, C. T., Murray, R. E.: Polym. Eng. Sci., 27(20), p. 1562 (1987)
219 Brauer, E., Gebauer, E., Wiegleb, H., Fuerling, W.: 31st IUPAC Macromol. Symp. (MACRO'87), Merseburg, V/Po/28, 1987
220 Clas, S.-D., McFaddin, K. E., Russell, K. E., Scammel-Bullock, M. V., Peat, I. R.: J. Polym. Sci., Part A: Polym. Chem. 25, p. 3105 (1987)
221 Mirabella, F. M., Jr., Ford, E. A.: J. Polym. Sci., Part B: Polym. Phys. 25, p. 777 (1987)
222 Bodor, G., Dalcomo, H. J., Schröter, O.: Colloid Polym. Sci. 267, p. 480 (1989)
223 Hazlitt, L. A., Moldovan, D. G.: US Pat. No. 4,798,081, 1989
224 Wild, L.: Adv. Polym. Sci. 98, p. 1 (1990)
225 Wilfong, D. L., Knight, G. W.: J. Polym. Sci., Part B: Polym. Phys. 28, p. 861 (1990)
226 Karbashewski, E., Kale, L., Rudin, A., Tchir, W. J., Cook, D. G., Pronovost, J. O.: J. Appl. Polym. Sci. 44, p. 425 (1992)
227 Karoglanian, S. A., Harrison, I. R.: Polym. Mater. Sci. Eng. 61, p. 748 (1989)
228 Wild, L., Ryle, T., Knobeloch, D.: Polym. Prepr. (Am. Chem. Soc. Div. Polym. Chem.) 23(2), p. 133 (1982)
229 Fujimoto, K., Ogawara, K., Emura, N.: Toyo Soda Kenkyu Hokoku 28(2), p. 89 (1984)
230 Kleintjens, L. A.: personal communication
231 Mathot, V., Pijpers, T., Bunge, W., in: Polym. Prepr. (Am. Chem. Soc. Div. Polym. Chem.), ACS Meeting on Recent Advances in Polyolefin Polymers, Washington, D.C., p. 143, 1992
232 Staub, R. B., in: LLDPE in Europe—World Perspectives and Developments, Proc. Conf. held in Madrid, November 1986, PRI, London, 1/1, 1986
233 Gray, A. P., Casey, K.: Polym. Lett. 2, p. 381 (1964)
234 Ford, R. W., Scott, R. A.: J. Appl. Polym. Sci. 11, p. 2325 (1967)
235 Mathot, V. B. F., Pijpers, M. F. J., in: Integration of Fundamental Polymer Science and Technology. Lemstra, P. J., Kleintjens, L. A. (Eds.), Vol. 2, p. 381, Elsevier Applied Science, Barking, UK, 1988
236 Deblieck, R. A. C., Mathot, V. B. F.: J. Mater. Sci. Lett. 7, p. 1276 (1988)
237 Mathot, V. B. F., Pijpers, M. F. J., in: Integration of Fundamental Polymer Science and Technology. Vol. 3, Lemstra, P. J., Kleintjens, L. A. (Eds.), p. 287, Elsevier, Amsterdam, 1989
238 Polyakov, A. V., Grigor'ev, V. A., Vinogradova, G. A., Korobova, I. A.: Int. Polym. Sci. Techn. 18(4), T/77 (1991)
239 Hunter, B. K., Russell, K. E., Scammell, M. V., Thompson, S. L.: J. Polym. Sci. Polym. Chem. Ed. 22, p. 1383 (1984)
240 Gutzler, F., Wegner, G.: Colloid Polym. Sci. 258, p. 776 (1980)
241 Vonk, C. G., Reynaers, H.: Polym. Commun. 31, p. 190 (1990)
242 Casey, K., Elston, C. T., Phibbs, M. K.: Polym. Lett. 2, p. 1053 (1964)
243 Bastien, I. J., Ford, R. W., Mak, H. D.: Polym. Lett. 4, p. 147 (1966)
244 Dohrer, K. K., Hazlitt, L. G., Whiteman, N. F.: J. Plast. Film Sheeting 4(3), p. 214 (1988)

245 *Baker, C. H., Mandelkern, L.:* Polymer 7, p. 71 (1966)

246 *Vonk, C. G.,* in: Proc. Golden Jubilee Conf. Polyethylenes, The Plastics and Rubber Institute. D2.1, Chameleon Press, London, 1983

247 *Baltá Calleja, F. J., Vonk, C. G.:* X-ray Scattering of Synthetic Polymers. Polymer Science Library 8, Elsevier, Amsterdam, 1989

248 *Hosoda, S., Nomura, H., Gotoh, Y., Kihara, H.:* Polymer 31, p. 1999 (1990)

249 *Bassett, D. C.,* in: Development in Crystalline Polymers. Vol. 1, *Bassett, D. C.* (Ed.), p. 115, Applied Science Publishers, New Jersey, 1982

250 *Seguela, R., Rietsch, F.:* J. Polym. Sci. Polym. Lett. Ed. 24, p. 29 (1986)

251 *Pérez, E., VanderHart, D. L., Crist, Jr., B., Howard, P. R.:* Macromol. 20, p. 78 (1987)

252 *Martinez de Salazar, J., Baltá Calleja, F. J.:* J. Cryst. Growth. 48, p. 283 (1979)

253 *France, C., Hendra, P. J., Maddams, W. F., Willis, H. A.:* Polymer 28, p. 710 (1987)

254 *Martinez-Salazar, J., Sánchez Cuesta, M., Baltá Calleja, F. J.:* Colloid Polym. Sci. 265, p. 239 (1987)

255 *Vonk, C. G.:* J. Polym. Sci., Part C 38, p. 429 (1972)

256 *Michajlov, L., Cantow, H.-J., Zugenmaier, P.:* Polymer 12, p. 70 (1971)

257 *Raine, H. C., Richards, R. B., Ryder, H.:* Trans. Faraday Soc. 41(2), p. 56 (1945)

258 *Kanig, G.:* Colloid Polym. Sci. 260(4), p. 356 (1982)

259 *Martinez-Salazar, J., Keller, A., Cagiao, M. E., Rueda, D. R., Baltá Calleja, F. J.:* Colloid Polym. Sci. 261, p. 412 (1983)

260 *Michler, G., Steinbach, H., Hoffmann, K.:* Acta Polymerica 34, p. 533 (1983)

261 *Baltá Calleja, F. J., Kilian, H. G.,* in: Integration of Fundamental Polymer Science and Technology. *Kleintjens, L. A., Lemstra, P. J.* (Eds.), p. 517, Elsevier Applied Science, Barking, UK, 1986

262 *Michler, G., Naumann, I.,* in: Morphology of Polymers. *Sedlácek, B.* (Ed.), p. 329, Walter de Gruyter, Berlin, New York, 1986

263 *Michler, G., Steinbach, H., Hoffmann, K.:* Acta Polymerica 37(5), p. 289 (1986)

264 *Voigt-Martin, I. G., Alamo, R., Mandelkern, L.:* J. Polym. Sci., Part B: Polym. Phys. 24, p. 1283 (1986)

265 *Schaper, A.:* Acta Polymerica 40(7), p. 479 (1989)

266 *Arakawa, T., Wunderlich, B.:* J. Polym. Sci., Part A2 4, p. 53 (1966)

267 *Bhateja, S. K., Pae, K. D.:* J. Macromol. Sci.-Revs. Macromol. Chem. C13(1), p. 77 (1975)

268 *Rastogi, S., Hikosaka, M., Kawabata, H., Keller, A.:* Macromol. 24, p. 6384 (1991)

269 *Vonk, C. G., Pijpers, A. P.:* J. Polym. Sci. Polym. Phys. Ed. 23, p. 2517 (1985)

270 *Wunderlich, B.:* J. Chem Phys. 29(6), p. 1395 (1958)

271 *Flory, P. J.:* J. Chem. Phys. 15(9), 684 (1947)

272 *Flory, P. J.:* Trans. Faraday Soc. 51, p. 848 (1955)

273 *Flory, P. J., Mandelkern, L.:* J. Polym. Sci. 21, p. 345 (1956)

274 *Juijn, J. A., Gisolf, J. H., de Jong, W. A.:* Kolloid-Z. u. Z. Polymere 251, p. 456 (1973)

275 *Scherrenberg, R.:* The Structural Aspects of Suspension Poly(Vinyl Chloride). Ph.D. Thesis, Katholieke Universiteit Leuven, Belgium, 1992

276 *Mathot, V. B. F., Pijpers, M. F. J.:* Thermochim. Acta 151, p. 241 (1989)

277 *Lee, Y. K., Jeong, Y. T., Kim, K. C., Jeong, H. M., Kim, B. K.:* Polym. Eng. Sci. 31(13), p. 944 (1991); *Kim, B. K., Kim, M. S., Kim, K. J., Jang, J. K.:* Makromol. Chem. 194, p. 91 (1992)

278 *Krigas, T. M., Carella, J. M., Struglinsky, M. J., Crist, B., Graessley, W. W., Schilling, F. C.:* J. Polym. Sci. Polym. Phys. Ed. 23, p. 509 (1985)

279 *Natta, G.:* J. Polym. Sci. 54, p. 411 (1961)

280 *Alamo, R. G., Mandelkern, L.,* in: Thermal Analysis and Calorimetry in Polymer Physics. *Mathot, V. B. F.* (Ed.), Special issue Thermochimica Acta, Elsevier Science Publishers, Amsterdam, 1994

281 *Mathot, V., Pijpers, M., Beulen, J., Graff, R., van der Velden, G.,* in: Proc. 2nd Eur. Symp. on Thermal Analysis 1981 (ESTA-2), *Dollimore, D.* (Ed.), p. 264, Heyden, London, 1981

282 *Mathot, V. B. F., Fabrie, Ch. C. M., Tiemersma-Thoone, G. P. J. M., van der Velden, G. P. M.,* in: Proc. Int. Rubber Conf. (IRC), Kyoto, October 15–18, 1985, p. 334, 1985

283 *Mathot, V. B. F., Fabrie, Ch. C. M.:* J. Polym. Sci., Part B: Polym. Phys. 28, p. 2487 (1990)

295

284 Mathot, V. B. F., Fabrie, Ch. C. M., Tiemersma-Thoone, G. P. J. M., van der Velden, G. P. M.: J. Polym. Sci., Part B: Polym. Phys. 28, p. 2509 (1990)
285 Cheng, H. N.: Macromol. 17, p. 1950 (1984)
286 Cheng, H. N., Bennett, M. A.: Makromol. Chem. 188, p. 135 (1987)
287 Mathot, V. B. F., Fabrie, Ch. C. M., Tiemersma-Thoone, G. P. J. M., van der Velden, G. P. M.: to be published
288 Cooper, W., in: Comprehensive Chemical Kinetics 15, Non-Radical Polymerization. Bamford, C. H., Tipper, C. F. H. (Eds.), Chapter 3, p. 133, Elsevier, Amsterdam, 1976
289 Burfield, D. R.: Macromol. 20, p. 3020 (1987)
290 Clegg, G. A., Gee, D. R., Melia, T. P.: Makromol. Chem., 116, p. 130 (1968)
291 Flisi, U., Valvassori, A., Novajra, G.: Rubber. Chem. Technol. 44, p. 1093 (1971)
292 Dole, M., Hettinger, Jr., W. P., Larson, N. R., Wethington, Jr., J. A.: J. Chem. Phys. 20(5), p. 781 (1952)
293 Dole, M.: J. Polym. Sci. 19, p. 347 (1956)
294 Coleman, B. D.: J. Polym. Sci. 31, p. 155 (1958)
295 Eby, R. K.: J. Appl. Phys. 34(8), p. 2442 (1963)
296 Baur, H.: Kolloid-Z. u. Z. Polymere 203(2), p. 97 (1965)
297 Baur, H.: Kolloid-Z. u. Z. Polymere 212(2), p. 97 (1966)
298 Miller, R. L., in: Encyclopedia of Polymer Science and Technology. Vol. 4, Bikales, N. M. (Ed.), p. 449, Interscience, New York, 1966
299 Ver Strate, G.: J. Appl. Polym. Sci. 14, p. 2509 (1970)
300 Helfand, E., Lauritzen, Jr., J. I.: Macromol. 6, p. 631 (1973)
301 Sanchez, I. C., Eby, R. K.: J. Res. NBS 77A(3), p. 353 (1973)
302 Sanchez, I. C., Eby, R. K.: Macromol. 8(5), p. 638 (1975)
303 Sanchez, I. C.: J. Polym. Sci. C59, p. 109 (1977)
304 Tanaka, N.: Sen-I Gakkaishi 39(10), p. 393 (1983)
305 Kilian, H.-G.: Progr. Colloid Polym. Sci. 72, p. 60 (1986): Kilian, H.-G., in: Thermal Analysis and Calorimetry in Polymer Physics. Mathot, V. B. F. (Ed.), Special issue Thermochimica Acta, Elsevier Science Publishers, Amsterdam, 1994
306 Goldbeck-Wood, G., Sadler, D. M.: Polym. Commun. 31, p. 143 (1990)
307 Alamo, R. G., Mandelkern, L.: Macromol. 24, p. 6480 (1991)
308 Goldbeck-Wood, G.: Polymer 33(4), p. 778 (1992)
309 Guillot, J., Rios, L.: Makromol. Chem. 180, p. 1049 (1979)
310 Goldbeck-Wood, G.: Polymer 31, p. 586 (1990)
311 Goldbeck-Wood, G.: Computer Simulation of Polymer Crystallization, to appear in Computer Simulation of Polymers, Polymer Science and Technology Series, Colbourn, E. A. (Ed.), Longman, London.
312 van Ruiten, J., van Dieren, F., Mathot, V. B. F.: in: Crystallization of Polymers, NATO ASI-C Series Mathematical and Physical Sciences. Dosière, M. (Ed.), p. 481–487, Kluwer Academic Publishers, 1993
313 Zachmann, H. G.: Kolloid-Z. u. Z. Polymere 216, 217, p. 180 (1967)
314 Baur, H.: Progr. Colloid Polym. Sci. 66, p. 1 (1979)
315 Lindenmeyer, P. H., Beumer, H., Hoseman, R.: Polym. Eng. Sci. 19(1), p. 40 (1979)
316 Baur, H.: Pure Appl. Chem. 52, p. 457 (1980)
317 Ke, B.: J. Polym. Sci. 61, p. 47 (1962)
318 Griskey, R. G., Foster, G. N.: J. Polym. Sci., Part A1 8, p. 1623 (1966)
319 Kimura, K., Yuasa, S., Maru, Y.: Polymer 25, p. 441 (1984)
320 Alamo, R., Mandelkern, L.: J. Polym. Sci., Part B: Polym. Phys. 24, p. 2087 (1986)
321 Nichols, M. E., Robertson, R. E.: J. Polym. Sci., Part B: Polym. Phys. 30, p. 305 (1992)
322 Cozewith, C., Ver Strate, G.: Macromol. 4, p. 482 (1971)
323 Ver Strate, G., Wilchinsky, Z. W.: J. Polym. Sci., Part A2 9, p. 127 (1971)
324 Baldwin, F. P., Ver Strate, G.: Rubber Chem. Technol. 45, p. 709 (1972)
325 Kissin, Y. V.: Adv. Polym. Sci. 15, p. 91 (1974)
326 Kissin, Y. V.: Isospecific Polymerization of Olefins with Heterogeneous Ziegler–Natta Catalysts. Springer Verlag, New York, 1985
327 Cozewith, C.: Macromol. 20, p. 1237 (1987)
328 Echevskaya, L. G., Bukatov, G. D., Zakharov, V. A.: Makromol. Chem. 188(11), p. 2573 (1987)

329 *Hayashi, T., Inoue, Y. Chûjô, R.:* Macromol. 21, p. 3139 (1988)
330 *Cheng, H. N.,* in: Computer Applications in Polymer Science II. ACS Symposium Series 404, p. 174, American Chemical Society, Washington, D.C., 1989
331 *de Cavalho, A. B., Gloor, P. E., Hamielec, A. E.:* Polymer 30, p. 280 (1989)
332 *Cheng, H. N.:* Polym. Bull. 23, p. 589 (1990)
333 *McAuley, K. B., MacGregor, J. F., Hamielec, A. E.:* Am. Inst. Chem. Engrs. 36(6), p. 837 (1990)
334 *Cheng, H. N., Kakugo, M.:* Macromol. 24, p. 1724 (1991)
335 *Cheng, H. N.:* Macromol. 25, p. 2351 (1992)
336 *Mathot, V. B. F., Pijpers, M. F. J.:* J. Therm. Anal. 28, p. 349 (1983)
337 *Mathot, V. B. F.:* Polymer 25, p. 579 (1984). Errata: Polymer 27, p. 969 (1986)
338 *Dole, M.:* Fortschr. Hochpolym.-Forsch. 2, p. 221 (1960)
339 *Gray, A. P.:* Thermochim. Acta 1, p. 563 (1970)
340 *Richardson, M. J.:* J. Polym. Sci., Part C 38, p. 251 (1972)
341 *Richardson, M. J.,* in: Developments in Polymer Characterization. Vol. 1, *Dawkins, J. V.* (Ed.), p. 205, Applied Science Publishers, London, 1978
342 *Richardson, M. J.,* in: Comprehensive Polymer Science, Vol. I, *Booth, C., Price, C.* (Eds.), p. 867, Pergamon, Oxford, 1989
343 *Runt, J. P.,* in: Encyclopedia of Polymer Science and Engineering. *Mark, H. F., Bikales, N. M.* (Eds.), Vol. 4, p. 482, Wiley, New York, 1986
344 *Sheldon, R. P.:* Polym. Lett. 1, p. 655 (1963)
345 *Clas, S.-D., Heyding, R. D., McFaddin, D. C., Russell, K. E., Scammell-Bullock, M. V., Kelusky, E. C., St-Cyr, D.:* J. Polym. Sci., Part B: Polym. Phys. 26, p. 1271 (1988)
346 *Mandelkern, L.:* Adv. Chem. Ser. 227, p. 377 (1990)
347 *Hagemann, H., Snyder, R. G., Peacock, A. J., Mandelkern, L.:* Macromol. 22, p. 3600 (1989)
348 *Axelson, D. E., Mandelkern, L., Popli, R., Mathieu, P.:* J. Polym. Sci. Polym. Phys. Ed. 21, p. 2319 (1983)
349 *Schouterden, P., Vandermarliere, M., Riekel, C., Koch, M. H. J., Groeninckx, G., Reynaers, H.:* Macromol. 22, p. 237 (1989)
350 *Bodor, G.:* Structural Investigation of Polymers, Series in Polymer Science and Technology, Ellis Horwood, Chichester, UK and Akadémiai Kiadó, Budapest, 1991
351 *Heink, M., Häberle, K.-D., Wilke, W.:* Colloid Polym. Sci. 269, p. 675 (1991)
352 *Kast, W.,* in: Die Physik der Hochpolymeren. *Stuart, H. A.* (Ed.), p. 232, Springer Verlag, Braunschweig, 1955
353 *Bergmann, K.:* J. Polym. Sci. Polym. Phys. Ed. 16, p. 1611 (1978)
354 *Bark, M., Schulze, C., Zachmann, H. G.:* Polym. Prepr. 31, p. 106 (1990)
355 *Hosoda, S., Kojima, K., Furuta, M.:* Makromol. Chem. 187, p. 1501 (1986)
356 *van Ruiten, J., Boode, J. W.:* Polymer 33(12), p. 2549 (1992)
357 *Marinow, S., May, M., Hoffmann, K.:* Plaste u. Kautschuk 12, p. 682 (1983)
358 *Mirabella, F. M., Jr., Westphal, S. P., Fernando, P. L., Ford, E. A., Williams, J. G.:* J. Polym. Sci., Part B: Polym. Phys. 26, p. 1995 (1988)
359 *Kanig, G.:* Progr. Colloid Polym. Sci. 57, p. 176 (1975)
360 *von Bassewitz, K., zur Nedden, K.:* Kautschuk Gummi Kunsstoffe 38, p. 42 (1985)
361 *Barham, P. J., Hill, M. J., Keller, A., Rosney, C. C. A.:* J. Mater. Sci. Lett. 7, p. 1271 (1988)
362 *Barham, P. J., Hill, M. J., Keller, A., Rosney, C. C. A.,* in: Integration of Fundamental Polymer Science and Technology. *Lemstra, P. J., Kleintjens, L. A.* (Eds.), Vol. 3, p. 291, Elsevier, Amsterdam, 1989
363 *Hill, M. J., Barham, P. J., Keller, A., Rosney, C. C. A.:* Polymer 32(8), p. 1384 (1991); *Hill, M. J., Organ, S. J., Barham, P. J.,* in: Thermal Analysis and Calorimetry in Polymer Physics. *Mathot, V. B. F.* (Ed.), Special issue Thermochimica Acta, Elsevier Science Publishers, Amsterdam, 1994
364 *van Ruiten, J., Pijpers, T.,* in: Polym. Prepr. (Am. Chem. Soc. Div. Polym. Chem.), ACS Meeting on Recent Advances in Polyolefin Polymers, Washington, D.C., 1992
365 *Nadkarni, J., Jog, J. P.,* in: Handbook of Polymer Science and Technology. Vol. 4, *Cheremisinoff, N. P.* (Ed.), p. 81, Marcel Dekker, New York, 1989
366 *Stafford, B. B.:* J. Appl. Polym. Sci. 9, p. 729 (1965)
367 *Clampitt, B. H.:* J. Polym. Sci., Part A3, p. 671 (1965)
368 *Sato, T., Takahashi, M.:* J. Appl. Polym. Sci. 13, p. 2665 (1969)

297

369 *Marinow, S.:* Plaste u. Kautschuk 4, p. 202 (1982)
370 *Radusch, H.-J.:* Plaste u. Kautschuk 4, p. 212 (1982)
371 *Hoffmann, K., Schmutzler, K.:* Plaste u. Kautschuk 2, p. 77 (1983)
372 *Marinow, S., Martin M., Hoffmann, K.:* Plaste u. Kautschuk 11, p. 620 (1983)
373 *Socher, M., Häußler, L., Müller, U., Stephan, M.:* Plaste u. Kautschuk 6, p. 310 (1983)
374 *Bailey, Jr., F. E., Walter, E. R.:* Polym. Eng. Sci. 15(12), p. 842 (1975)
375 *Wild, L., Chang, S., Shankernarayanan, M. J.:* Polym. Prepr. (Am. Chem. Soc. Div. Polym. Chem.) 31(1), p. 270 (1990)
376 *Okabe, M., Mitsui, K., Sasai, F., Matsuda, H.:* Polym. J. 21(4), p. 313 (1989)
377 *Parikh, D. R., Childress, B. S., Knight, G. W.,* in: Proc. Annual Technical Conf. (Antec) '91, p. 1543, 1991
378 *Phuong-Nguyen, H., Delmas, G.:* Macromol. 25, p. 414 (1992)
379 *See, B., Smith, T. G.:* J. Appl. Polym. Sci. 10, p. 1625 (1966)
380 *Blackadder, D. A., Schleinitz, H. M.:* Polymer 7, p. 603 (1966)
381 *Koenig, J. L., Carrano, A. J.:* Polymer 9, p. 359 (1968)
382 *Koenig, J. L., Carrano, A. J.:* Polymer 9, p. 401 (1968)
383 *Schreiber, H. P.:* J. Appl. Polym. Sci. 16, p. 539 (1972)
384 *Schreiber, H. P.:* Polymer 13, p. 78 (1972)
385 *Ogawa, T., Tanaka, S., Inaba, T.:* J. Appl. Polym. Sci. 17, p. 319 (1973)
386 *Motooka, M.:* Jpn. Pat. Appl. Discl. No. 59-68307, 1984

Nomenclature

A	area belonging to a DSC curve
B	number of end- and side-chain groups per 1000 C
B_e	number of end-groups per 1000 C
C_p	heat capacity at constant pressure
D	density
dQ	amount of heat transferred
G	free enthalpy or Gibbs free energy
H	enthalpy
l	lamellar thickness
$l_{extr \to T}$	extrapolation of melt part of DSC curve to T
m	mass
M	molar mass
M^*	apparent molar mass as determined by SEC in the case of branching
s	sequence length
S	entropy
S_c, S_h	scanning rates in cooling and heating respectively
T	temperature
t	time
T_c	(DSC) crystallization (peak) temperature
$T_{c,h}$	(DSC) highest crystallization temperature
T_d	dissolution temperature
T_g	glass transition temperature
T_m	(DSC) melting peak temperature
$T_{m,h}$	(DSC) highest melting temperature
T_m^o	equilibrium crystal–melt transition temperature
T^*	point of intersection of DSC curve with extrapolation from the melt part of the curve
w	mass fraction
W	mass percentage
$w^c(T)$	crystallinity
x	mole fraction
X	mole percentage, number of C atoms in the main chain
Δ	difference between reference states
$\Delta g(T)$	specific driving force of melting at T, specific free enthalpy differential function
$-\Delta g(T)$	specific driving force of crystallization

ΔT	supercooling
σ_e	end-surface free enthalpy
σ_s	side-surface free enthalpy

(Specific quantities are indicated by lower-case letters.)

Abbreviations

A	analytical
CSD	compact semi-crystalline domain
DE	direct extraction
DSC	differential scanning calorimetry/calorimeter
EO	ethylene-1-octene copolymer
EP	ethylene propylene copolymer
EPDM	ethylene propylene diene terpolymer
ESL(D)	ethylene sequence length (distribution)
HDPE	high-density polyethylene (low-pressure process)
LAM	longitudinal acoustic mode as determined by Raman spectroscopy
LCB(D)	long chain branching (distribution)
LDPE	low-density polyethylene (high-pressure process)
LLDPE	linear low-density polyethylene
LPE	linear polyethylene
MFI	melt flow index
MMD	molar mass distribution
P	preparative
P_{ij}	probability of addition of a j unit to a chain ending in an i unit
P set	set of chain propagation probabilities of (co)mononomer incorporation in the chain
PVC	polyvinyl chloride
r set	set of reactivity ratios
SCB(D)	short chain branching (distribution)
SEC	size exclusion chromatography or gel permation chromatography
TREF	temperature rising elution fractionation
UHMWPE	ultra-high molecular weight polyethylene
ULDPE	ultra-low-density polyethylene or VLDPE
VLDPE	very low-density polyethylene

Subscripts and superscripts

a	amorphous
b	base line
c	crystal, comonomer
cc	cooling curve
i, j	numbers: $i, j = 1, 2$ or 3
e	excess, ethylene
g	glass
hc	heating curve
m	melt, monomer
n	number-average
p	propylene
R	Raman
s	solid
s:e	ethylene sequence of length s
s:p	propylene sequence of length s
S	sample
v	viscosity-average
w	weight-average
z	z-average
4	1-butene as comonomer
8	1-octene as comonomer
*	initial

299

Chapter 10

Microcalorimetry

Luc Benoist, SETARAM, 7, rue de l'Oratoire, 69300 Caluire, France

Contents

10.1 Introduction

Microcalorimeters using the fluxmeter principle were developed by Calvet, following the work done by A. Tian. An innovation was to measure the heat exchanged and not the temperature. This technique has many advantages, for example it enables direct access to the heat of reaction and presents the possibility of investigating both fast and slow reactions. The temperature of the calorimeter and consequently that of the sample is pre-set by the operator.

The design of the Calvet microcalorimeters includes a space surrounded by heat-flux transducers (Fig. 10.1). The top of this cylindrical space is open; this allows the operator to introduce and remove any vessels or set-ups. These must have the correct external diameter and their height must not exceed that of the heat-flux transducer. The open top is also used to connect the system under investigation to the outside of the calorimeter. The link is optical (beam–material interaction), electrical (wires to a probe), flow (liquid and gas through pipes) or mechanical (stirring etc.). Any type of set-up can be introduced into the space surrounded by the heat-flux transducers in the Calvet microcalorimeters. The temperature is controlled and the heat transfer is monitored with a high degree of accuracy.

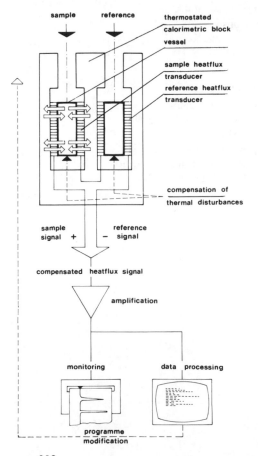

Fig. 10.1. Calorimeter: schematic diagram

Calorimeters are instruments that determine the behavior of a material and its reaction in a specific environment in which the pressure, the temperature and the surrounding atmosphere can be controlled. The requirements of scientists have led Setaram to develop a range of calorimeters whose designs have been optimized for some of the following criteria:
- versatility
- high-sensitivity investigations
- high-temperature studies
- low-temperature tests
- large samples to be investigated
- gas–solid interaction
- liquid–liquid and liquid–solid interaction
- scanning facility
- simultaneous thermogravimetric and calorimetric investigation
- heat conductivity of liquids and gases
- vapor pressure determination and enthalpies of evaporation.

The optimization of the calorimeters has resulted in the following instruments.
- The high-temperature microcalorimeter (HT 1000) (ambient to 1000 °C) designed to perform investigations at high temperatures with a high sensitivity. The crucibles have a capacity of 15 cm^3. There is a slow scanning facility.
- The low temperature microcalorimeter (BT 215) (-196 °C to $+200$ °C) designed to work at low temperatures with a high sensitivity. Its temperature can be varied by scanning (Fig. 10.2).

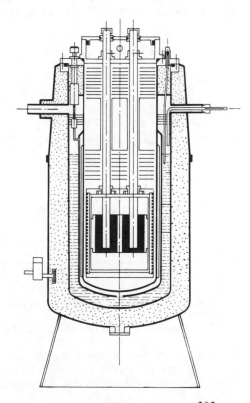

Fig. 10.2. BT 215: cross-section

- The standard microcalorimeter (MS 80) (ambient to 200 °C) designed for high-sensitivity research with samples of volumes up to 100 cm³. Used mainly under isothermal conditions.
- The micro-DSC batch and flow calorimeter designed for high sensitivity of heat-flow measurement under both isothermal (0–85 °C) and temperature scanning conditions (−20 °C to +120 °C). Vessel capacity up to 1 cm³.
- The C 80 calorimeter, the most versatile calorimeter with an optimized temperature range (ambient to 300 °C) and the widest range of vessels and set-ups (standard, high-pressure, batch and flow mixing, heat conductivity of liquids and gases, enthalpies of evaporation, vapor pressure).
- The HV 108 calorimeter designed for large samples. This type of calorimeter can have a volume of up to 1 l and energy transfer can be measured down to 10 µW in the temperature range 10–50 °C.

10.2 BT 215 microcalorimeter
10.2.1 Glass transition of elastomers (Fig. 10.3)

The (de) vitrification of a polymer determines its product application. Elastomers have very low glass transition temperatures. Glass transition is characterized by a change in heat capacity, which can be observed during upward or downward scanning. Even though it uses low scanning rates, the low-temperature BT 215 microcalorimeter can be used for such determinations, particularly for revealing small heat effects that can occur below (at low temperatures) or above the glass transition temperature. The example (Fig. 10.3) shows the glass transition of a polyurethane in heating, followed by an endothermic transformation.

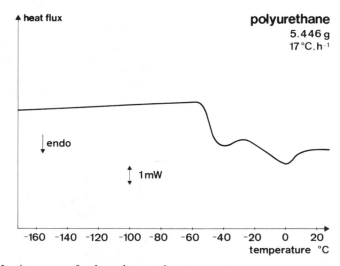

Fig. 10.3. Heating curve of polyurethane at low temperatures

10.2.2 Ice formation in porous materials (Fig. 10.4)

The 'low-temperature' microcalorimeter has been useful in the study of the behavior of cement under freezing conditions [1]. Samples of hardened cement paste are saturated in water before the test. The recordings obtained during the cooling of the

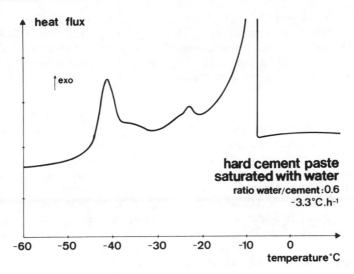

Fig. 10.4. Cooling curve of hard cement paste at low temperatures

sample show various crystallization peaks, which correspond to the amount of water held in the differently sized pores. The distribution of pore sizes in the cement can be deduced from the diagram thus obtained. With the aid of low-temperature calorimetry, the behavior of cement subjected to low temperatures at various degrees of humidity can be investigated.

10.2.3 Wax crystallization in mineral oils (Fig. 10.5)

The presence of wax in oils and fuels used in engines is a very serious problem, especially when these engines are operated at low temperatures. The sensitivity of the low-temperature calorimeter allows the study of wax crystallization at a low rate of cooling for the purpose of simulating natural temperature variation during the night.

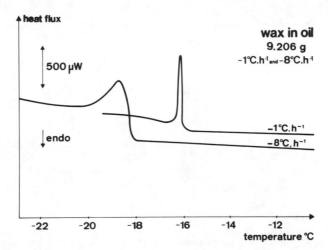

Fig. 10.5. Wax crystallization in oil

305

It should be noted that when the cooling rate varies from 8 °C/h to 1 °C/h, a considerable difference (of more than 2 °C) in the temperature of crystallization is observed. Depending on the climatic conditions under which these oils will be used, the corresponding cooling rate must be chosen so that the wax crystallization temperature can be accurately determined.

10.2.4 Thermoporometry

The size of a material's pores largely determines its properties. Various methods are used for pore measurement: mercury porosimetry, 'desorption nitrogen', etc. Thermoporometry can use the curves produced by the 'low-temperature' microcalorimeter when a pure substance in the porous material under investigation changes state.

The temperature of melting or crystallization of this substance depends on the size of the pores under investigation: it is higher when the pores are large, and lower when they are small. Irregular pore sizes lead to various melting points or crystallization of the pure substance: the curve reveals a change of state over a wide temperature range. The measurement is carried out after immersion in liquid: the medium used is of important when the porous structure itself is affected by immersion, as in the case with polymers.

The method applies to melting curves (e.g. melting of benzene in the pores of a styrene–divinyl benzene resin) (Fig. 10.6) or crystallization curves (e.g. crystallization of benzene in the pores of alumina). Two heat effects can be observed in either the cooling stage or the heating stage: a sharp peak due to the transformation of the excess liquid, and a gradual peak caused by the change of state of the substance in the pores. With the aid of this second peak it is possible to determine the pore sizes and their total volume at each temperature [2].

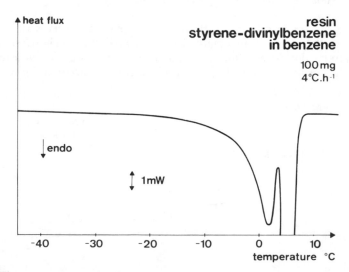

Fig. 10.6. Thermoporometry: heating curve

10.3 C 80 calorimeter

Figures 10.7 and 10.8 show a cross-section of this calorimeter. Various experimental vessels are used, depending on the application: the closed standard vessel for

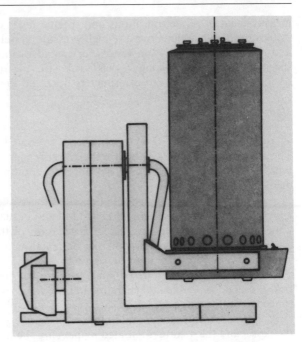

Fig. 10.7. C 80 calorimeter

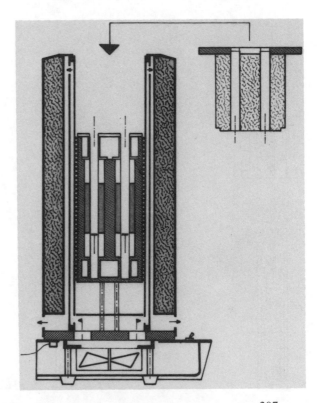

Fig. 10.8. C 80 calorimeter:
cross-section

measurement of phase transitions and specific heats of solids, the vacuum vessel for evaporation and sublimation, the high-pressure vessel (100 bars) for pressure studies, and gas or liquid circulation vessels for gas adsorption, solid–liquid interaction and specific liquids. The mixing vessels used with the rotating device have a different design from those used for calorimeters operated in the static mode.

10.3.1 Mixing vessels

These are composed of two compartments separated by either a lid or a membrane. The samples are introduced before the vessel is placed in the calorimeter. Two main types of mixing vessel have been developed: a rotating mixing vessel with a lid and a membrane mixing vessel.

In the rotating mixing vessel (Fig. 10.9) the two compartments are separated by a metallic lid. A drop of mercury around the lid ensures a seal between the lower and upper parts. The samples are mixed by rotation of the calorimeter. There are many applications of this technique: mixing or reaction of two liquids or a liquid and a solid in dissolution, hydration or wetting.

The rotating mixing vessel is not suitable for some applications because of the risk of reactions between the mercury and the samples or because a mechanical stirrer is needed for viscous samples. In such cases the membrane mixing vessel must be used (Fig. 10.9). A membrane of aluminum or PTFE separates the two compartments of the vessel and the mixing is initiated by piercing the membrane with a rod, which then also acts as a stirrer. For some applications the stirring can be motorized. Both

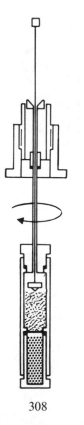
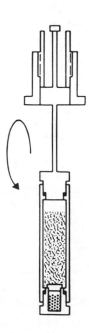

Fig. 10.9. Mixing vessels

vessels can be used for either isothermal or temperature-programmed experiments. The following examples illustrate the versatility of a calorimeter employing mixing vessels and the rotating system.

10.3.2 Dissolution of potassium chloride (Fig. 10.10)

The calorimeter is calibrated electrically using the Joule effect. The sample of potassium chloride dissolved in water is used to check the accuracy and precision of the calorimeter. The heat of dissolution determined with the aid of the calibration curve is 17 680 J/mol, which is very close to the mean value of 17 520 J/mol found in the literature.

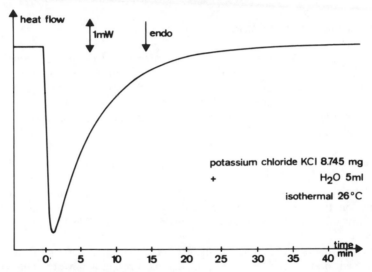

Fig. 10.10. Dissolution of KCl in water

10.3.3 Heat flow monitoring during mixing

The symmetrical design allows the detection of very small quantities of heat, even during the rotation of the calorimeter. For example, in the dissolution of the organic compounds pyrazole and pyrazolone derivatives in a buffer solution (pH 7) at 29 °C, this design shows its sensitivity. A test run with a buffer solution in both the measurement and reference vessels shows that the disturbance due to the rotation of the calorimeter is very small. The heat of dissolution is 0.135 J for pyrazole derivatives and 0.085 J for pyrazolone derivatives, for a sample mass of 1.85 mg.

10.3.4 Heat of mixing of polymer alloys: PPO/PS

Direct measurement of the heat of mixing (interaction) of polymers is not feasible due to their high viscosity. The indirect approach described here is a heat of solution method, in which a Hess's Law cycle is used to determine the heat of mixing from the individual heats of solution of the alloy and its constituent polymers in a common solvent [3].

Polymer A + Solvent $\xrightarrow{\Delta H_s^A}$ Solution of A

    +   +

Polymer B + Solvent $\xrightarrow{\Delta H_s^B}$ Solution of B

  $\Big\downarrow \Delta H_m$ $\Big\downarrow \Delta H_s^S$

Alloy AB + Solvent $\xrightarrow{\Delta H_s^{AB}}$ Solution of AB

$$\Delta H_s^A + \Delta H_s^B + \Delta H_s^S = \Delta H_m + \Delta H_s^{AB}$$

With this method the specific heat of mixing of a 75/25 m/m% poly(dimethyl) phenylene oxide–polystyrene alloy has been determined as 4.9 ± 0.2 J/g.

10.3.5 Mixing of viscous products

A membrane mixing vessel is used for the investigation of the polycondensation of a polyisocyanate (264 mg) and a polyol (292 mg) resulting in a polyurethane (Fig. 10.11). The test is run isothermally at 100 °C. The two monomers are mixed by piercing the membrane with the rod, which is then used to stir the mixture. With the aid of the recorded exothermic peak it is possible to determine the reticulation rate as a function of time.

10.3.6 Flow vessel for liquid adsorption on powders (Fig. 10.12)

The applications of many powdered compounds (catalysts, oxides, coals, etc.) are dependent on their surfaces. By measuring the physical or chemical interactions between any solute and the surface of powders it is possible to characterize their reactivity.

Flow calorimetry is suitable for such an investigation. A special flow vessel has been designed to allow the percolation of liquids through a powder. The solutes are introduced in a carrier liquid. Adsorption and desorption reactions can be investigated, such as hydrocarbons on a coal powder: the carrier liquid n-heptane flows, at a rate of 0.5 ml/min, through the powdered coal (164.0 mg) in order to wet it. Pure n-heptane is then substituted by a solution of butanol. An exothermic effect corresponding to butanol adsorption is recorded at 32 °C:

$$q = 4.39 \text{ J/g of coal}$$

After the substitution of the butanol solution by pure n-heptane, the butanol is desorbed and the absolute value of the heat is found to be the same as during the adsorption process. The butanol is completely desorbed, which is not the case with the $C_{32}H_{66}$ solution. The specific heat of adsorption (18.2 J/g) is greater than the specific heat of desorption (8.5 J/g).

10.3.7 Vapor pressure and enthalpy of evaporation of liquids

A special set-up has been designed to measure the vapor pressure and heats of evaporation of liquids up to 200 °C. It fits inside the C 80 calorimeter (Fig. 10.13).

The principle is to evaporate a liquid in a container at a constant rate. This is achieved by creating a vacuum, with the aid of a capillary. Depending on the vapor

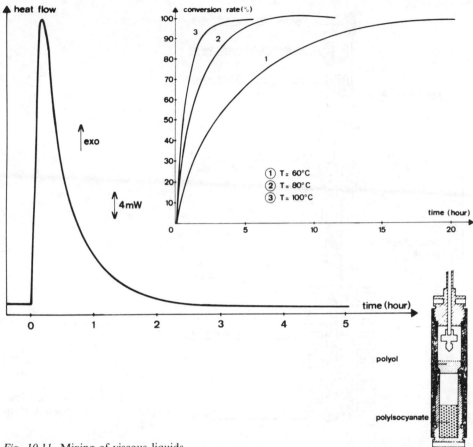

Fig. 10.11. Mixing of viscous liquids

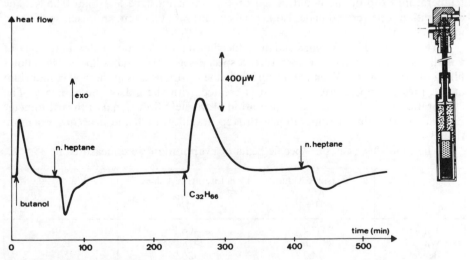

Fig. 10.12. Liquid adsorption

311

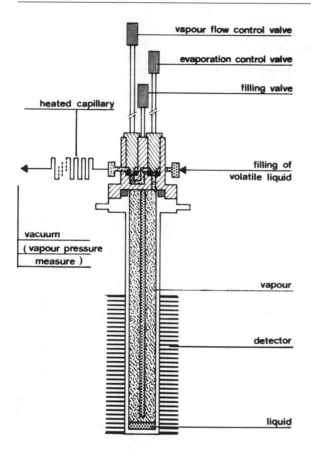

vapour flow control valve

evaporation control valve

filling valve

heated capillary

filling of
volatile liquid

vacuum
(vapour pressure
measure)

vapour

detector

Fig. 10.13. Set-up for the determination of evaporation enthalpy

liquid

pressure, different types of capillaries have to be used in order to minimize the
temperature drop of the liquid in the vessel, which must be less than 0.05 K. The
evaporation is performed under isothermal conditions at a temperature chosen for the
measurement.

The vessel is filled, weighed and introduced into the C 80 calorimeter. In the case of
volatile liquids at ambient temperature, a special set of inlet tubes and valves allows
filling under pressure. When the set-up inside the calorimeter is in thermal equilibrium
at the preset temperature, a vacuum is created with the aid of the capillary. The
evaporation takes 1–3 h and the integration of the heat-flow signal of the calorimeter
provides the enthalpy of evaporation with a repeatability of 1 % and an accuracy of more
than 3 %.

The following values of specific heats of evaporation were measured at 35.5 °C.

	Measured values Δh_{exp} J/g	Literature values Δh_{th} J/g	$\dfrac{\Delta h_{exp} - \Delta h_{th}}{\Delta h_{th}}$ %
Toluene	397.4	406.1	2.14
Water	2375.852	2418.045	1.75
Benzene	414.2	426.0	2.77

The vapor pressure can also be measured by applying a small counter-pressure of helium at the outlet of the capillary. This pressure is monitored by an accurate gauge (0.01 bar) (Figs. 10.14, 10.15). The pressure is slowly increased stepwise until the

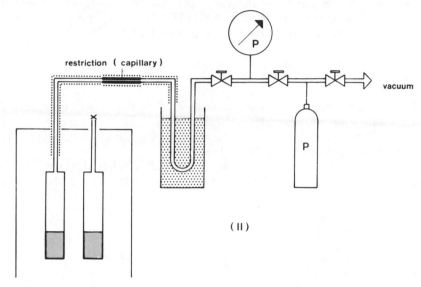

Fig. 10.14. Vapor pressure determination

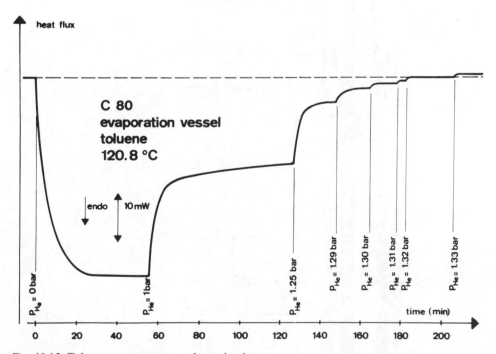

Fig. 10.15. Toluene: vapor pressure determination

evaporation is stopped, i.e. the heat-flow signal returns to the base line. The vapor pressure of the liquid at the preset temperature is equal to the counter-pressure of the helium which stopped its evaporation. The repeatability of the measurement is of the order of 0.5 % and the accuracy is 1 %.

By following this procedure the vapor pressure of toluene was measured at 120.8 °C (Fig. 10.15) and was found to be 0.135 MPa (1.33 bars) which is in very good agreement with the value found in the literature [3] (1.326 bar).

10.3.8 Thermal conductivity of liquids and gases
10.3.8.1 Principle

The thermal conductivity of a material characterizes its ability to transfer heat. This is an important property, especially in chemical engineering, but it cannot always be found in the literature. It can now easily be measured by means of a new set-up which fits inside the C 80 calorimeter (Fig. 10.16).

The design of the set-up is based on the coaxial cylinders principle. The fluid, whose conductivity is measured, is located between two cylinders: a container and an

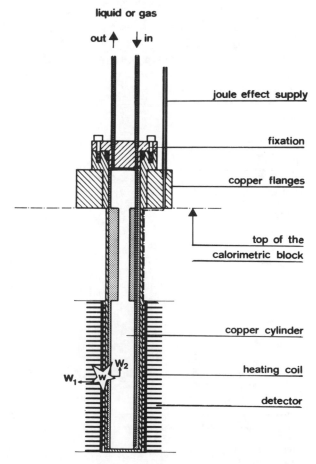

liquid or gas

out ↑   ↓ in

joule effect supply

fixation

copper flanges

top of the
calorimetric block

copper cylinder

heating coil

detector

W_2
W_1 W

Fig. 10.16. Set-up for thermal conductivity determination

314

inserted cylinder. The latter is made of copper, which shows a high degree of heat conductivity. The former, of stainless steel, fits into the space surrounded by the heat-flux transducer.

Inside the wall of the container there is a heating coil in which a constant power is generated by the Joule effect. The power W will flow from the heating coil into the heat sink of the calorimeter by two paths. One part, W_1, will flow through the heat-flux transducer and the other part, W_2, through the fluid into the inserted cylinder to the heat sink. W_1 is monitored by the heat-flux transducer (Fig. 10.16).

$$W = W_1 + W_2$$

$$W_1 = W \frac{\dfrac{1}{\Gamma'} + \dfrac{1}{\Upsilon}}{\dfrac{1}{\Gamma'} + \dfrac{1}{\Gamma} + \dfrac{1}{\Upsilon}}$$

where Υ is the thermal conductance of the fluid between the cylinders

$$\Upsilon = k \times \lambda \qquad (\lambda \text{ is the thermal conductivity of the fluid})$$

Γ is the thermal conductance of the heat-flux transducer, Γ' is the thermal conductance of the copper insert, W is the power generated in the coil by the Joule effect, W_1 is the power monitored by the heat-flux transducer, W_2 is the power transferred by the copper insert.

10.3.8.2 Calibration and measurement (Fig. 10.17)

The power W generates a calorimetric signal S at a given temperature with a fluid of thermal conductivity of λ. From the above equation, the following one is calculated:

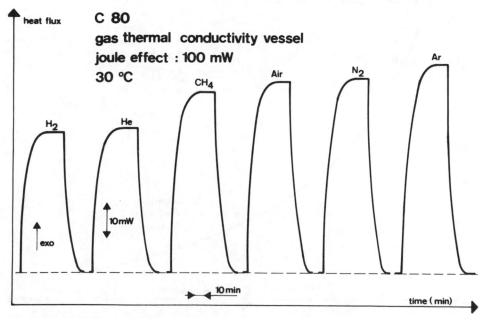

Fig. 10.17. Thermal conductivity of gases

315

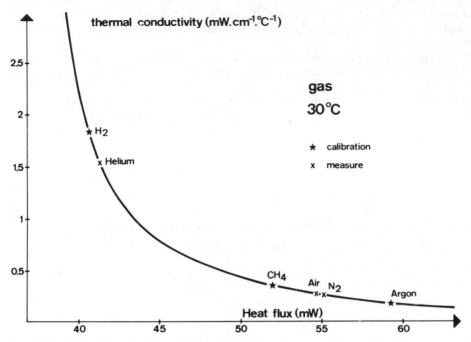

Fig. 10.18. Thermal conductivity of gases: calibration curve

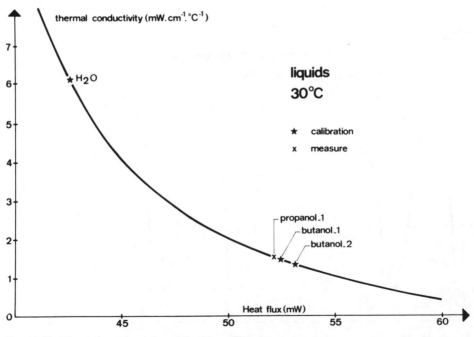

Fig. 10.19. Thermal conductivity of liquids: calibration curve

$$\lambda = \frac{-S + A}{BS + C}$$

where A, B and C are constants.

Three pairs of (λ, S) values are required to determine the curve at a given temperature. For the determination of the thermal conductivity of a gas, hydrogen, methane and argon are chosen as standards (Fig. 10.18). A calibration curve is drawn and the thermal conductivity coefficients of helium, air and nitrogen are calculated. The difference between these values and those given in the literature is $\sim 1\%$, and the repeatability of the measurements is better than 1%. With the aid of this method it is possible to differentiate between the thermal conductivity coefficients of air and nitrogen, which are very similar.

For liquids, water, butanol 1 and butanol 2 are chosen as standards for the determination of the curve, which is then used to measure the thermal conductivity of propanol 1 (Fig. 10.19). The repeatability of this method is better than 1% and its accuracy is 1%.

References

1 *Sellevold, E. J., Barger, D. H.*, in: Durability of Building Materials and Components. *Sereda, P. J., Litvan, G. G.* (Eds.) STP 691, American Society for Testing and Materials, Philadelphia, PA, p. 439, 1980
2 *Brun, M., Quinson, J. F., Eyraud, C.:* L'Actualité Chimique, p. 21, Oct. 1979
3 *Aukett, P. N., Brown, C. S.:* J. Therm. Anal. 33, p. 1079 (1988)
4 *Riddick, J. A., Bunger, W. B..:* Organic Solvents: Physical Properties and Methods of Purification, Third Edition, Vol. II, Wiley Interscience, 1970

Nomenclature

ΔH_m	enthalpy of mixing
ΔH_s^A	heat of solution of A
ΔH_s^B	heat of solution of B
ΔH_s^{AB}	heat of solution of alloy AB
W	total power generated in the coil
W_1	power monitored by the transducer
W_2	power transferred by the copper insert
Γ	thermal conductance of the transducer
Γ'	thermal conductance of the copper insert
Υ	thermal conductance of the fluid
λ	thermal conductivity of the fluid

Chapter 11

Reaction Calorimetry for Polymerization Studies

Rudolf Riesen, Mettler-Toledo AG Analytical, CH-8603 Schwerzenbach, Switzerland

Contents

11.1 Introduction to reaction calorimetry
11.1.1 The application field of reaction calorimetry

The chemical industry is currently undergoing some major changes as far as products are concerned. Many companies are moving away from the more commodity types of chemicals or polymers into producing highly specialized compounds. As a result, many new processes are being developed, all of which have to be optimized. This means maximizing product performance and quality and minimizing processing costs. To achieve this, the end-products must be characterized fully, accurately and quickly in terms of quality and performance.

Thermoanalytical methods can be used for this, as discussed below. But not only is the testing of end-products important, the production process also must be known in detail, i.e. the way in which reactants are converted to products and the optimization of the processes with respect to time and costs, e.g. yield, quality, environmental consequences and chemical hazards. For this, quantitative calorimetric information is needed, such as:

- thermophysical properties of reaction mixtures
- heat generation rates due to reactions
- thermodynamic data of chemical reactions
- kinetic analysis data
- heat-transfer properties.

To shorten process development, reliable scale-up calculations have to be performed. The necessary data must be determined under conditions representative of the actual industrial process configuration. A recent approach is to conduct the measurements in a bench-scale reaction calorimeter system, e.g. in the Mettler RC1. In this system, typical process parameters that can be controlled in the fully automated mode are reaction temperature, stirring speed, feed rates of reagents in semibatch operation, and pressure, pH and other selectable variables.

All these process parameters are monitored, e.g. displayed, printed and stored, and lead in a synergistic way, together with possible visual observations, to a detailed qualitative analysis of the on-going process. The calorimetric data from these records give quantitative information on the heat production rate, heat of reaction, specific heat and heat-transfer coefficient, and hence may provide information on the following topics for process design:

- necessary cooling power
- rate of reaction (global reaction kinetics)
- reactant accumulation and potential adiabatic temperature increase (safety implications)
- heat-transfer coefficient for scale-up studies.

These data are basic to many fields of application, from research and development to production of chemicals and safety investigation.

11.1.2 Illustrative example (nitration)

Every exothermic reaction that shows a potential adiabatic temperature rise sufficiently high to initiate evaporation or decomposition of the reaction product is hazardous. Typical reactions of this type are nitrations, diazotizations, etc., but also polymerizations (see sections 11.3 and 11.4).

In the following example, the safety of a production procedure for the nitration of a substituted nitrobenzene had to be improved [1]. An isothermal measurement in

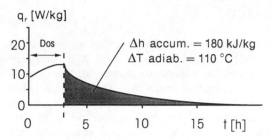

Fig. 11.1. Nitration of a substituted nitrobenzene. Reproduction of the existing plant process by an isothermal reaction calorimetry experiment at 80 °C. The feed of mixed acid is stopped after addition of an equimolar amount. The heat production rate due to reaction is monitored as a function of time. The heat evolved during the feed period amounts to 90 J/g of reaction mixture. The black area represents the accumulation of unreacted product after the feed is stopped

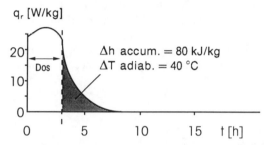

Fig. 11.2. Nitration of a substituted nitrobenzene at the modified process temperature of 100 °C (see Fig. 11.1): the heat evolved during the feed period amounts to 190 J/g of reaction mixture

a reaction calorimeter simulating the production process showed a significant accumulation of unreacted material after dosing of the reactants (Fig. 11.1). This could lead to a runaway reaction within an hour in the case of loss of stirring or cooling power. Under such adiabatic conditions heat accumulation would raise the temperature to 190 °C. At this temperature a very strong exothermic hazardous decomposition might start within hours, as DSC experiments revealed. Thus, to prevent excessive reactant accumulation the reaction rate should be increased with respect to the dosing rate. The optimum conditions found are shown in Fig. 11.2, demonstrating that the higher temperature is safer due to the faster reaction, as is apparent from the higher reaction power. The adiabatic temperature rise of only 40 °C would lead to a temperature of 140 °C, far below the rapid decomposition point.

However, the high reaction rate of the new chemical process might exceed the cooling power of the plant, with the result that the reactor would go out of isothermal control. Here, the accurate measurement of the heat production rate helped to find the optimum feed rate of the mixed acid in such a way that the nitration was dosing controlled. Not only was the safety of the process increased, but also the yield. In the new process the reaction time is about half that of the old one: productivity has been doubled by applying the reaction calorimetry method.

321

11.2 Instrumentation (RC1)

A reaction calorimeter should be operated under conditions that resemble the plant-scale as closely as possible. Hence, the RC1 is a computer-controlled batch reactor developed from practical experience to carry out isothermal and adiabatic as well as linear heated or cooled experiments. The principle of the RC1 reaction calorimeter is based on the continuous measurement of the temperature difference between the reactor contents and the heat carrier oil in the external jacket. An extremely efficient thermostat ensures rapid and exact adjustment of the jacket temperature to the required value (Fig. 11.3) [2 4]. The RC1 reaction calorimeter incorporates the following features:

– various glass or metal reactors—construction similar to industrial vessels (effective volume 0.25–2 l), for easy visual observation of physical or chemical changes
– reactor cover with standard taper joints for widely different external fittings, e.g. reflux condenser
– reactors for pressures up to 60 bar
– powerful stirrer motor with adjustable speed control for various types of stirrer
– temperature range from $-20\,°C$ to $200\,°C$ (depending on the cooling facilities) with a high temperature stability
– calorimetric sensitivity (peak-to-peak noise) of 0.2 W in a range of 0–500 W
– versatile RD10 dosing controller for additional measurement and control purpose
– manual and automatic operation.

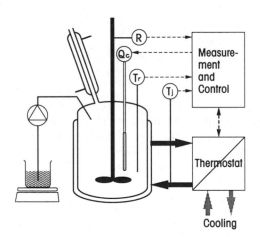

Fig. 11.3. Schematic illustration of the Mettler RC1 reaction calorimeter system, showing the combination of heat-flux calorimetry, using a fast thermostatic unit, and realistic process operations (e.g. dosing, stirring or refluxing) in a bench-scale reactor

The heat-flow balance around the reactor includes the following individual heat-flow terms [3]. The heat generation rate of the reactor content Q_r, the heat flow through the reactor wall Q_{flow}, the calibration power Q_c, heat effects by dosing Q_{dos}, the cooling power in the reflux condenser Q_{reflux} and heat accumulation due to temperature changes Q_{accum}:

$$Q_r = Q_{flow} + Q_{accum} + Q_{dos} + Q_{reflux} - Q_c$$

Q_b is the so-called base line, and is therefore equal to Q_r if no chemical reaction or physical transition is present. (In case of specific quantities, lower-case letters are used).

11.3 Case study 1: homo-polymerization of styrene

During process development, a change in concentration of one of the starting materials caused a rapid and violent exothermic polymerization with an exponential increase in the reaction temperature and, finally, partial venting of reactor contents. This qualitative statement in the report represents standard manual lab bench operation in a reaction flask. Detailed analysis and quantitative conclusions are not possible due to lack of precise and recorded data.

Reproduction of the same experiment in the reaction calorimeter showed the same behavior, but 'post-mortem' analysis was possible: the temperature profile and the heat generation rate are shown in Fig. 11.4. To start the reaction, catalyst was added. During the first minute, a constant high heat generation rate of 500 W was observed. This exceeded the cooling capacity of the system at the low temperature of 21 °C, and led to a temperature rise which in turn increased the rate of reaction and started the expected runaway. To stop this, a maximum temperature of 80 °C was specified in the safety program. (When this limit is exceeded an emergency program will be activated which will automatically cool down the reaction mass as fast as possible to the lowest possible temperature.) As the temperature curve of the experiment shows, only 82.5 °C was reached and the residual product was rapidly cooled down to 50 °C within 75 s. Even though the run was aborted by emergency cooling, the global reaction kinetics (Fig. 11.5) and the heat of reaction as well as the specific heat and the heat transfer coefficient could be evaluated. Figure 11.5 shows the profile of the thermal conversion, which runs parallel to the chemical conversion. From the

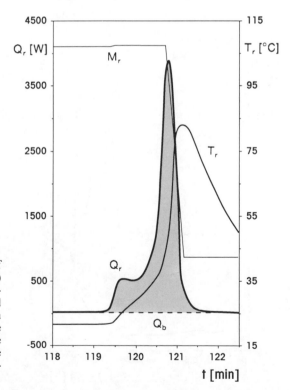

Fig. 11.4. Homo-polymerization of styrene: heat generation rate Q_r (W) with base line Q_b (W), reactor contents temperature T_r (°C) and total mass M_r (kg). The small increase in M_r at time $t = 119.3$ min shows the catalyst addition, the large decrease shows the 'spill'. Cooling, at the fastest possible range, is automatically started when T_r reaches 80 °C

323

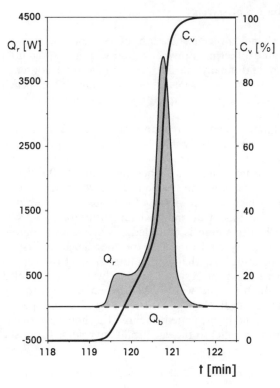

Fig. 11.5. Thermal conversion plot (curve C_v) that reflects the overall chemical kinetics of the reaction described in Fig. 11.4

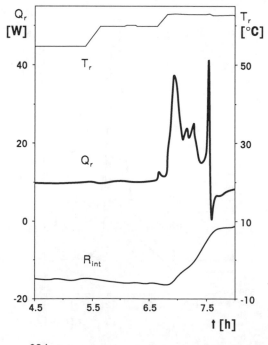

Fig. 11.6. Solution polymerization of acrylamide containing initiator, batch A. Temperature T_r was increased stepwise from 55 °C to 60 °C and 63 °C, and the heat production rate Q_r was monitored. The viscosity changes were recorded by a proportional value R_{int} (arbitrary units). After Q_r reached 40 W, the reaction was quenched by the addition of sodium sulfate solution

shape of the curve, two reaction steps can be distinguished: first, after a short initialization period, a zero-order reaction releasing approximately constant heat flow; second, an exponential rate increase to 100 % conversion within 2 min. The heat of the reaction determined as the area between the reaction power Q_r and the base line Q_b in Fig. 11.5 amounts to 146 kJ/batch, and the specific heat capacity before catalyst addition is 1.750 J/g per °C.

11.4 Case study 2: solution polymerization of acrylamide

In this example the RC1 system was used to check the test-batching possibilities for assessing the reactivity and quality of the monomers before plant production. In the solution polymerization, a water-soluble initiator was used and the reaction was started by increasing the temperature. Various grades of monomers and purities and various initiator types and concentrations were studied.

    With batch A, containing normal monomer and initiator, the temperature was increased stepwise from 20 °C to 65 °C in steps of 5 °C. The temperature was kept constant when the reaction started; this moment was determined by the heat production rate. Figure 11.6 displays this rate, the reaction power Q_r; the temperature T_r was increased in steps from 55 °C via 60 °C to 63 °C, where the reaction started. When 40 W was reached the reaction was quenched automatically by feeding of sodium sulfate solution. The on-going polymerization increased the viscosity and hence the torque required to maintain the stirring speed. A proportional value R_{int} for this is also shown in Fig. 11.6.

    In batch B, containing a different monomer quality, the reaction had already started at 60 °C. In Fig. 11.7, the corresponding signals are displayed in the same way

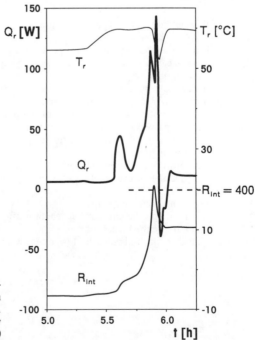

Fig. 11.7. Solution polymerization of acrylamide containing initiator, batch B, with a different monomer to batch A (Fig. 11.6): the reaction was automatically quenched after R_{int} reached a value of 400

as in Fig. 11.6. Again the reaction was quenched automatically when a high viscosity, i.e. $R_{int} > 400$, was reached.

Comparing the two batches, the influence of monomer quality can easily be seen in the recorded curves (Figs. 11.6 and 11.7). For example, the Q_r curve profile indicates a strongly different initiation behavior followed by a fast reaction. In these cases only the start of the polymerization had to be monitored.

11.5 Conclusions

Examples from practice demonstrate how easily reaction calorimetry characterizes chemical processes in terms of qualitative description of the reactions to be studied. The simultaneous and continuous monitoring of important process parameters such as temperature, heat production rate, stirring speed and torque help in a synergistic way to interpret the reaction steps. For optimization of scale-up calculations and plant design, the quantitative calorimetric information, e.g. heat of reaction, cooling power and heat transfer coefficients, can be determined from the same data. The results show how the reaction calorimeter RC1, as a pilot plant on a bench, can be used as an efficient tool for screening and test-batching the various grades of starting materials for polymerization.

References

1 *Fierz, H., Finck, P., Giger, G., Gygax, R.:* The Chemical Engineer, February, p. 9 (1984)
2 *Grob, B., Riesen, R., Vogel, K.:* Thermochim. Acta 114, p. 83 (1987)
3 *Riesen, R.:* Thermochim. Acta 119, p. 219 (1987)
4 *Riesen, R., Grob, B.:* Swiss Chem 7, No. 5b, p. 39 (1985)

Nomenclature

C_v	thermal conversion (partial integral of Q_r)
M_r	mass of the reactor content
Q_{accum}	heat accumulation due to temperature change
Q_b	base line, equal to Q_r if no chemical reaction or physical transition is present
Q_c	calibration power
Q_{dos}	heat effects by dosing
Q_{flow}	heat flow through the reactor wall
Q_r	heat generation rate of the reactor content
Q_{reflux}	cooling power in the reflux condenser
R_{int}	proportional value to the stirrer torque
t	experiment time
T_j	temperature of the jacket fluid
T_r	temperature of the reactor contents

Δh	reaction specific enthalpy
Δh_{accum}	potential specific enthalpy of reaction after dosing period
ΔT_{adiab}	adiabatic temperature increase

(Specific quantities are indicated by lower-case letters)

Chapter 12

Coupled Techniques: Thermogravimetry, Differential Thermal Analysis and Mass Spectrometry

Mark Wingfield, Polymer Laboratories, Robert-Bosch-Straße 30–34, D-69190 Walldorf, Germany

Contents

12.1 Introduction

The coupling of a thermal analysis technique with mass spectrometry provides a powerful analytical tool in which the usefulness of each technique is enhanced by the other. For instance, in the case of coupled thermogravimetry, differential thermal analysis and mass spectrometry (TG–DTA–MS), the interpretation of TG–DTA data is made easier by mass spectrometric data, while TG data allow a quantification of the mass spectrometric data. TG–DTA–MS data have therefore proved a valuable aid for the elucidation of reaction mechanisms and for the determination of the chemical composition of unknown materials [1, 2].

Presented here are the results of our experience with an apparatus where TG, DTA and MS are coupled. This has been developed by Polymer Laboratories [3] in combination with VG Quadrupoles [4].

12.2 The apparatus

The TG–DTA–MS technique requires :
– that the reaction gases are representatively transferred from the thermal analyzer to the mass spectrometer
– that the transfer time for reaction gases be short, thus allowing a real-time correlation of TG–DTA and MS data.

In order to meet these demands, a stainless steel glass-lined capillary has been built into the furnace of the STA 1500 apparatus (Fig. 12.1). The capillary terminates on

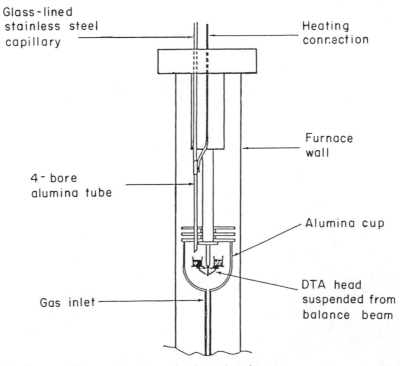

Glass-lined stainless steel capillary

Heating connection

4-bore alumina tube

Furnace wall

Alumina cup

Gas inlet

DTA head suspended from balance beam

Fig. 12.1. Gas sampling arrangement at the thermal analyzer

328

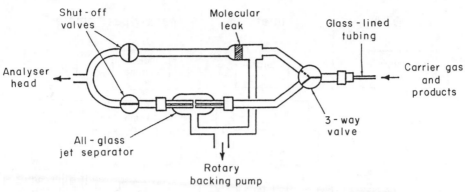

Fig. 12.2. Dual molecular leak–jet separator inlet system

reaching the hot zone of the furnace into a short length of aluminum oxide capillary, which protrudes inside the small chamber surrounding the sample. Gas is withdrawn via the capillary at a rate of 20 l/s by a two-stage rotary pump and transferred into a capillary with a by-pass inlet system where a small but representative sample is sampled into the mass spectrometer (Fig. 12.2). The capillary inlet system and mass spectrometer housing are heated to 150 °C to avoid condensation. In this way gases are transferred from the thermal analyzer to the mass spectrometer within 150 ms. Moreover, the gas mixture is unaltered up to the capillary with the by-pass inlet.

An alternative jet separator inlet (Fig. 12.2) is intended for work with helium purge gas where it enriches the gas stream with evolved gases. The jet separator exploits molecular mass-dependent diffusion. Gas is drawn down the capillary from the thermal analyzer and through a nozzle into a glass bulb by a rotary pump. Sampling into the mass spectrometer is via a capillary, which is precisely aligned with the nozzle. As the gas comes through the nozzle, it diffuses. Molecules of lower molecular mass diffuse away from the line of axis and are preferentially pumped away. Heavier molecules maintain a straighter course and are preferentially sampled into the mass spectrometer. The mass spectrometric analysis is achieved with a VG Micromass Quadrupole mass spectrometer of mass range 1–300 amu. The quadrupole mass spectrometer is favored because of its low cost and ease of use. Suitable quadrupole mass spectrometers are available for a molecular mass range up to 800 amu. The mass spectrometer analyzer is equipped with an enclosed ion source [5]. This means that levels of residual CO and CO_2 are up to 100 times lower than in the case of the conventional RGA ion source, and hence the sensitivity of this ion source for these gases is up to 100 times higher than in the case of the conventional TGA ion source. The system for TG–DTA–MS is summarized in Fig. 12.3.

12.3 Scope of the technique

Although the quadrupole mass spectrometer serves as a relatively inexpensive tool for detection of evolved gases, it cannot be straightforwardly used for quantitative gas analysis. Its ability to analyze complex mixtures—even in a purely qualitative fashion—is restricted by its low resolution. This manifests itself in two ways.

(a) Small peaks in the region of a large peak can be hidden under the tail of the large peak. This means that the purge gas must be chosen carefully. For instance with

329

TG – DTA | Inlet | Mass Spectrometer

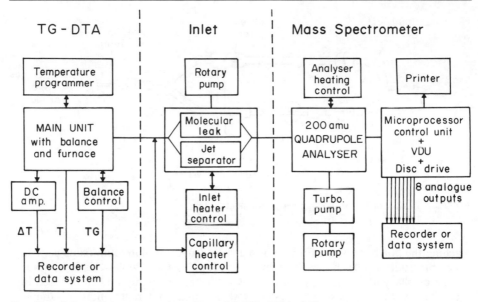

Fig. 12.3. Schematic representation of coupled TG–DTA–MS system

air or nitrogen as purge gas, peaks at 27, 28, 29 and 30 can be monitored and the monitoring of peaks at 13, 14 and 15 is difficult.

(b) The quadrupole cannot distinguish between ions of the same nominal m/e ratio. Hence, for instance, CO and N_2 cannot be distinguished at m/e 28 peak (although they can readily be distinguished from the cracking patterns). This makes the investigation of nitrogen-containing polymers, for instance, very difficult. For such an investigation NO, N_2O and NO_2 together with carbon monoxide, carbon dioxide and unburnt hydrocarbons would be of interest. Such an investigation is practically impossible, as can be seen from Table 12.1.

Another example is the study of decomposition or combustion of polyvinylchloride or chlorine-containing polymers. This is rendered impossible by the use of argon or argon-containing air as purge gas, because argon has isotope peaks at 36 (0.3 %) and 38 (0.6 %). These are present in similar ratios to the isotopes of chlorine.

The sensitivity of the quadrupole is not usually determined by the sensitivity of the detection system, but varies from gas to gas depending on how identifiable the corresponding ion is, and also on the residual gas levels within the vacuum system of

Table 12.1. Cracking patterns of commonly encountered combustion products

Gas	Peak 1 (intensity)	Peak 2 (intensity)	Peak 3 (intensity)
NO	30 (100 %)		
N_2O	30	44 ($\sim$1 %)	
NO_2	30	46 ($\sim$1 %)	
CO	28 (100 %)	12 (5 %)	16 (2 %)
CO_2	44 (100 %)	16 (9 %)	28 (8 %)
Low molecular mass hydrocarbons	29 (100 %)	27 ($\sim$30 %)	28, 29, 30 (5 %)

the mass spectrometer. Peaks at m/e 2 (hydrogen), 16 (oxygen), 17, 18 (water), 14, 28 (nitrogen), 16, 32 (oxygen), 28 (carbon monoxide), 40 (argon) and 44 (carbon dioxide) belong to the typical spectrum of residual gases in a vacuum system. Levels of these gases can be reduced by the following means:

- the use of an enclosed ion source results in significantly (up to 100 times) lower background levels of CO and CO_2
- baking the system before use and operating the mass spectrometer at elevated temperatures results in significantly lower levels of all the residual gases typically found in the vacuum system
- the inclusion of an ion pump (in addition to the turbo-molecular pump) in the vacuum system of the mass spectrometer results in significantly lower levels of hydrogen within the system
- the use of a jet separator inlet enhances detection limits for gases in the 1–100 amu range by up to a factor of 10: this is of limited usefulness, however, because it necessitates the use of helium as a purge gas.

12.4 Applications

The limitations of a quadrupole must be borne in mind when considering it for a particular application. Complex gaseous mixtures of reaction products present a problem, particularly when the gases are of low molecular mass (less than 49 amu). Despite this, a combined TAMS system can be applied to the study of a wide range of materials.

Figure 12.4 shows TG–DTA–MS curves of a partially cured phenol–formaldehyde resin. Water ($m/e = 18$), ammonia ($m/e = 17$—from the urea filler) and

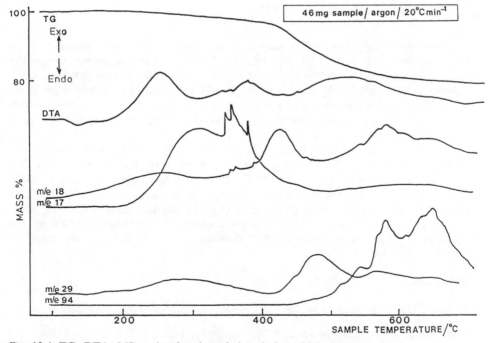

Fig. 12.4. TG–DTA–MS study of curing of phenol–formaldehyde composite

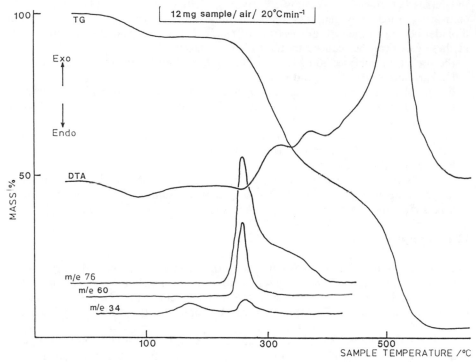

Fig. 12.5. TG–DTA–MS investigation of formation of reduction products during the combustion of wool

formaldehyde ($m/e = 94$) were monitored while heating from room temperature to 800 °C at 20 K/min. A broad exotherm in the DTA curve arising from curing of the resin coincides with evolution of water and formaldehyde. Decomposition of the polymer is accompanied by evolution of formaldehyde and phenol. Sharp peaks on the ammonia curve correspond to similar events on the water curve and DTA curve. These were assigned after investigation with hot-stage microscopy to the bursting of bubbles on the surface of the resin.

In a TG–DTA–MS investigation of the combustion of wool, water ($m/e = 18$), carbon dioxide ($m/e = 44$), sulfur dioxide ($m/e = 64$), hydrogen sulfide ($m/e = 34$), carbonylsulfide (O=C=S, $m/e = 60$) and CS_2 ($m/e = 76$) were monitored as a function of temperature. The most striking feature was the formation of numerous sulfur-containing species prior to the combustion of carbonaceous sulfur-containing material (Fig. 12.5) while also curves corresponding to the oxidation products of the combustion could be measured. An interesting new application is the study of pyrotechnique systems [6].

References

1 *Brown, M. E.:* Introduction to Thermal Analysis, Techniques and Applications. Chapman and Hall, London, 1988
2 *Wendlandt, W. W.:* Thermal Analysis. 4th Edn., Wiley Interscience, New York, 1986
3 Polymer Laboratories (Epsom) Ltd., Kiln Lane, Epsom, Surrey KT17 1JF, UK

4 V.G. Quadrupoles, Aston Way, Holmes Chapel Road, Middlewich, Cheshire CW10 0HS, UK

5 *Dawson, P. H.:* Quadrupole Mass Spectrometry and Applications. Elsevier, Barking, UK, 1976

6 *Charlsley, T., Warrington, S. S., Griffiths, T. T., Quay, J.:* Role of Simultaneous Thermal Analysis, Mass Spectrometry in the Study of Pyrotechnic Systems. Proc. 14th Int. Pyrotechnic Seminar, Jersey, p. 763, RARDE, 1989

Nomenclature

amu	atomic mass unit
DTA	differential thermal analysis
m/e	mass-to-charge ratio
MS	mass spectrometry
TG	thermogravimetry

Chapter 13

Evolved Gas Analysis of Polymers by Coupled Gas Chromatography, Fourier Transform Infrared Spectroscopy and Mass Spectrometry

Jan A. J. Jansen, Philips, Competence Centre Plastics B.V., P.O. Box 518, NL-5600 MD Eindhoven, The Netherlands

Contents

13.1 Introduction

Evolved gas analysis is a powerful technique that yields much information about polymeric materials. The thermal events that occur when a material is heated are related to the structure and thermal behavior of the material [1]. By feeding the evolved volatile components from the thermal system into a Fourier transform infrared spectrometer (FTIR) [2] and/or a mass spectrometer (MS) [3], these molecular fragments can be detected and identified. This information provides insight into the composition of the material and the chemistry of the thermal processes. The thermal events often result in complex gaseous mixtures, which complicate the interpretation of FTIR and/or MS spectra. Therefore, a separation method, coupled between the thermal system and the FTIR and/or MS, is necessary. Gas chromatography (GC) is a suitable separation method for this purpose. GC–FTIR [4] and/or MS [5] are well-known techniques for the separation of gaseous components and their subsequent detection and identification.

We developed a characterization technique capable of on-line temperature-controlled outgassing by thermal desorption (TD) followed by gas chromatographic separation of the volatiles and subsequent detection and identification by means of a parallel FTIR and MS configuration [6, 7]. The combination of FTIR and MS constitutes a powerful means of detailed characterization, with detection limits at the sub-ppb level. In addition, using off-line sampling techniques with preconcentration via Tenax adsorption cartridges, the components evolved during thermogravimetric (TG) experiments can be sampled for subsequent analysis via TD–GC–FTIR–MS.

Curie-point pyrolysis (PY) [8] of the sample and subsequent separation, detection and identification of the volatile pyrolysis products by means of GC–FTIR–MS provides a useful means of characterizing (co)polymers and blends. In Curie-point pyrolysis the sample is heated very quickly to a precisely defined temperature to obtain a reproducible and well-defined degradation process. The instrumentation, experimental set-up and some representative results are presented and discussed in this chapter.

13.2 Experimental details

The GC–FTIR–MS equipment configuration is shown schematically in Fig. 13.1. This configuration has been extended with on-line facilities for thermal desorption or Curie-point pyrolysis. The five basic components of the on-line system together with the off-line thermogravimetric system are explained in more detail below.

13.2.1 Thermal desorption

For thermal desorption experiments a thermal desorption cold trap injector (Chrompack, TCT) [9], schematically shown in Fig. 13.2, is used to perform temperature-controlled outgassing of polymeric materials. A finely cut sample is transferred to the glass tube in the desorption oven and heated via an isothermal or dynamic temperature program with a maximum temperature of 350 °C. With a helium carrier gas stream the resulting volatile components from the sample are transferred to the cold trap, which is cooled with liquid nitrogen. The components are preconcentrated at low temperatures (lower than 120 °C) in a fused silica capillary tube. After the desorption period the cold-fused silica trap is heated quickly, so that the trapped components are injected on-column into the analytical capillary column. In the design, great care was taken to avoid cold spots, and all flow paths are glass-lined.

336

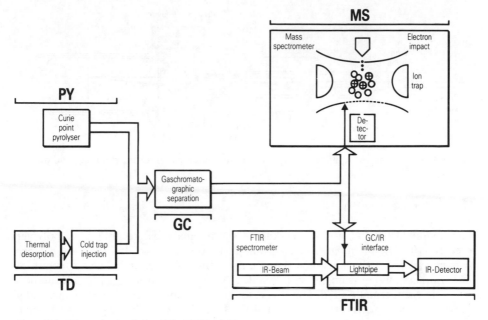

Fig. 13.1. Flow chart of the GC–FTIR–MS system

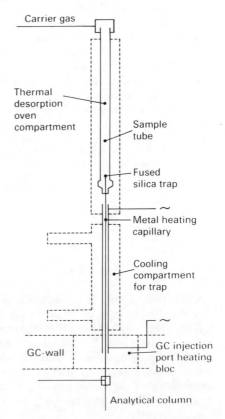

Fig. 13.2. Thermal desorption cold-trap injector

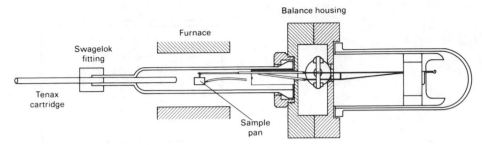

Fig. 13.3. TG equipment equipped with a Tenax adsorption cartridge for sampling and off-line analysis

13.2.2 Thermogravimetry

Thermogravimetry (TG) is a well-known thermoanalytical technique in polymer research [10]. The thermogravimetric equipment consists basically of a microbalance by means of which the mass of a material is continuously registered as a function of time (isothermally) or temperature (dynamically). This provides mass-loss information related to the thermal degradation processes of the sample. Figure 13.3 schematically shows the thermogravimetric equipment (Du Pont TG 951/990). A platinum sample cup is suspended from the arm of the microbalance, a thermocouple is mounted close to the sample, and the assembly is covered with a glass tube which is flushed with a constant gas flow. Provisions are made for working under different atmospheres, which can also be altered during the run. However, TG provides no information about the chemical nature of the mass loss. Therefore, off-line sampling of the evolved volatile components is performed via Tenax adsorption cartridges (Chrompack), which afterwards are desorbed in the thermal desorption cold-trap injector. The Tenax cartridges are connected to the oven of the TG equipment via a Swagelok fitting.

The thermogravimetric technique has several advantages over the thermal desorption cold-trap injector, as shown in Table 13.1.

Table 13.1. Some specifications of the TCT and TG equipment

	TCT	TG
Maximum temperature	350 °C	1200 °C
Maximum duration	1 h	No restrictions
Gas atmosphere	Helium	Various

13.2.3 Curie-point pyrolysis

A Curie-point pyrolyzer (Philips Scientific, 795050), shown schematically in Fig. 13.4, was used for the pyrolysis experiments. The operating principle of this apparatus is as follows. The polymer sample is deposited onto a ferromagnetic wire, which is placed in the pyrolysis unit. The wire is surrounded by an induction coil, which generates an oscillating magnetic flux within the wire. This leads to movement of the magnetic domains, as a result of which the wire is rapidly heated. When the Curie-point temperature of the wire is reached, a transition to paramagnetism occurs. The energy intake drops and, in principle, the Curie-point temperature is maintained. The

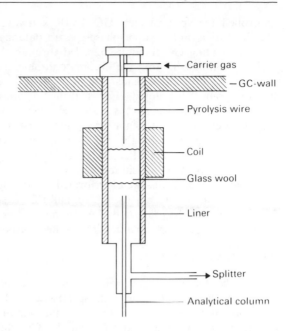

Labels on figure: Carrier gas, GC-wall, Pyrolysis wire, Coil, Glass wool, Liner, Splitter, Analytical column

Fig. 13.4. Curie-point pyrolyzer

Curie point of a pyrolysis wire is determined by its composition, and each pyrolysis temperature requires an alloy of different composition [11]. In our experiments pure iron was used, with a Curie point of 770 °C, a temperature rise-time of 1.6 s and a pyrolysis time of 8 s.

13.2.4 Gas chromatography

The volatiles were injected into a gas chromatograph (Philips Scientific, PU 4500/11) for separation. The chromatograph was equipped with a wall-coated open tubular fused silica capillary column (25 m, 0.32 mm internal diameter, 1.2 µm CpSil5CB, Chrompack), temperature-programming control and a splitter for coupling with both FTIR and MS. Figure 13.1 shows the parallel FTIR–MS configuration. By use of a dead volume splitter (Gerstel, Graphpack 3D), make-up gas and suitable capillary tubes, ~0.5 % of the column effluent was fed into the MS and the remainder into the FTIR. Parallel FTIR–MS operation is preferred to tandem FTIR–MS for several reasons [12, 13]. First, the rather large volume of the FTIR light pipe causes peak broadening and thus loss of resolution in the MS detection. Second, water is a frequently observed component in outgassing studies of plastics and rubbers, causing deterioration of the KBr windows of the FTIR light pipe. For this reason, an open-ended light pipe was used. Third, the unmatched sensitivities of FTIR and MS result in overloading of the MS ion source.

13.2.5 Fourier transform infrared spectroscopy

The separated components were detected by means of an FTIR spectrometer (Nicolet, 20 SXB) with GC interface. The interface was equipped with a gold-covered capillary light pipe (15 cm, 1.5 mm internal diameter) and a liquid-nitrogen-cooled mercury cadmium telluride A (MCT-A) detector. A minicomputer (Nicolet 1280)

339

controlled the spectrometer. GC–FTIR software was available, including library search routines and a vapor-phase spectra database.

The eluting GC fractions were led through the light pipe and IR spectra were recorded on-line. During the chromatographic separation, in an interval of 4 s, a number of consecutive scans (16) with 4 cm^{-1} resolution were co-added and the interferogram was stored. For real-time monitoring, high-speed low-resolution (16 cm^{-1}) Fourier transforms were performed and the resulting IR spectrum was displayed on the monitor in a three-dimensional plot. Simultaneously, the integrated absorbances in five selectable wavelength windows, representing specific group frequencies, were printed on the plotter, giving five spectral traces of the chromatogram. At the end of the GC-elution, full-resolution (4 cm^{-1}) Fourier transforms and absorbance calculations were performed on the stored co-added interferograms. A chromatogram was reconstructed from all the IR spectra: this is called a chemigram. The positions of the peaks in this chemigram were calculated and qualitative analyses, i.e. library search and identification of these peaks, were performed.

13.2.6 Mass spectrometry

Parallel to the FTIR spectrometer, detection of the separated components was performed with an ion-trap mass spectrometer (Finnigan Mat, ITD 800) [14]. The ITD provides positive ion detection while scanning a mass range from 20 to 650 atomic mass units (amu) throughout the GC run with unit mass resolution. The components were ionized inside the ITD by conventional electron-impact ionization (EI) with a filament [5]. Automatic gain control software enhances high-quality EI spectra in a large concentration range. For molecular mass confirmation, chemical ionization (CI) was used, providing a soft ionization capability via charge transfer of a reagent gas, e.g. methane [5]. A personal computer (IBM, PC-AT) controlled the ITD equipment and the data acquisition, including library search routines.

The gas chromatographic capillary column was coupled directly to the mass spectrometer via a heated transfer line. The helium carrier gas flow of about 1 ml/min was pumped down by a turbomolecular pump.

Data acquisition was similar to that of the FTIR. Four spectra were co-added in time intervals of 1 s and stored. On a monitor the real-time chromatogram was displayed with the total ion current on the intensity scale. At the end of the GC run, the mass spectra of the peaks in the chromatogram could be analyzed, with library search routines as a useful interpretation aid.

13.3 Results and discussion

The GC–FTIR–MS system has been operational in our laboratory for several years. In this period a variety of polymer samples have been thermally analyzed, including thermoplastics, thermosets, elastomers and lacquers [6, 7]. A number of examples are given below of how the thermal desorption unit, the thermogravimetric analyzer and the pyrolysis unit can help to predict product performance and to solve problems in this field.

13.3.1 Thermal desorption

During processing and in product applications of polymeric materials, especially at elevated temperatures, outgassing of initially present low molar mass products, possibly accompanied by degradation products, may occur, causing deterioration of the properties of the material. Outgassing phenomena may also be related to product lifetime, toxicological aspects, mold and environmental contamination, suitability for finishing processes, etc. [6, 7].

In addition, the study of outgassing phenomena has been found to be very useful for identification of components that are responsible for environmental stress cracking of plastic products. When a plastic material is stressed or strained in air below its yield point, stress cracking may occur after a certain period of time. The stresses may be internal or external or a combination of both. Simultaneous exposure to a chemical medium may result in a dramatic shortening of the time to failure. This phenomenon is referred to as environmental stress cracking (ESC) [15].

In thermoplastics, what appear to be cracks are often load-bearing entities; the walls of the 'cracks' are not quite apart but are connected by stretched micofibrils. Since they are not true cracks, these entities, with dimensions in the nanometer to micrometer range, are called crazes. Crazes may penetrate through the entire thickness of the material, separating it into two or more pieces, or may be arrested when they reach regions of lower stress or different material morphology, e.g. in rubber- modified thermoplastics. Craze growth can be accelerated by temperature and environment. ESC is a major problem in the long-term service behavior of plastic products.

Via outgassing experiments using thermal desorption–GC–FTIR–MS, harmful components causing ESC can be detected and identified, and preventive measures can be taken. Two examples are discussed below.

13.3.1.1 Example 1: ESC of an acrylonitrile–butadiene–styrene product during lacquering

In a lacquering process of an acrylonitrile–butadiene–styrene (ABS) product, a large number of the products broke spontaneously. A small ABS sample was taken at a fracture surface and subjected to an outgassing treatment of 5 min at 300 °C. Figure 13.5 shows the resulting chromatogram. Besides the degradation products of ABS

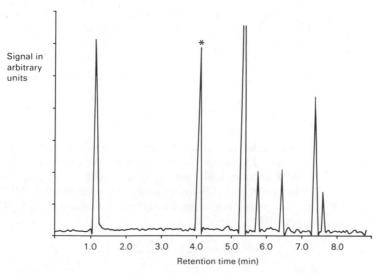

Fig. 13.5. Chromatogram of the volatiles after 5 min at 300 °C originating from ∼10 mg of a failed ABS product

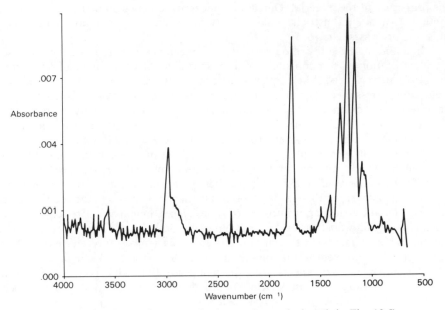

Fig. 13.6. IR spectrum of ethyl lactate (referring to the marked peak in Fig. 13.5)

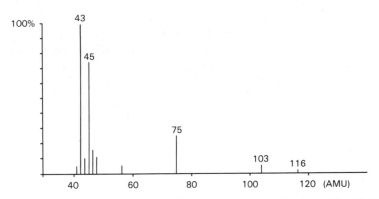

Fig. 13.7. MS spectrum of ethyl lactate (referring to the marked peak in Fig. 13.5)

(styrene, acetophenone etc.), ethyl lactate was identified. This component was used in the thinner. Figures 13.6 and 13.7 show the IR and MS spectra of ethyl lactate. A subsequent test of the ABS samples under an external stress [16] in an environment of ethyl lactate confirmed their poor ESC resistance. The problem was solved by a small change in the mold construction to reduce the internal stress and by optimization of the thinner composition.

13.3.1.2 Example 2: ESC of polycarbonate optical components

During lifetime tests of polycarbonate (PC) optical components premature failure was observed, possibly due to deterioration or degradation of the PC. In the product

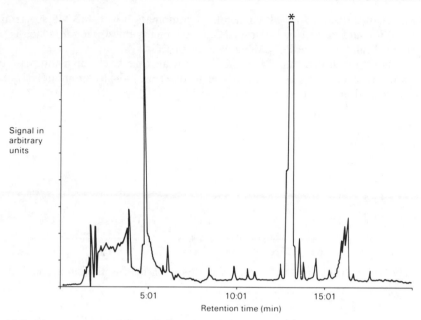

Fig. 13.8. Chromatogram of the volatiles after 5 min at 300 °C originating from ∼10 mg of the degraded surface of PC optical components

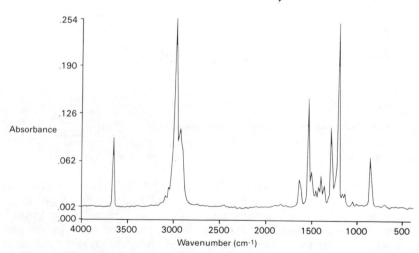

Fig. 13.9. IR spectrum of p-(1,1,3,3-tetra-methylbutyl)phenol (referring to the marked peak in Fig. 13.8)

application the optical components were glued with an epoxy glue containing an amine hardener. The nature of this phenomenon was investigated by reflectance-optics FTIR [17]. Amide groups were detected on the surface of the PC components. A small amount of the degraded surface and a small amount of the bulk of the components were studied via outgassing for 5 min at 300 °C. Compared with the chromatogram of the bulk material, the chromatogram of the surface material shows an additional large peak, as can be seen in Fig. 13.8. Figure 13.9 shows the IR spectrum of this peak, which

343

could be characterized as a *p*-alkylphenol. In combination with the MS spectrum of Fig. 13.10 it could be specified as *p*-(1,1,3,3-tetra-methylbutyl)phenol, which is used as a chain regulator in the PC polymerization process [18].

It was demonstrated that PC products with an excessively high internal stress level show stress cracking phenomena due to deterioriation by ammonolysis, especially at the end-groups as shown in Fig. 13.11.

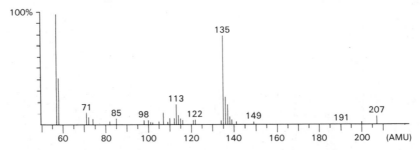

Fig. 13.10. MS spectrum of *p*-(1,1,3,3-tetra-methylbutyl)phenol (referring to the marked peak in Fig. 13.8)

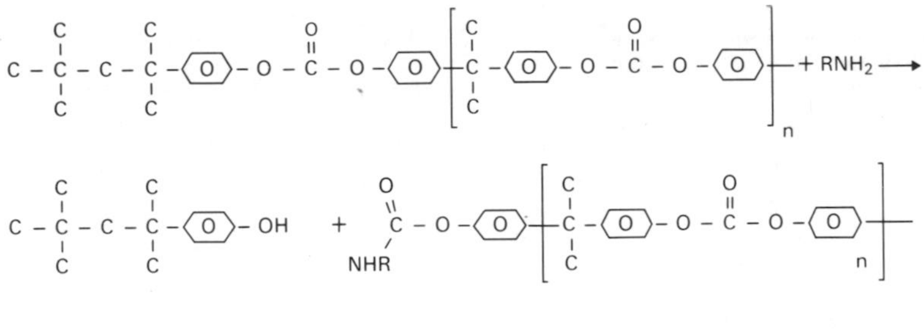

R = hydrogen or alkyl group

Fig. 13.11. Ammonolysis reaction of PC at the end-groups

13.3.2 Thermogravimetry

In thermogravimetric analysis, the mass of a substance as a function of time and temperature is used to assess the thermal stability and degradation of polymers, which includes the generation of kinetic data and lifetime predictions [19]. In general, thermogravimetry is a convenient and relatively simple method with a high sensitivity. A disadvantage is that no information is obtained about the chemical nature of the mass loss.

For this reason, we developed an off-line method in which the volatiles liberated during the thermogravimetric run are sampled by Tenax adsorption cartridges, which can be desorbed afterwards in the thermal desorption unit. Two examples are discussed below.

13.3.2.1 Example 1: composition of a styrene–butadiene rubber

For optimization and batch-to-batch control of rubber recipes, TG is a useful technique [20]. Figure 13.12 shows the TG curve (i.e. mass *vs.* temperature) of a

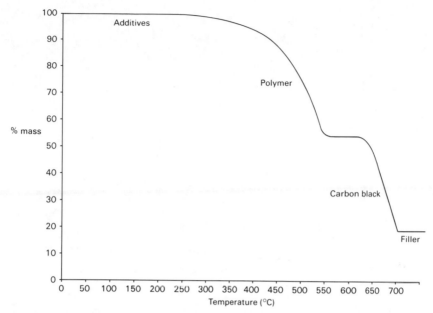

Fig. 13.12. Thermal decomposition of 22.5 mg of SBR rubber in the TG analyzer at a heating rate of 5 °C/min, the analyzer being flushed with a nitrogen flow of 50 ml/min: at 550 °C the flow was switched to air

styrene–butadiene rubber (SBR) in a dynamic run at a heating rate of 5 °C/min. The mass of the sample as a function of temperature and time provides information about its composition, including polymers, additives, carbon black and inorganic fillers. At about 250 °C the evolved gases were sampled for 10 min by a Tenax adsorption cartridge. The cartridge was desorbed afterwards in the thermal desorption unit for 10 min at 300 °C, and the volatile components were separated, detected and identified. Figure 13.13 shows the resulting chromatogram and Table 13.2 lists the identified components and their origins.

13.3.2.2 Example 2: lifetime prediction of polybutyleneterephthalate products

TG is widely applied in the study of polymer degradation reactions, which includes the generation of kinetic data and lifetime predictions. In these studies it is assumed that temperature is the only accelerating factor from service conditions to test conditions, with the activation energy representing the temperature dependence of the degradation process. Among others, FLYNN and WALL [21] have developed a method to predict the lifetime of polymers. This method was based on the evaluation of several mass-loss curves obtained at different heating rates.

A number of polymers have been successfully studied with this method. We encountered some anomalies in the mass-loss curves obtained by thermogravimetric analysis when comparing two types of flame-retardant polybutyleneterephthalate (PBT), which resulted in large differences in lifetime predictions for these materials. It was speculated that the higher mass loss of one PBT depended on the specific type of flame-retardant additive. To verify this, the two PBTs were heated for 30 min

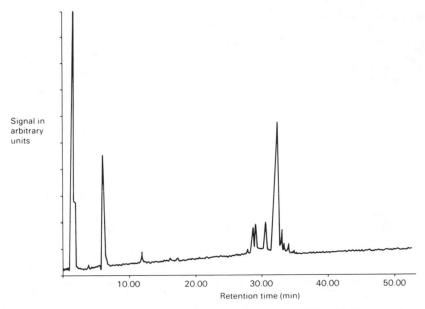

Fig. 13.13. Chromatogram of the volatiles of the SBR rubber of Fig. 13.12 trapped at about 250 °C for 10 min by a Tenax adsorption cartridge

Table 13.2. Volatile components evolved at about 250 °C in a dynamic TG run from an SBR rubber (referring to Fig. 13.13)

Retention time min	Components
2.5	Water
2.5	Carbon disulfide[1]
8.2	Morpholine
32.8	Aromatic amine[2]
28–35	Hydrocarbons (>C20)[3]

[1]Accelerator 2-(4-morpholinylthio)benzothiazol
[2]Antioxidant *N*-isopropyl-*N'*-phenyl-*p*-phenylene-diamine
[3]Plasticizer

at 250 °C in the TG equipment while the liberated volatiles were sampled with Tenax adsorption tubes. Mass losses after this period amounted to 0.1 % and 1.5 % respectively, as shown in Fig. 13.14. Subsequent analysis of the Tenax tubes via TD–GC–FTIR–MS for both materials revealed the presence of water and tetrahydrofuran in a total quantity of about 0.1 %. The missing 1.4 % of one sample appeared as a cold spot deposit in the TG equipment, which, by infrared analysis, was analyzed as a brominated diphenylether with a molar mass of about 800 Daltons, as shown in Fig. 13.15. This means that the higher mass loss for one type of PBT must be attributed to the flame-retardant system. This experiment also shows that the results of the TG–GC–FTIR–MS analysis must be evaluated with care.

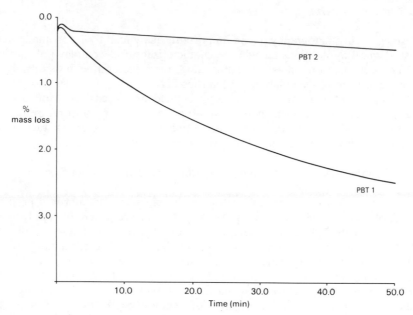

Fig. 13.14. Mass-loss curves of two flame-retardant PBTs covering a period of 30 min at 250 °C in the TG equipment, which was flushed with an air flow of 50 ml/min

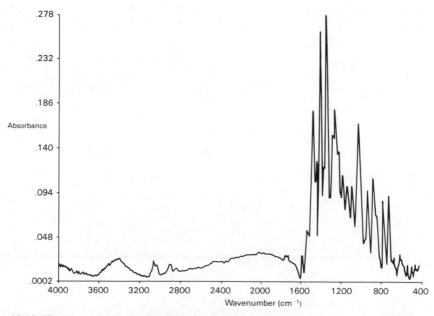

Fig. 13.15. IR spectrum of a brominated diphenylether with a molar mass of about 800 Daltons deposited in the TG equipment during the experiment (referring to Fig. 13.14)

347

13.3.3 *Curie-point pyrolysis*

Pyrolysis gas chromatography is an important analytical tool for the polymer chemist. Typical applications include the characterization of complex polymer mixtures, the assessment of copolymer composition, the structural analysis of polymers and the elucidation of degradation mechanisms [22].

Pyrolysis conditions must be precisely defined and accurately controlled for proper and reproducible characterization of polymers. The temperature-rise-time must be very rapid to ensure that the pyrolysis products obtained are characteristic of the chosen pyrolysis temperature [11]. Two examples are discussed below.

13.3.3.1 Example 1: monomeric composition of polyalkyl(meth)acrylate

An increasing variety of modified polyalkyl(meth)acrylates with improved properties are available for optical applications. These polymers are generally copolymers of acrylates and methacrylates. For identification and classification purposes, an analysis of the monomeric composition is therefore desirable. The chromatograms obtained from the pyrolysis of these materials indicate that the most important pyrolysis fragments were the monomeric entities [23]. Figure 13.16 shows the pyrolysis chromatogram of such a polyalkylmethacrylate, pyrolyzed for 8 s at 770 °C. Peak A was identified as methylmethacrylate. Figure 13.17 shows the IR spectrum of peak C, which was characterized as a cycloalkylmethacrylate. With the help of the MS spectrum of Fig. 13.18, it was specified as tricyclododecanemethacrylate. Peak B was identified as the tricyclododecane fragment.

13.3.3.2 Example 2: polybutadiene content of high-impact polystyrene

The amount of polybutadiene (PB) in high-impact polystyrene (HIPS) is an important quantity, since it is related to the impact properties of this polymer. For the

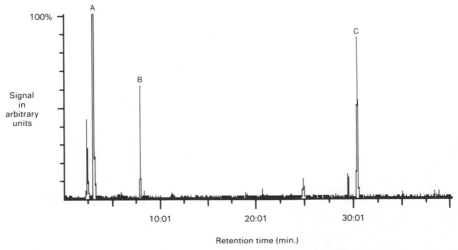

Fig. 13.16. Chromatogram of the pyrolysis products of a polyalkylmethacrylate pyrolyzed for 8 s at 770 °C

348

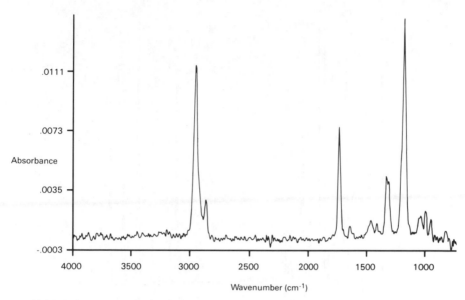

Fig. 13.17. IR spectrum of tricyclodecanemethacrylate (referring to peak C in Fig. 13.16)

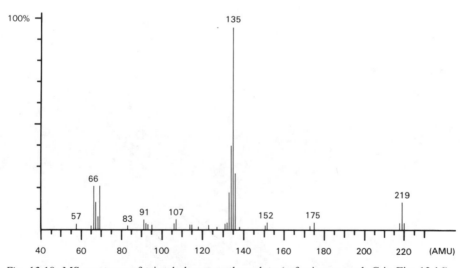

Fig. 13.18. MS spectrum of tricyclodecanemethacrylate (referring to peak C in Fig. 13.16)

determination of the PB content, two analytical methods are available, namely the time-consuming wet-chemical WIJS method [24] and an infrared spectroscopic method [25]. The latter is relatively simple, but is suitable only for trans-PB content determination.

A quantitative pyrolysis–GC–MS method was developed for HIPS types with a PB content of 0 %–7 %. In this method about 0.1–0.2 mg of the polymer sample was attached to a pyrolysis wire. The Curie-point temperature was 770 °C; the

349

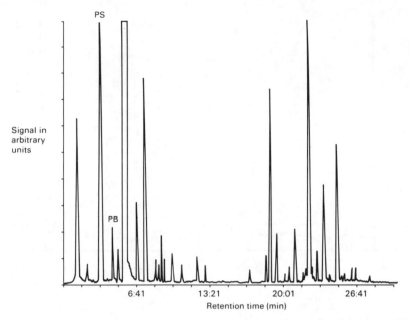

Fig. 13.19. Chromatogram of the pyrolysis products of a HIPS pyrolyzed for 8 s at 770 °C

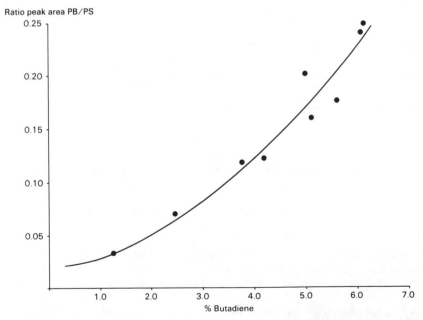

Fig. 13.20. Calibration curve of the ratio of the peak areas of the dimer of butadiene (PB) and toluene (PS) obtained from the pyrolysis chromatograms of HIPS *vs.* the polybutadiene content

pyrolysis time was 8 s. The pyrolysis products were gas chromatographically separated in a standard temperature program and detected by the mass spectrometer. Figure 13.19 shows a typical pyrolysis chromatogram in which peaks originating from butadiene and styrene compounds can be identified. Samples of known PB content from several suppliers were used as standards. Figure 13.20 shows the calibration curve of the ratio of the peak areas of the dimer of butadiene (PB) and toluene (PS) as a function of the PB content. Some scattering was observed, which may be due to different polymerization processes. Nevertheless, it was seen that a quantitative analysis of the PB content in HIPS was possible with an accuracy of 0.5 %.

13.4 Conclusions

The coupling of thermal and chemical analytical techniques is an extremely useful means to obtain insight into the composition of polymeric materials and the chemistry of the thermal processes. The combination of FTIR and MS is a powerful means of detailed characterization, with detection limits at the sub-ppb level.

In the context of product applications and problems, a variety of polymer samples, including thermoplastics, thermosets, elastomers and lacquers were analyzed using the thermal desorption unit, the thermogravimetric analyzer and the pyrolysis unit.

References

1 *Turi, E. A.:* Thermal Characterization of Polymeric Materials. Academic Press, New York, 1981
2 *Cassel, B., McCure, G.:* Int. Lab. 19, p. 32 (1989)
3 *Whiting, L. F., Langvardt, P. W.:* Anal. Chem. 56, p. 1755 (1984)
4 *Herres, W.:* Capillary Gas Chromatography Fourier Transform Infrared Spectroscopy. Huethig, Heidelberg, 1988
5 *Chapman, J. R.:* Practical Organic Mass Spectrometry. John Wiley, Chichester, UK, 1985
6 *Jansen, J. A. J., Haas, W. E.:* Anal. Chim. Acta 196, p. 69 (1987)
7 *Jansen, J. A. J., Haas, W. E., Neutkens, H. G. M., Leenen, A. J. H.:* Thermochim. Acta 134, p. 307 (1988)
8 *Liebman, S. A., Levy, E. J.:* Pyrolysis and GC in Polymer Analysis. Marcel Dekker, New York, 1985
9 *Rijks, J. A., Drozd, J., Novak, J.:* J. Chromatogr. 186, p. 167 (1979)
10 *Flynn, J. H.,* in: Degradation and Stabilization of Polymers. *Jellinek, H. H. G.* (Ed.), Elsevier, Amsterdam, 1983
11 *Oguri, N., Kim, P.:* Int. Lab. 19, p. 58 (1989)
12 *Demirgian, J. C.:* Trends in Anal. Chem. 6, p. 58 (1987)
13 *Leibrand, R. J., Duncan, W. P.:* Int. Lab. 19, p. 46 (1989)
14 *Stafford, G. C., Kelley, P. E., Syka, J. E. P., Reynolds, W. E., Todd, J. F. J.:* Int. J. Mass Spectrom. & Ion Proces. 60, p. 85 (1984)
15 *Brostow, W., Corneliussen, R. D.:* Failure of Plastics. Carl Hanser Verlag, München, 1986
16 *ISO 4599:* Plastics—Determination of Resistance to Environmental Stress Cracking—Bent Strip Method. International Organization for Standardization, 1986
17 *Jansen, J. A. J., Haas, W. E.:* Polym. Commun. 29, p. 77 (1988)
18 *Bailly, C., Daoust, D., Legras, R., Mercier, J., De Valck, M.:* Polymer 27, p. 776 (1986)
19 *Flynn, J. H.:* Polym. Eng. Sci. 20, p. 675 (1980)
20 *Staub, F.:* Int. Lab. 16, p. 55 (1986)
21 *Flynn, J. H., Wall, L. A.:* J. Polym. Sci. Polym. Lett. 4, p. 323 (1966)
22 *Hammond, T., Lehrle, R. S.:* Br. Polym. J. 21, p. 23 (1989)
23 *Herres, W., Fölster, U.:* Farbe und Lack 91, p. 6 (1985)

24 *Schröder, E., Franz, J., Hagen, E.:* Ausgewahlte Methoden der Plastanalytik. Akademie Verlag, Berlin, 1976
25 *Haslam, J., Willis, H. A., Squirrell, D. C. M.:* Identification and Analysis of Plastics. Iliffe Books, London, 1972

Nomenclature

ABS	poly(acrylonitrile–butadiene–styrene)
amu	atomic mass unit
CI	chemical ionization
EI	electron-impact ionization
ESC	environmental stress cracking
(FT)IR	(Fourier transform) infrared spectroscopy
GC	gas chromatography
(HI)PS	(high-impact) polystyrene
ITD	ion-trap detector
MS	mass spectrometry
PB	polybutadiene
PBT	polybutyleneterephthalate
PC	polycarbonate
PY	pyrolysis
SBR	styrene–butadiene rubber
TCT	thermal desorption cold-trap injector
TD	thermal desorption
TG	thermogravimetry

Index

355

357